Compressive Sensing for Wireless Networks

Compressive sensing is a new signal-processing paradigm that aims to encode sparse signals by using far lower sampling rates than those in the traditional Nyquist approach. It helps acquire, store, fuse and process large data sets efficiently and accurately. This method, which links data acquisition, compression, dimensionality reduction, and optimization, has attracted significant attention from researchers and engineers in various areas. This comprehensive reference develops a unified view on how to incorporate efficiently the idea of compressive sensing over assorted wireless network scenarios, interweaving concepts from signal processing, optimization, information theory, communications, and networking to address the issues in question from an engineering perspective. It enables students, researchers, and communications engineers to develop a working knowledge of compressive sensing, including background on the basics of compressive sensing theory, an understanding of its benefits and limitations, and the skills needed to take advantage of compressive sensing in wireless networks.

Zhu Han is an Associate Professor in the Electrical and Computer Engineering Department at the University of Houston, Texas. He received an NSF CAREER award in 2010 and the IEEE Fred W. Ellersick Prize in 2011. He has co-authored papers that won the best paper award at the IEEE International Conference on Communications 2009, the 7th International Symposium on Modeling and Optimization in Mobile, Ad Hoc, and Wireless Networks (WiOpt09), the IEEE Wireless Communication and Networking Conference, 2012 and IEEE Smartgridcomm Conference, 2012.

Husheng Li is an Assistant Professor in the Electrical and Computer Engineering Department at the University of Tennessee. He received the Best Paper Award of the EURASIP Journal on Wireless Communications and Networking in 2005 (together with his PhD advisor, Professor H. V. Poor), the Best Demo Award of IEEE Globecom in 2010, and the Best Paper Award at IEEE ICC in 2011.

Wotao Yin is an Associate Professor at the Department of Computational and Applied Mathematics at Rice University. He won an NSF CAREER award in 2008 and an Alfred P. Sloan Research Fellowship in 2009.

Compressive Sensing for Wireless Networks

ZHU HAN
University of Houston, USA

HUSHENG LI
University of Tennessee, USA

WOTAO YIN
Rice University, USA

CAMBRIDGE UNIVERSITY PRESS
Cambridge, New York, Melbourne, Madrid, Cape Town,
Singapore, São Paulo, Delhi, Mexico City

Cambridge University Press
The Edinburgh Building, Cambridge CB2 8RU, UK

Published in the United States of America by Cambridge University Press, New York

www.cambridge.org
Information on this title: www.cambridge.org/9781107018839

First published 2013

Printed and bound by CPI Group (UK) Ltd, Croydon CR0 4YY

A catalogue record for this publication is available from the British Library

Library of Congress Cataloguing in Publication data
Han, Zhu, 1974–
Compressive sensing for wireless networks / Zhu Han, University of Houston, USA, Husheng Li, University of Tennessee, USA, Wotao Yin, Rice University, USA.
pages cm
Includes bibliographical references and index.
ISBN 978-1-107-01883-9 (hardback)
1. Coding theory. 2. Data compression (Telecommunication) 3. Signal processing – Digital techniques. 4. Sampling (Statistics) I. Li, Husheng, 1975– II. Yin, Wotao. III. Title.
TK5102.92.H355 2013
621.39′81 – dc23 2013000272

ISBN 978-1-107-01883-9 Hardback

For the people I met in the Barneo ice camp, North Pole, who showed me the bravery to conquer any difficulty, which encouraged me to finish this challenging book **Zhu Han**

To my wife, Min Duan, and my son, Siyi Li **Husheng Li**

To those who advocate for intellectual honesty and defend academic integrity **Wotao Yin**

Contents

Preface

Over the past few decades, wireless communications and networking have witnessed an unprecedented growth, and have become pervasive much sooner than anyone could have predicted. For example, cellular wireless networks are expected to become the dominant and ubiquitous telecommunication means in the next few decades. The widespread success of cellular and WLAN systems prompts the development of advanced wireless systems to provide access to information services beyond voice such as telecommuting, video conferencing, interactive media, real-time internet gaming, and so on, anytime and anywhere. The enormous potential demands for these wireless services require a careful design of the future networks. Many technical challenges remain to be addressed such as limited resources, adverse natures of wireless channels, interference, etc.

Today, with the increasing demand of higher resolution and increasing number of modalities, the traditional wireless signal processing hardware and software are facing significant challenges since the Nyquist rate, which is part of the dogma for signal acquisition and processing, has become too high in many wireless applications. How to acquire, store, fuse, and process these data efficiently becomes a critical problem. The most current solution to this problem is to compress after sensing densely. However, this oversampling-then-discarding procedure wastes time, energy, and/or other precious resources.

A new paradigm of signal acquisition and processing, named compressive sensing (CS), has emerged since 2004. Starting with the publication of "Compressed sensing" by D. Donoho, and a few seminal works by E. J. Candès, J. Romberg, and T. Tao, the CS theory, which integrates data acquisition, compression, dimensionality reduction, and optimization, has attracted lots of research attention. The CS theory consists of three key components: signal sparsity, incoherent sensing, and signal recovery. It claims that, as long as the signal to be measured is sparse or can become sparse under a certain transform or dictionary, the information in the signal can be encoded in a small number of incoherent measurements, and the signal can be faithfully recovered by tractable computation.

Since CS is so new a tool bearing a large number of potential applications in engineering, there is not yet a published book for the engineers. However, the applications of CS in wireless communication are very important and have the potential to revolutionize certain traditional design concepts. This produces the foremost motivation of this book: to equip engineers with the fundamental knowledge of CS and demonstrate its strong potential in wireless networking fields. Secondly, understanding a large portion of the existing CS results in the literature requires a good mathematical background, but this book is written at a level for the engineers. Most parts of this book are suitable for readers who want to broaden their views, and it is also very useful for engineers and researchers in applied fields who deal with sampling problems in their work.

We would like to thank Drs. Richard Baraniuk, Stephen Boyd, Rick Chartrand, Ekram Hossain, Kevin Kelly, Yingying Li, Lanchao Liu, Jia Meng, Lijun Qian, Stanley Osher, Zaiwen Wen, Zhiqiang Wu, Ming Yan, and Yin Zhang for their support and encouragement. We also would like to thank Lanchao Liu, Nam Nguyen, Ming Yan, and Hui Zhang for their assistance and Mr. Ray Hardesty for text editing. Finally, we would like to acknowledge NSF support (ECCS-1028782), ARL and ARO grant W911NF-09-1-0383 and NSF grant DMS-0748839.

ZHU HAN
HUSHENG LI
WOTAO YIN

1 Introduction

Sampling is not only a beautiful research topic with an interesting history, but also a subject with high practical impact, at the heart of signal processing and communications and their applications. Conventional approaches to sample signals or images follow Shannon's celebrated theorem: the sampling rate must be at least twice the maximum frequency present in the signal (the so-called Nyquist rate) has been to some extent accepted and widely used ever since the sampling theorem was implied by the work of Harry Nyquist in 1928 ("Certain topics in telegraph transmission theory") and was proved by Claude E. Shannon in 1949 ("Communication in the presence of noise"). However, with the increasing demand for higher resolutions and an increasing number of modalities, the traditional signal-processing hardware and software are facing significant challenges. This is especially true for wireless communications.

The compressive sensing (CS) theory is a new technology emerging in the interdisciplinary area of signal processing, statistics, optimization, as well as many application areas including wireless communications. By utilizing the fact that a signal is sparse or compressible in some transform domain, CS can acquire a signal from a small set of incoherent measurements with a sampling rate much lower than the Nyquist rate. As more and more experimental evidence suggests that many kinds of signals in wireless applications are sparse, CS has become an important component in the design of next-generation wireless networks.

This book aims at developing a unified view on how to efficiently incorporate the idea of CS over assorted wireless network scenarios. This book is interdisciplinary in that it covers materials in signal processing, optimization, information theory, communications, and networking to address the issues in question. The primary goal of this book is to enable engineers and researchers to understand the fundamentals of CS theory and tools and to apply them in wireless networking and other areas. Additional important goals are to review some up-to-date and state-of-the-art techniques for CS, as well as for industrial engineers to obtain new perspectives on wireless communications.

1.1 Motivation and objectives

CS is a new signal-processing paradigm and aims to encode sparse signals by using far fewer measurements than those in the Nyquist setup. It has attracted a great amount of attention from researchers and engineers because of its potential to revolutionize many sensing modalities. For example, in a cognitive radio system, to increase the efficiency of the utility of spectrum, it is necessary to separate occupied spectrum and unoccupied spectrum first, which becomes a spectrum sensing problem and can leverage CS techniques. However, as with many great techniques, there is a gap between the theoretical breakthrough of CS and its practical applications, in particular the applications in wireless networking. This motivates us to write a book to narrow this gap by presenting the theory, models, algorithms, and applications in one place. The book was written with two main objectives. The first one is to introduce the basic concepts and typical steps of CS. The second one is to demonstrate its effective applications, which will hopefully inspire future applications.

1.2 Outline

In order to achieve the objectives, the book first presents an introduction to the basics of wireless networks. The book is written in two parts as follows: the first part studies the CS framework, and the second part discusses its applications in wireless networks by presenting several existing implementations. Let us summarize the remaining chapters of this book as follows:

Chapter 2 Overview of wireless networks
Different wireless network technologies such as, cellular wireless, WLAN, WMAN, WPAN, WRAN technologies, and the related standards are reviewed. The review includes the basic components, features, and potential applications. Furthermore, advanced wireless technologies such as cooperative communications, network coding, and cognitive radio are discussed. Some typical wireless networks such as ad hoc/sensor networks, mesh networks, and vehicular networks are also studied. The research challenges related to the practical implementations at the different layers of the protocol stack are discussed.

Part I: Compressive sensing framework

Before we discuss how to employ CS in different wireless network problems, the choice of a design technique is crucial and must be studied. In this context, this part presents different CS techniques, which are applied to the design, analysis, and optimization of wireless networks. We introduce the basic concepts, theorems, and applications of CS schemes. Both theoretical analysis and numerical algorithms are discussed and CS examples are given. Finally, we discuss the current state-of-the-art for CS-based analog-to-digital converters.

Chapter 3 Compressive sensing framework
This chapter overviews the basic concepts, steps, and theoretical results of CS. It is a methodology using incoherent linear measurements to recover sparse signals. The preliminaries and notation are set up for further usage. This chapter also

presents the elements of a typical CS process. The conditions that guarantee successful CS encoding and decoding are presented.

Chapter 4 Sparse optimization algorithms

There are a collection of various algorithms for recovering sparse solutions, as well as low-rank matrices, from their linear measurements. Generally, they can be classified into optimization models and algorithms and non-optimization ones. The chapter gives more emphasis on the first class and briefly discusses the second one. When presenting algorithms, the big picture is focused, and some detailed analyses are omitted and referred to related papers. The advantages and disadvantages of presented algorithms are discussed to help the reader pick appropriate ones to solve their own problems.

Chapter 5 CS Analog-to-digital converter

Wideband analog signals push contemporary analog-to-digital conversion systems to their performance limits. In many applications, however, sampling at the Nyquist rate is inefficient because the signals of interest contain only a small number of significant frequencies relative to the limited band, though the locations of the frequencies may not be known a priori. In this chapter, we discuss several possible strategies in the literature. First, we study the CS-based ADC and its applications to 60 GHz communication. Then we describe the random demodulator, which demodulates the signal by multiplying it with a high-rate pseudonoise sequence and smears the tones across the entire spectrum. Next, we study the modulated wideband converter, which first multiplies the analog signal by a bank of periodic waveforms. The product is then low-pass filtered and sampled uniformly at a low rate, which is orders of magnitude smaller than Nyquist. Perfect recovery from the proposed samples is achieved under certain necessary and sufficient conditions. Finally, we study Xampling, a design methodology for analog CS in which analog band-limited signals are sampled at rates far lower than Nyquist without loss of information.

Part II: Compressive Sensing Applications in Wireless Networks

To exploit CS in wireless communication, many applications using CS are given in detail. However, CS has more places to be adopted and emphasized. Because of the authors' limited time and effort, we have only contributed those listed related to wireless networking in the book, but hope this can motivate the readers to discover more in the future. The process of designing a suitable model for CS and problem formulation is also described to help engineers who are also interested in using the new technology – CS – in their research.

Chapter 6 Compressed channel estimation

In communications, CS is largely accepted for sparse channel estimation and its variants. In this chapter, we highlight the fundamental concepts of CS channel estimation with the fact that multipath channels are sparse in their equivalent baseband representation. Popular channels such as OFDM and MIMO are investigated by use of CS. Then, a belief-propagation-based channel estimation scheme is used with a standard bit-interleaved coded OFDM transmitter, which performs joint sparse-channel estimation and data decoding. Next, blind channel estimation is studied to show how to use the CS and matrix completion. Finally, a special channel, the underwater acoustic channel, is investigated from the aspect of CS channel estimation.

Chapter 7 Ultra-wideband systems

Ultra-wideband (UWB) has been heavily studied due to its wide applications like short-range communications and localizing. However, it suffers from the extremely narrow impulse width that makes the design of the receiver difficult. Meanwhile, the narrow impulse width and low duty cycle also provides the sparsity in the time domain that facilitates the application of CS. In this chapter, we will provide a brief model of UWB signals. Then, we will review different approaches of applying CS to enhance the reception of UWB signals for general purposes. The waveform template-based approach and Bayesian CS method will be explained as two case studies.

Chapter 8 Positioning

Precise positioning (e.g., of the order of centimeters or millimeters) is useful in many applications like robot surgery. Usually it is achieved by analyzing the narrow pulses sent from the object and received at multiple base stations. The precision requirement places a pressing demand on the timing acquisition of the received pulses. In this chapter, we discuss the precision positioning using UWB impulses. In contrast to the previous chapter, this chapter is focused on the CS with the correlated signals received at the base stations. We will first introduce the general models and approaches of positioning. Then, we will introduce the framework of Bayesian CS and explain the principle of using the a priori distribution to convey the correlated information. Moreover, we will introduce the general principle of how to integrate CS with the subsequent positioning algorithm like the Time Difference Arrival (TDOA) approach, which can further improve the precision of positioning.

Chapter 9 Multiple access

In wireless communications, an important task is the multiple access that resolves the collision of the signals sent from multiple users. Traditional studies assume that all users are active and thus the technique of multiuser detection can be applied. However, in many practical systems like wireless sensor networks, only a random and small fraction of users send signals simultaneously. In this chapter, we study the multiple access with sparse data traffic, in which the task is to recover the data packets and the identities of active users. We will formulate the general problem as a CS one due to the sparsity of active users. The algorithm of reconstructing the above information will be described. In particular, the feature of discrete unknowns will be incorporated into the reconstruction algorithm. The CS-based multiple access scheme will further be integrated with the channel coding. Finally, we will describe the application in the advanced metering infrastructure (AMI) in a smart grid using real measurement data.

Chapter 10 Cognitive radio networks

In wideband cognitive radio (CR) networks, spectrum sensing is an essential task for enabling dynamic spectrum sharing. For sensor networks, the event detection is critical for the whole network performance. But it entails several major technical challenges: very high sampling rates required for wideband processing, limited power and computing resources per CR or sensor, frequency-selective wireless fading, possible failure of reporting to the fusion center, and interference due to signal leakage from other coexisting CRs or sensor transmitters. The algorithms in the literature using CS, joint sparsity recovery, and matrix completion are reviewed. The dynamic for such a system is also investigated. Then the distributed solution is studied based on decentralized consensus optimization algorithms. Next, by utilizing a Bayesian CS framework, the sampling reduction advantage of CS can be achieved with significantly less computational complexity. Moreover, the CR or sensor does not have to reconstruct the entire signal because it is only interested in detecting the presence of Primary Users, which can also reduce the complexity. Finally, the joint spectrum sensing and localization problem is introduced.

In conclusion, this book focuses on teaching engineers to use CS and connecting engineer research and CS. On the other hand, it helps mathematicians to obtain feedback from engineers in their design and problem solving. This connection will help fill the gap and benefit both sides. This book will serve this purpose by explaining CS in engineering language and concentrating on the applications in wireless communication.

2 Overview of wireless networks

A wireless network refers to a telecommunications network that interconnects between nodes that are implemented without the use of wires. Wireless networks have experienced unprecedented growth over the past few decades, and they are expected to continue to evolve in the future. Seamless mobility and coverage ensure that various types of wireless connections can be made anytime, anywhere. In this chapter, we introduce some basic types of wireless networks and provide the reader with some necessary background on the state-of-the-art developments.

Wireless networks use electromagnetic waves, such as radio waves, for carrying the information. Therefore, their performance is greatly affected by the randomly fluctuating wireless channels. To develop an understanding of channels, in Section 2.1 we will study the radio frequency band first, then the existing wireless channel models used for different network scenarios, and finally the interference channel.

There exist many wireless standards, and we describe them according to the order of coverage area, starting with cellular wireless networks. In Section 2.2.1, we provide an overview of the key elements and technologies of the third-generation (3G) wireless cellular network standards. WiMax, based on the IEEE 802.16 standard for the wireless metropolitan area network, is discussed in Section 2.2.2. In Section 2.2.3, we study a wireless local area network (WLAN or WiFi), which is a network in which a mobile user can connect to a local area network through a wireless connection. A wireless personal area network (WPAN) is a personal area network for wireless interconnecting devices centered around an individual person's workspace. IEEE 802.15 standards specify some technologies used in Bluetooth, ZigBee, and ultra-wideband, and these are investigated in Section 2.2.4. Networks without any infrastructure, such as ad hoc and sensor networks, are also discussed in Sections 2.2.5 and 2.2.6, respectively.

Finally, we briefly discuss various advanced wireless technologies in Section 2.3, such as OFDM, MIMO (space-time coding, beamforming, etc.), cognitive radios, localization, scheduling, and multiple access. The motivations for deploying such techniques, the design challenges to maintain basic functionality, and recent developments in real implementation are explained in detail.

2.1 Wireless channel models

2.1.1 Radio propagation

Unlike wired channels, which are stationary and predictable, wireless channels are extremely random and hard to analyze. Modeling wireless channels is one of the most

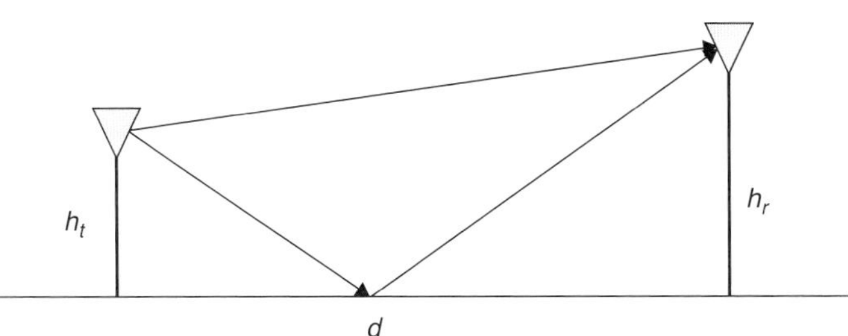

Figure 2.1 Reflection and two-ray model.

challenging tasks encountered in wireless network design. Wireless channel models can be classified as large-scale propagation models and small-scale propagation models relative to the wavelength.

Large-scale models predict behavior averaged over distances much longer than the wavelength. The models are usually functions of distance and significant environmental features and are roughly frequency independent. The large-scale models are useful for modeling the range of a radio system and rough capacity planning. Some large-scale theoretical models (e.g., first four) and large-scale experimental models (the rest) are listed as follows:

- Free space model
  Path loss is a measure of attenuation based only on the distance from the transmitter to the receiver. The free space model is only valid in the far-field and only if there is no interference and obstruction. The received power $P_r(d)$ of the free space model as a function of distance d can be written as

$$P_r(d) = \frac{P_t G_t G_r \lambda^2}{(4\pi)^2 d^2 L}, \tag{2.1}$$

  where P_t is the transmit power, G_t is the transmitter antenna gain, G_r is the receiver antenna gain, λ is the wavelength, and L is the system loss factor not related to propagation. Path loss models typically define a "close-in" point d_0 and reference other points from that point. The received power in dB form can be written as

$$P_d(d) \text{ dBm} = 10 \log_{10}\left[\frac{P_r(d_0)}{0.001W}\right] + 20 \log_{10}\left(\frac{d_0}{d}\right). \tag{2.2}$$

- Reflection model
  Reflection is the change in the direction of a wave front at an interface between two different media so that the wave front returns to the medium from which it originated. A radio propagation wave impinges on an object that is large compared with the wavelength, e.g., the surface of the Earth, buildings, or walls.

  A two-ray model is one of the most important reflection models for wireless channels. An example of a reflection and the two-ray model is shown in Figure 2.1. The two-ray model considers a model in which the receiving antenna sees a direct path signal as well as a signal reflected off the ground. Specular reflection, much

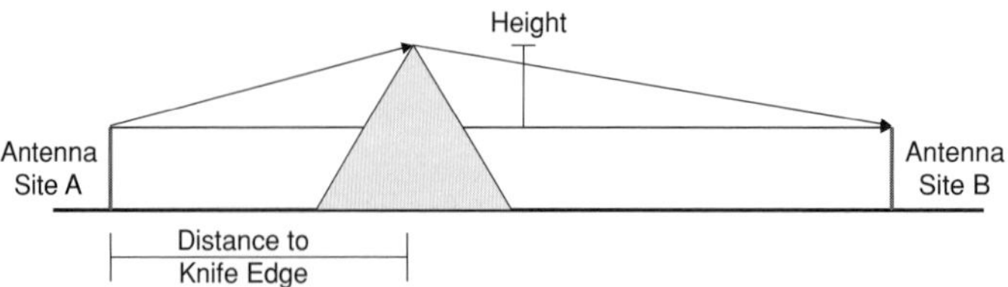

Figure 2.2 Diffraction and knife edge model.

like light off a mirror, is assumed and to a very close approximation. The specular reflection arrives with strength equal to that of the direct path signal (i.e., without loss in strength by reflection). The reflected signal shows up with a delay relative to the direct path signal and as a consequence, may add constructively (in phase) or destructively (out of phase). The received power of the two-ray model can be written as

$$P_r = P_t G_t G_r \frac{h_t^2 h_r^2}{d^4}, \tag{2.3}$$

where h_t and h_r are the transmitter height and receiver height, respectively, and d is the distance between the two antennas.

- Diffraction model
  Diffraction occurs when the radio path between the transmitter and receiver is obstructed by a surface with sharp, irregular edges. Radio waves bend around the obstacle even when a line of sight (LOS) does not exist. In Figure 2.2, we show a knife edge diffraction model in which the radio wave of the diffraction path from the knife edge and the radio wave of LOS are combined at the receiver. Similar to the reflection, the radio waves might add constructively or destructively.
- Scattering model
  Scattering is a general physical process whereby the radio waves are forced to deviate from a straight trajectory by one or more localized non-uniformities in the medium through which they pass. In conventional use, this also includes deviation of reflected radiation from the angle predicted by the law of reflection. The obstructing objects are smaller than the wavelength of the propagation wave, e.g., foliage, street signs, and lampposts. One scattering example is shown in Figure 2.3.
- Log scale propagation model and log-normal shadowing model
  From the experimental measurement, the received signal power decreases logarithmically with distance. However, because of the variety of factors, the decreasing speed is very random. To characterize the mean and variance of this randomness, the log scale propagation model and log-normal shadowing model are used, respectively.

  The log scale propagation model generalizes path loss to account for other environmental factors. The model chooses a distance d_0 in the far-field and measures the path loss $PL(d_0)$. The propagation path loss factor α indicates the rate at which the path loss increases with distance. The path loss of the log scale propagation model is given

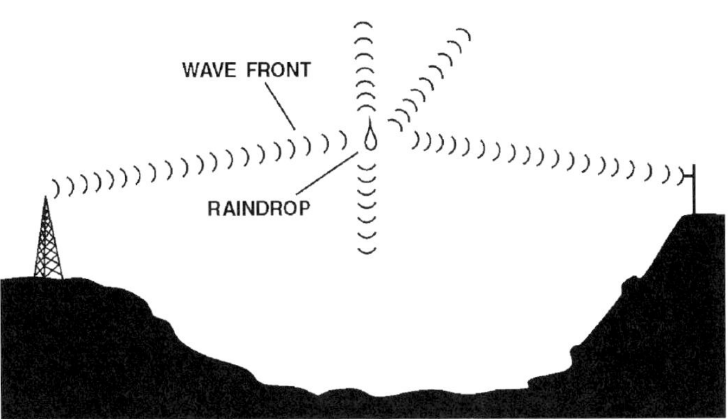

Figure 2.3 Scattering.

by

$$PL(d)\,(\text{dB}) = PL(d_0) + 10\alpha \log_{10}\left(\frac{d}{d_0}\right). \tag{2.4}$$

In the free space propagation model, the path loss factor α equals 2.

Shadowing occurs when objects block the LOS between the transmitter and the receiver. A simple statistical model can account for unpredictable "shadowing" as

$$PL(d)\,(\text{dB}) = PL(d) + X_0, \tag{2.5}$$

where X_0 is a 0-mean Gaussian random variable with variance typically from 3 to 12. The propagation factor and the variance of log-normal shadowing are usually determined by experimental measurement.

- Outdoor propagation models
  In the outdoor models, the terrain profile of a particular area needs to be taken into account for estimating the path loss. Most of the following models are based on a systematic interpretation of measurement data obtained in the service area. Some typical outdoor propagation models are Longley–Rice Model, ITU Terrain Model, Durkins Model, Okumura Model, Hatas Model, PCS Extension of the Hata Model, Walfisch and Bertoni Model, and Wideband PCS Microcell Model [1].
- Indoor propagation models
  For indoor applications, the distances are much shorter than those in the outdoor models. The variability of the environment is much greater, and key variables are layout of the building, construction materials, building type, and antenna location. In general, indoor channels may be classified either as LOS or obstruction with varying degrees of clutter. The losses between floors of a building are determined by the external dimensions and materials of the building, as well as the type of construction used to create the floors and the external surroundings. Some of the available indoor propagation models are Ericsson multiple breakpoint model, ITU model for indoor attenuation, log distance path loss model, attenuation factor model, and Devasirvathams model.

Small-scale (fading) models describe signal variability on a scale of wavelength. In fading, multipath and Doppler effects dominate. Fading is frequency dependent and time variant. The focus is on modeling "fading," which is the rapid change in signal strength over a short distance or length of time.

Multipath fading is caused by interference between two or more versions of the transmitted signal that arrive at slightly different times. The multipath fading causes rapid changes in signal strength over a small travel distance or time interval, random frequency modulation due to varying Doppler shifts on different multipath signals, and time dispersion caused by multipath propagation delays. The response of a multipath fading channel can be written as

$$h(t) = \sum_k a_k \delta(t - \tau_k), \tag{2.6}$$

where τ_k is the delay of the kth multipath and a_k is its corresponding amplitude.

To measure the time dispersion of multiple paths, power delay profile and root mean square (RMS) are the most important parameters. Power delay profiles are generally represented as plots of relative received power as a function of excess delay with respect to a fixed time delay reference. The mean excess delay is the first moment of the power delay profile and is defined as $\bar{\tau} = \frac{\sum_k a_k^2 \tau_k}{\sum_k a_k^2}$. The RMS is the square root of the second central moment of the power delay profile, defined as $\sigma_\tau = \sqrt{\bar{\tau^2} - (\tau)^2}$, where $\bar{\tau^2} = \frac{\sum_k a_k^2 \tau_k^2}{\sum_k a_k^2}$. Typical values of RMS delay spread are on the order of microseconds in outdoor mobile radio channels and on the order of nanoseconds in indoor radio channels.

Analogous to the delay spread parameters in the time domain, coherent bandwidth is used to characterize the channel in the frequency domain. Coherent bandwidth is the range of frequencies over which two frequency components have a strong potential for amplitude correlation. If the frequency correlation between two multipaths is above 0.9, then the coherent bandwidth is [1] $B_c = \frac{1}{50\sigma}$. If the correlation is above 0.5, this coherent bandwidth is $B_c = \frac{1}{5\sigma}$. Coherent bandwidth is a statistical measure of the range of frequencies over which the channel can be considered flat.

Delay spread and coherent bandwidth describe the time-dispersive nature of the channel in a local area, but they do not offer information about the time-varying nature of the channel caused by the relative motion of the transmitter and the receiver. Next, we define Doppler spread and coherence time, which describe the time-varying nature of the channel in a small-scale region.

Doppler frequency shift is due to the movement of the mobile users. Frequency shift is positive when a mobile moves toward the source; otherwise, the frequency shift is negative. In a multipath environment, the frequency shift for each ray may be different, leading to a spread of received frequencies. Doppler spread is defined as the maximum Doppler shift $f_m = \frac{v}{\lambda}$, where v is the mobile user's speed and λ is the wavelength. If we assume that signals arrive from all angles in the horizontal plane, the Doppler spectrum can be modeled as Clarke's Model [1].

Coherence time is the time duration over which the channel impulse response is essentially invariant. Coherence time is defined as $T_c = \frac{C}{f_m}$, where C is a constant [1]. This definition of a coherence time implies that two signals arriving with a time

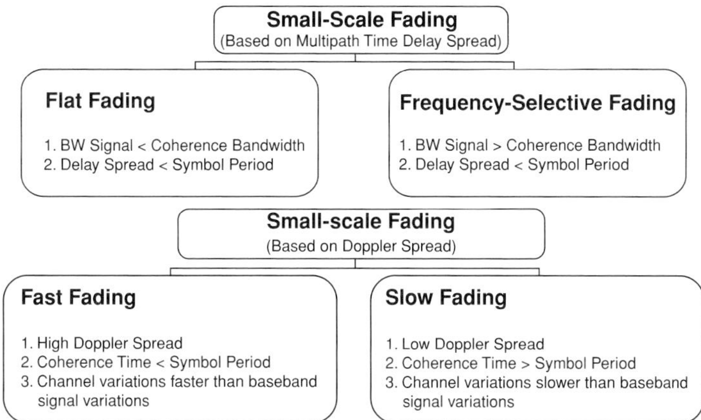

Figure 2.4 Classification of small fading.

separation greater than T_c are affected differently by the channel. If the symbol period of the baseband signal (reciprocal of the baseband signal bandwidth) is greater than the coherence time, then the signal will distort, since the channel will change during the transmission of the signal.

Based on the transmit signal's bandwidth and symbol period relative to the multipath RMS and coherent bandwidth, the small-scale fading can be classified as flat fading and frequency-selective fading. This classification means that a band-limited transmit signal sees a flat frequency channel or a frequency-selective channel. Based on coherence time due to Doppler spread, the small-scale fading can be classified as fast fading or slow fading. This classification is according to whether the channel changes during each signal symbol. The details are shown in Figure 2.4.

Multipath and Doppler effects describe the time and frequency characteristics of wireless channels. But further analysis is necessary for statistical characterization of the amplitudes. Rayleigh distributions describe the received signal envelope distribution for channels, where all the components are non-LOS. Ricean distributions describe the received signal envelope distribution for channel where one of the multipath components is the LOS component. Nakagami distributions are used to model dense scatters, and can be reduced to Rayleigh distributions, but this provides more control over the extent of the fading.

2.1.2 Interference channel

Since networks accommodate an increasing number of users and because bandwidth is limited, radio frequencies are reused beyond a certain distance, which leads to co-channel interference. In this subsection, we study the interference channel. The system model for an interference channel is shown in Figure 2.5. The received signal vector $\mathbf{y}$ can be written as:

$$\mathbf{y} = \mathbf{G}\,\mathbf{x} + \mathbf{z}, \tag{2.7}$$

where $\mathbf{x}$ is the transmitted signal vector, $\mathbf{z}$ is the noise vector, and $\mathbf{G}$ is the channel gain matrix with elements $G_{k,n}$. Here, k is the transmitter index and n is the receiver index.

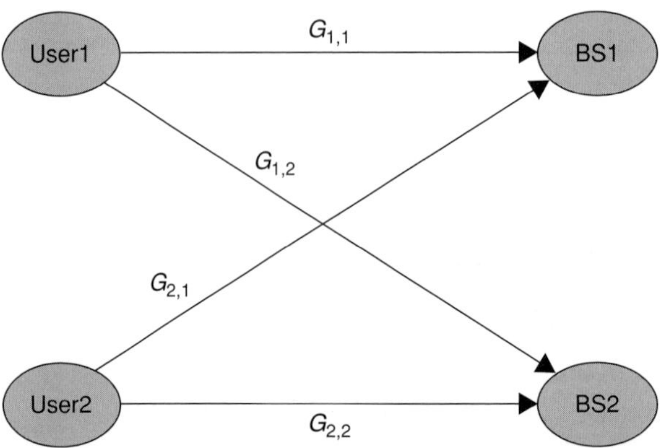

Figure 2.5 2-user interference channel.

For an interference channel, the interference from other users is generally considered as noise. This assumption leads to optimal rates for weak and medium interference. So instead of simply using SNR (signal-to-noise ratio, given by $\frac{P_k G_{k,k}}{\sigma^2}$), we consider the SINR (signal-to-interference-and-noise ratio) to calculate the capacity of the network. Therefore, the capacity of user k, R_k is given by:

$$R_k = \log_2 \left(1 + \frac{P_k G_{k,k}}{\sum_{i \neq k}^{n} P_i G_{i,k} + \sigma^2} \right), \tag{2.8}$$

where P_k is the transmit power of the kth user, $G_{i,k}$ is the channel gain from user i to base station k, and the term $\sum_{i \neq k} P_i G_{i,k}$ represents the interference caused by other users to user k. Without loss of generality, we consider the variance of additive Gaussian noise as a constant σ^2 for all subcarriers.

The spectrum management problem defines the objective of the network and the various constraints that are to be applied depending on the network capabilities. One example of a spectrum management problem has the following objective and limitations:

- Objective – To maximize the overall rate of the network.
- Constraints – Limited transmit power to achieve the minimum data rate while causing the least interference to other users.

Mathematically, by defining w_k as the weight factor, the problem of finding the capacity region in an interference channel can be expressed as:

$$\max_{P_k^n \geq 0, \forall n,k} \sum_{k=1}^{K} w_k \sum_{n=1}^{N} \log_2 \left(1 + \frac{P_k^n G_{k,k}^n}{\sum_{i \neq k} P_i^n G_{i,k}^n + \sigma^2} \right) \qquad \text{s.t.} \sum_n P_k^n \leq P_k^{max}. \tag{2.9}$$

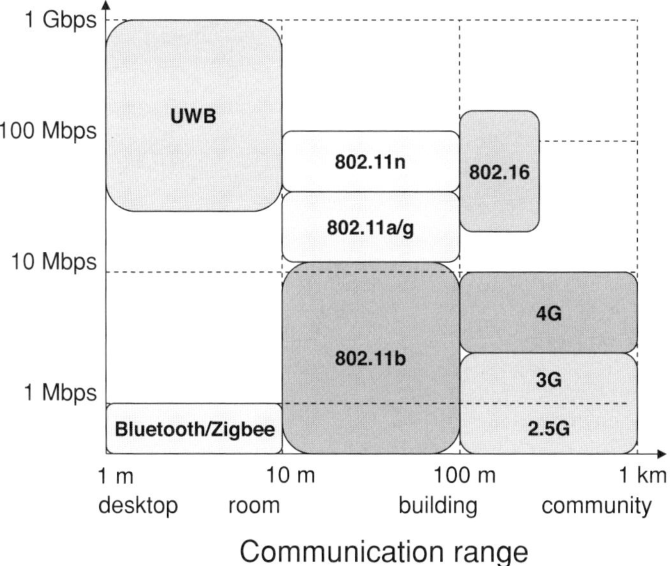

Figure 2.6 Standards comparison.

The capacity region of an interference channel is still an open problem. Once the goals of the network have been tied down, there are various algorithms proposed in the literature (such as Iterative Water-filling [2], Optimal Spectral Balancing [3], Iterative Spectral Balancing [4], SCALE [5], Autonomous Spectral Balancing [6], and Band Preference [7, 8]), all of which try to achieve the maximum capacity region possible, while adhering to the constraints of maximum transmitter power and minimum target rate of each user.

2.2 Categorization of wireless networks

We list different standards in Figure 2.6 and Figure 2.7 for different communication rates and different communication ranges. Those standards will fit different needs of various applications, and we will also discuss the techniques that can utilize the multiple standards in different situations, so that the connection can be made anytime, anywhere. In the following, we categorize wireless networks and provide some specifics.

2.2.1 3G cellular networks and beyond

Third-generation (3G) mobile communication systems based on the wideband code-division multiple access (WCDMA) and CDMA2000 radio access technologies have seen widespread deployment around the world. The applications supported by these commercial systems range from circuit-switch services such as voice and video telephony to packet-switched services such as video streaming, email, and file transfer. As

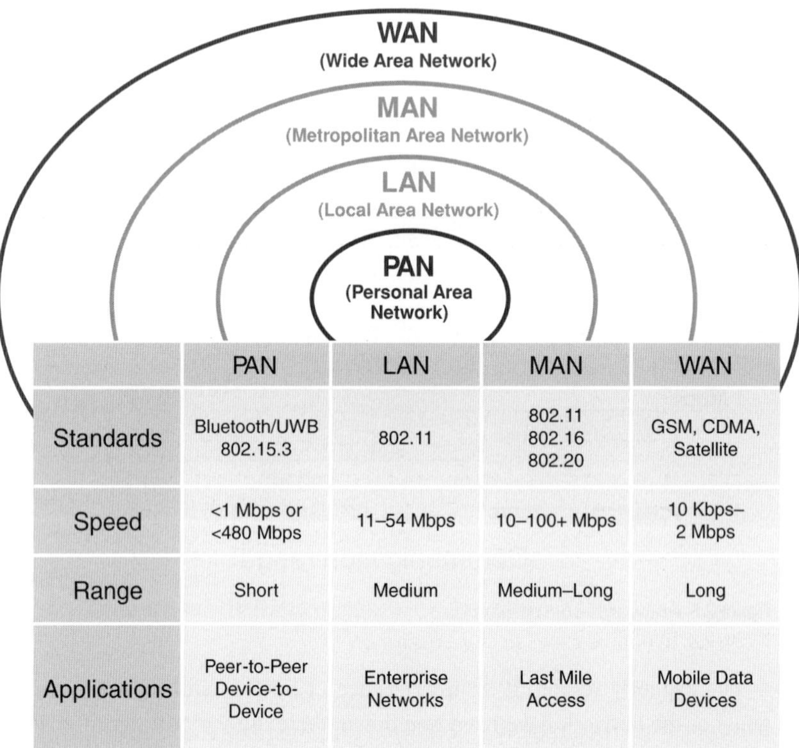

	PAN	LAN	MAN	WAN
Standards	Bluetooth/UWB 802.15.3	802.11	802.11 802.16 802.20	GSM, CDMA, Satellite
Speed	<1 Mbps or <480 Mbps	11–54 Mbps	10–100+ Mbps	10 Kbps– 2 Mbps
Range	Short	Medium	Medium–Long	Long
Applications	Peer-to-Peer Device-to-Device	Enterprise Networks	Last Mile Access	Mobile Data Devices

Figure 2.7 Comparison of different wireless networks.

more packet-based applications are developed and put into service, there is increased need for better support for different quality of service (QoS), higher spectral efficiency, and higher data rate for packet-switched services, in order to further enhance user experience while maintaining efficient use of the system resources. This need has resulted in the evolution of 3G standards as shown in Figure 2.8. For 3G cellular systems, there are two camps: 3G Partnership Project (3GPP) [9] and 3G Partnership Project 2 (3GPP2) [10], which is based on different 2G technologies.

The development of 3G will follow a few key trends, and the evolution following these trends will continue as long as the physical limitations or backward compatibility requirements do not force the development to move from evolution to revolution. The key trends include:

- Voice services will continue to be important in the foreseeable future, which means that capacity optimization for voice services will continue.
- Along with increasing use of IP-based applications, the importance of data as well as simultaneous voice and data will increase.
- Increased need for data means that the efficiency of data services needs to be improved.
- When more and more attractive multimedia terminals emerge in the markets, the usage of such terminals will spread from office, homes, and airports to roads, and finally

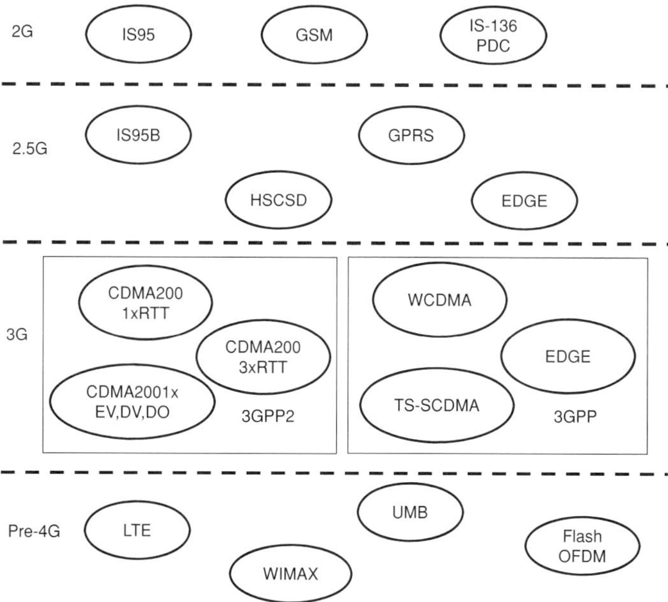

Figure 2.8 Evolution of 2G to 3G cellular networks.

everywhere. This means that high-quality high-data-rate applications will be needed everywhere as well.

- When the volume of data increases, the cost per transmitted bit needs to decrease in order to make new services and applications affordable for everybody. The other current trend is that in the 3G evolution path very high data rates are achieved in hot spots with WLAN rather than cellular-based standards.

CDMA2000

The CDMA family of cellular networks grew out of work undertaken by Qualcomm, a California-based company. Working on direct sequence spread spectrum techniques, by using different spreading codes, a large number of users could occupy the same channel at the same time, which could provide a multiple-access scheme for cellular telecommunications. The first standard was the IS-95 and the first network was launched in Hong Kong in 1996 under the brand name CDMAOne.

The CDMA system also has the following standards in its developmental stages as shown in Figure 2.8: IS-95, IS-95A, IS-95B, and CDMA2000 (1x/EV-DO, 1xEV-DV, 1xRTT, and 3xRTT). The first version of system standard IS-95 has never been launched for commercial purposes due to its prematurity. The IS-95A has been applied for businesses since then and is still used widely nowadays. IS-95B was a short version since the CDMA2000 standard was announced six months after it came out. The original IS-95A standard only allowed for circuit-switched data at 14.4 kbit/s, and IS-95B provided up to 64 kbit/s data rates as well as a number of additional services. A major step improvement came later with the development of 3G services. The first 3G standard was known as CDMA2000 1x, which initially provided data rates up to 144 kbit/s. With further developments, the systems hold the promise of allowing a maximum data rate of 307 kbits/s.

WCDMA/UMTS

WCDMA was developed by NTT DoCoMo as the air interface for their 3G network FOMA. Later, NTT DoCoMo submitted the specification to the International Telecommunication Union (ITU) as a candidate for the international 3G standard known as IMT-2000. The ITU eventually accepted WCDMA as part of the IMT-2000 family of 3G standards, as an alternative to CDMA2000, EDGE, and the short-range DECT system. Later, WCDMA was selected as the air interface for the Universal Mobile Telecommunications System (UMTS), the 3G successor to GSM.

TD-SCDMA

Transmit diversity (TD) is one of the key contributing technologies to defining the ITU-endorsed 3G systems WCDMA and CDMA2000. Spatial diversity is introduced into the signal by transmitting through multiple antennas. The antennas are spaced far enough apart[1] that the signals emanating from them can be assumed to undergo independent fading. In addition to diversity gain, antenna gain can also be incorporated through channel state feedback. This leads to the categorization of TD methods into open-loop and closed-loop methods. Several methods of transmit diversity in the forward link have been either under consideration or adopted for the various 3G standards.

4G and beyond

Looking at the development of the internet and applications, it is clear that the complexity of the transferred content is rapidly increasing and will increase further in the future. Generally, it can be said that the more bandwidth that is available, the more bandwidth applications will consume. In order to justify the need for a new air interface, goals need to be set high enough to ensure that the system will be able to serve us long into the future. A reasonable approach would be to aim at 100 Mb/s full-mobility wide-area coverage and 1 Gb/s low-mobility local-area coverage with a next-generation cellular system. Also, the future application and service requirements will bring new requirements to the air interface and new emphasis on air-interface design. One such issue, which already strongly impacts the 3G revolution, is the need to support IP and IP-based multimedia. If both technology and spectrum to meet such requirements cannot be found, the whole discussion of 4G may become obsolete. In Table 2.1, we compare key parameters of 4G with 3G. There are also some pre-4G systems such as long-term evolution (LTE), WiMAX, Ultra-Mobile Broadband (UMB, formerly EV-DO Rev. C), and Flash-OFDM, as shown in Figure 2.8.

[1] A typical spacing is half the wavelength.

Table 2.1 Comparison of 3G and 4G

	3G	4G
Major Requirement	Voice Driven	Data/Voice
Driving Architecture	Data Add On	Over IP
Network Architecture	Wide Area Cell Based	Hybrid with WiFi and WPAN
Speed	384 kbps to 2 Mbps	20 to 10 Mbps
Frequency Band	1.8–2.4 GHz	2–8 GHz
Bandwidth	1.25–5–20 MHz	100 MHz
Switching Design	Circuit and Packet	Packet
Access	DS-CDMA	OFDM/MC-CDMA
FEC	Convolution/Turbo Code	Concatenated Coding
Component Design	Antenna, Multiband Adapter	Smart Antennas, Software Radios

2.2.2 WiMAX networks

Wireless Metropolitan Area Network (WMAN) technology is a relatively new field that began in 1998. From that time, a new standard has emerged to handle the implementation, IEEE 802.16. The equivalent of 802.16 in Europe is HIPERMAN. The WiMAX Forum is working to ensure that 802.16 and HIPERMAN interoperate seamlessly. This standard has helped to pave the way for WMAN technology globally, and since its first inception it has received six expansions onto the standards. WMAN differs from other wireless technologies in that it is designed for a broader audience, such as a large corporation or an entire city.

The 802.16 MAC uses a scheduling algorithm for which the subscriber station needs to compete once (for initial entry into the network). After the competition, the subscribe station is allocated an access slot by the base station. The time-slot can enlarge and contract, but it remains assigned to the subscriber stations, which means that other subscribers cannot use it. The 802.16 scheduling algorithm is stable under overload and oversubscription (unlike 802.11). It can also be more bandwidth efficient. The scheduling algorithm also allows the base station to control QoS parameters by balancing the time-slot assignments among the application needs of the subscriber stations. Moreover, the MAC layer is also in charge of protocol data unit (PDU) assembly and disassembly. A detailed illustration for different layers' protocols of 802.16 is shown in Figure 2.9.

The operation standards for WMANs are regulated under IEEE standard 802.16 [11], and the WMANs are allowed the operating frequency range of 10–66 GHz. With such a broad spectrum to work with, WMANs have the ability to transmit over previous wireless frequencies such as IEEE 802.11b/g, causing less interference with other wireless products. The only downside to using such high frequencies is that WMAN needs a line of sight between the transmitters and receivers, much like a directional antenna. Line of sight, however, decreases multipath distortion, allowing higher bandwidths to be achieved, and it can attain up to 75 Mbps for both uplink and downlink on a

Table 2.2 Comparison of 802.16 Standards

	802.16	802.16a/802.16d	802.16e
Date	Dec. 2001	Jan. 2003/Q3 2004	Q3, 2004
Spectrum	10–66 GHz	<11 GHz	<6 GHz
Channels	Line of sight only	Non-line of sight	Non-line of sight
Modulation	QPSK,16QAM, 64QAM	OFDM256, QPSK, 16QAM, 64QAM	Same as 802.11a
Mobility	Fixed	Fixed	Pedestrian Mobility regional roaming
Bandwidth	20, 25 or 28 MHz	1.25–20 MHz	Same as 802.16a
Throughput	Up to 75 Mbps	Up to 75 Mbps	Up to 30 Mbps
Cell radius	1–3 miles	3–5 miles	1–3 miles

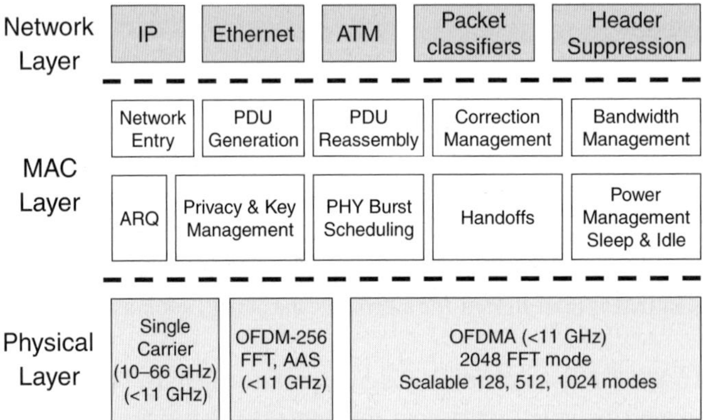

Figure 2.9 WIMAX protocol stacks.

single channel [12]. Some extensions of 802.16 standards are listed as follows and in Table 2.2.

- IEEE 802.16a: The IEEE has developed 802.16a for use in licensed and license-exempt frequencies from 2 to 11 GHz. Most commercial interest in IEEE 802.16 is in these lower frequency ranges. In the lower ranges, the signals can penetrate barriers and, thus, do not require a line of sight between transceiver and antenna. This enables more flexible WiMax implementations while maintaining the technology's data rate and transmission range. IEEE 802.16a supports mesh deployment, in which transceivers can pass a single communication on to other transceivers, thereby extending the basic 802.16s transmission range.
- IEEE 802.16b: This extension increases the spectrum the technology can use in the 5 and 6 GHz frequency bands and improves quality of service. WiMax provides QoS to ensure priority transmission for real-time voice and video and to offer differentiated service levels for different traffic types.

- IEEE 802.16c: IEEE 802.16c represents a 10 to 66 GHz system profile that standardizes more details of the technology. This encourages more consistent implementation and, therefore, interoperability.
- IEEE 802.16d: IEEE 802.16d includes minor improvements and fixes to 802.16a. This extension also creates system profiles for compliance testing of 802.16a devices.
- IEEE 802.16e: This technology will standardize networking between carriers' base stations and mobile devices, rather than just between base stations and fixed recipients. IEEE 802.16e would enable the high-speed signal handoffs necessary for communications with users moving at vehicular speeds.

In addition to IEEE 802.16, IEEE 802.20 (IEEE802.20) and the Mobile Broadband Wireless Access (MBWA) Working Group aim to prepare a formal specification for a packet-based air interface designed for IP-based services. The goal is to create an interface that will allow the creation of low-cost, always-on, and truly mobile broadband wireless networks, nicknamed Mobile-Fi. IEEE 802.20 will be specified according to a layered architecture, which is consistent with other IEEE 802 specifications. The scope of the working group consists of the following layers: the physical (PHY), medium access control (MAC), and logical link control (LLC) layers. The air interface will operate in bands below 3.5 GHz and with a peak data rate of more than 1 Mbit/s. The goals of 802.20 and 802.16e, the so-called mobile WiMAX, are similar.

WiMAX can be viewed as "last mile" connectivity at high data rates. This could result in lower pricing for both home and business customers as competition lowers prices. In areas without pre-existing physical cable or telephone networks, WiMAX may be a viable alternative for broadband access that has been economically unavailable. Prior to WiMAX, many operators have been using proprietary, fixed wireless technologies for broadband services. For this reason, WiMAX has its significant markets in rural areas and developing countries.

2.2.3 WiFi networks

IEEE 802.11 denotes a set of Wireless Local Area Network (WLAN) standards developed by working group 11 of the IEEE LAN/MAN Standards Committee (IEEE 802). WiFi is a brand originally licensed by the WiFi Alliance to describe the underlying technology of WLAN based on the IEEE 802.11 specifications. It was developed to be used for mobile computing devices, such as laptops, in LANs, but it is now increasingly used for more services, including internet and VoIP phone access, gaming, and basic connectivity of consumer electronics such as televisions, DVD players, and digital cameras.

In the physical layer, 802.11b operates within the 2.4 GHz industrial, scientific, and medical (ISM) band. The original 802.11b defines data rates of 1 Mbps and 2 Mbps via radio waves using frequency hopping spread spectrum (FHSS) or direct sequence spread spectrum (DSSS). For FHSS, the 2.4 GHz band is divided into 75 1-MHz subchannels. The sender and receiver agree on a hopping pattern, and data are sent over a sequence of the subchannels. Each conversation within the 802.11 network occurs over a different

Table 2.3 802.11b rates

Data Rate	Code Length	Modulation	Symbol Rate	Bits/Symbol
1 Mbps	11(DSSS)	BPSK	1 MSps	1
2 Mbps	11(DSSS)	QPSK	1 MSps	2
5.5 Mbps	8(CCK)	QPSK	1.375 MSps	4
11 Mbps	8(CCK)	QPSK	1.375 MSps	8

hopping pattern. Because of Federal Communications Commission (FCC) regulations that restrict subchannel bandwidth to 1 MHz, FHSS techniques are limited to speeds of no higher than 2 Mbps. DSSS divides the 2.4 GHz band into 14 22-MHz channels. Adjacent channels overlap one another partially, with three of the 14 being completely non-overlapping. The spreading code is an 11-bit Barker sequence. Binary Phase Shift Keying (BPSK) and Quadrature Phase Shift Keying (QPSK) are used to provide different rates.

To increase the data rate to 5.5 Mbps and 11 Mpbs in the 802.11b standard, an advanced coding technique, Complementary Code Keying (CCK), is employed. A complementary code contains a pair of finite bit sequences of equal length, in which the number of pairs of identical elements (1 or 0) with any given separation in one sequence is equal to the number of pairs of unlike elements having the same separation in the other sequence. A network using CCK can transfer more data per unit time for a given signal bandwidth than a network using the Barker code, because CCK makes more efficient use of the bit sequences. CCK consists of a set of 64 8-bit code words. The 5.5 Mbps rate uses CCK to encode 4 bits per carrier, while the 11 Mbps rate encodes 8 bits per carrier. Both speeds use QPSK as the modulation technique and signal at 1.375 MSps. Table 2.3 shows the different rates for 802.11b.

Standard 802.11a adopts orthogonal frequency division multiplexing (OFDM) at 5.15–5.25 GHz, 5.25–5.35 GHz, and 5.725–5.825 GHz to support multiple data rates up to 54 Mbps. Standard 802.11g utilizes the 2.4 GHz band with OFDM modulation and is also backward compatible with 802.11b. For OFDM, the FFT has 64 subcarriers. There are 48 data subcarriers and 4 carrier pilot subcarriers, for a total of 52 non-zero subcarriers defined in IEEE 802.11a, plus 12 guard subcarriers. The IEEE 802.11a/g physical layer provides eight PHY modes with different modulation schemes and different convolutional coding rates, and it can offer various data rates.

To achieve higher data rates in the PHY layer, in January 2004, IEEE announced that it had formed a new 802.11 Task Group (TGn) to develop a new amendment to the 802.11 standard for wireless local-area networks. 802.11n builds upon previous 802.11 standards by adding MIMO (multiple-input multiple-output). MIMO uses multiple transmitter and receiver antennas to allow for increased data throughput through spatial multiplexing and increased range by exploiting the spatial diversity. There are several proposal groups named TGnSync, WWiSE (short for "World-Wide Spectrum Efficiency"), and MITMOT ("Mac and mImo Technologies for More Throughput"). All proposals occupy frequency band 2.5 GHz with 20 MHz or 40 MHz bandwidth so as to support the communication speed of more than 200 Mbps. 802.11n is backward

Table 2.4 Comparison of 802.11 Standards

	802.11b	802.11a/g	802.11n	802.11ac
Air Rate	11 Mbps	54 Mbps	200+ Mbps	433 Mbps–6.93 Gbps
Range	30 m	30 m	50 m	
Frequency	2.4 G	5.25, 5.6, 5.8 G/2.4 G	2.4 G	5 G band
Bandwidth	20 M	20 M	20 M or 40 M	80 M/160 M
Modulation	DSSS/CCK	DSSS/CCK/OFDM	DSSS/CCK/OFDM with MIMO	256 QAM Multi-user MIMO

compatible with 802.11b and 802.11g. Recently, the 802.11ac standard has been under development. In Table 2.4, we compare various parameters of the 802.11 standards.

In order to take full advantage of the future market opportunity for WiFi, several key challenges must be overcome. In the following, we list some near-future design topics and their possible solutions.

- Security
  Most concentration for WiFi is on the free public access. However, the eavesdroppers and hackers can fully take advantage of the WiFi system. Currently, all 802.11a, b, and g devices support WEP (Wired Equivalent Privacy) encryption, which contains flaws.

  IEEE 802.11i, also known as WiFi Protected Access 2 (WPA2), is an amendment to the 802.11 standard specifying security mechanisms for wireless networks. The 802.11i specification defines two classes of security algorithms: Robust Security Network Association (RSNA), and Pre-RSNA. Pre-RSNA security consists of Wired Equivalent Privacy (WEP) and 802.11 entity authentication. RSNA provides two data confidentiality protocols, called the Temporal Key Integrity Protocol (TKIP) and the Counter-mode/CBC-MAC Protocol (CCMP). The RSNA establishment procedure includes 802.1X authentication and key management protocols. Beyond IEEE 802.11i, it is worth mentioning that WAPI (WLAN Authentication and Privacy Infrastructure) is a Chinese National Standard for Wireless LAN (GB 15629.11-2003).
- Mobility
  Mobility is an important attribute of wireless networks. Current Wireless LAN standards provide mobility through roaming capabilities. IEEE 802.11p also referred to as Wireless Access for the Vehicular Environment (WAVE) defines enhancements to 802.11 required to support Intelligent Transportation Systems (ITS) applications. This includes data exchange between high-speed vehicles and between the vehicles and the roadside infrastructure in the licensed ITS band of 5.9 GHz (5.85–5.925 GHz). 802.11p will be used as the groundwork for DSRC (Dedicated Short–Range Communications), a U.S. Department of Transportation project that will be emulated elsewhere looking at vehicle-based communication networks, particularly for applications such as toll collection, vehicle safety services, and commerce transactions via cars. The ultimate vision is a nationwide network that enables communications between vehicles and roadside access points or other vehicles.

- QoS support
  802.11e is the first wireless standard that spans home and business environments. It adds quality-of-service (QoS) features and multimedia support to the existing 802.11 wireless standards, while maintaining full backward compatibility with these standards. QoS and multimedia support are critical to wireless home networks where voice, video, and audio will be delivered. Broadband service providers view QoS and multimedia-capable home networks as an essential ingredient for offering residential customers video on demand, audio on demand, voice over IP, and high-speed internet access.

  802.11e introduces two enhancements, Enhanced DCF (EDCF) and Hybrid Coordination Function (HCF). In EDCF, a station with high-priority traffic waits a little less before it sends its packet, on average, than a station with low-priority traffic, so that high-priority traffic has a higher chance of being sent than low-priority traffic. In addition, each priority level is assigned a Transmit Opportunity (TXOP), which is a bounded time interval during which a station can send as many frames as possible. HCF works more like PCF. With the PCF, QoS can be configured with great precision. QoS-enabled stations have the ability to request specific transmission parameters (data rate, jitter, etc.), which should allow advanced applications like VoIP and video streaming to work more effectively on a WiFi network.
- Integration of 3G and WLAN
  The third-generation cellular networks and 802.11 local area wireless networks possess complementary characteristics. 3G cellular networks promise to offer always-on, ubiquitous connectivity, and mobility with relatively low data rates. 802.11 offers much higher data rates, comparable to the cellular networks, but can cover only smaller areas without mobility, suitable for hot-spot applications in hotels and airports. The performance and flexibility of wireless data services would be dramatically improved if users could seamlessly roam across the two networks. By offering integrated 802.11/3G services, 3G operators and Wireless Internet Service Providers (WISP) can attract a wider user base and ultimately facilitate the ubiquitous introduction of high-speed wireless services. Users can also benefit from the enhanced performance and lower overall cost of such a combined service. For a network node changing the type of connectivity between 3G cellular phone and WLAN, the concept of vertical handoff will be discussed in the later chapters.

2.2.4 Wireless personal area networks

A wireless personal area network (WPAN) is a computer network used for wireless communication among devices (including telephones and personal digital assistants) close to a person. The reach of a WPAN is typically a few meters. WPANs can be used for communication among the personal devices themselves (intrapersonal communication), or for connecting to a higher-level network and the internet (an uplink). 802.15 is a communications specification that was approved in early 2002 by the IEEE Standards Association (IEEE-SA) for wireless personal area networks (WPANs). Specifically, we list the following three sub-standards:

1. The IEEE Standard 802.15.1 was approved as a new standard for Bluetooth by the IEEE-SA Standards Board on April 15, 2002. The Bluetooth standard enables wireless communication between multiple electronic devices within 10 meters of each other. Bluetooth devices are organized in piconets, which include one master device and up to seven slave devices. The Bluetooth devices communicate in the 2.4 GHz radio frequency band, enabling devices to communicate without line-of-sight spacing, such as through walls or through a person's body. Bluetooth piconets utilize frequency-hopping spread spectrum in 79 1-MHz bands, reducing the likelihood of interference with other Bluetooth piconets.
2. 802.15.3 is the IEEE standard for high data rate WPAN designed to provide Quality of Service (QoS) for real-time distribution of multimedia content, like video and music. It is ideally suited for a home multimedia wireless network. The original standard uses a "traditional" carrier-based 2.4 GHz radio as the physical layer (PHY). A follow-on standard, 802.15.3a, defines an alternative PHY, based on UWB, that will provide in excess of 110 Mbps at a 10 m distance and 480 Mbps at 2 m. This will allow applications requiring streaming of high-definition video between media servers and flat-screen HD monitors and extremely fast transfer of media files between media servers and portable media devices.
3. IEEE 802.15.4-2003 (Low Rate WPAN) deals with low data rate but very long battery life (months or even years) and very low complexity. The first edition of the 802.15.4 standard was released in May 2003. In March 2004, after forming Task Group 4b, Task Group 4 put itself in hibernation. The ZigBee set of high-level communication protocols is based upon the specification produced by the IEEE 802.15.4 task group.

Bluetooth/ZigBee

Bluetooth is a standard for wireless communications that uses short-range radio frequencies to enable communications among multiple electronic devices. Bluetooth technology is envisioned as a replacement for the interconnection cables between personal devices such as notebook computers, cellular phones, personal digital assistants, and digital cameras. If widely adopted, Bluetooth would enable a uniform interface for accessing data services. Thus, calendars, address books, and business cards stored in personal devices could be automatically synchronized using push-button synchronization and proximity operation.

The Bluetooth Special Interest Group was founded by Ericsson, IBM, Intel, Nokia, and Toshiba in February 1998 to develop an open specification for short-range wireless connectivity. The Special Interest Group offered all of the intellectual property explicitly included in the Bluetooth specification royalty-free to adopter members to facilitate the widespread acceptance of the technology. The Special Interest Group now includes thousands of companies. To use the intellectual property in the Bluetooth specification, adopter members must qualify any Bluetooth products they intend to bring to market through the Bluetooth qualification program. The Bluetooth qualification program includes radio and protocol conformance testing, profile conformance testing, and interoperability testing.

Bluetooth is a radio-frequency technology utilizing the unlicensed 2.4 GHz ISM band. It enables wireless connections up to 10 meters under standard transmitter power, and due to the use of radio frequencies, devices need not be within line of sight of each other and may connect through walls or other non-metal objects. In active mode, Bluetooth devices typically consume 0.1 W of active power for class 1 with range of 100 m, 2.5 mW for class 2 with range 10 m, and 1 mW for class 3 with range 1 m. The modulation technique utilized in Bluetooth technology is binary Gaussian frequency shift-keying, and the baud rate is 1 Msymbol/s. Thus, the bit time is 1 μs and the raw transmission speed is 1 Mb/s.

The baseband signals used in Bluetooth devices, which are typically 1 MHz in bandwidth, cannot be directly transmitted on the wireless medium. Modulation of the 1 MHz baseband signals into the 2.4 GHz band is difficult to achieve in one step because CMOS transistors do not operate at these frequencies. Bluetooth radio devices solve this problem by modulating the baseband signal onto an intermediate frequency, such as 3 MHz, and then they use a frequency mixer to increase the frequency of the signal to the 2.4 GHz band.

Because the unlicensed ISM band in which Bluetooth operates is often cluttered with signals from other devices, such as garage-door openers, baby monitors, cordless phones, and microwave ovens, Bluetooth utilizes frequency hopping spread spectrum for security and to avoid interference with the signals from other devices. Frequency hopping also allows multiple piconets to exist within range of each other with minimal interference. Frequency hopping typically involves generating a frequency shift-keyed signal, and then shifting the frequency of the frequency shift-keyed signal by an amount determined by a pseudonoise code. The pseudonoise code is random in that it appears to be unpredictable to an outsider, but it is generated by deterministic means. The pseudonoise code is unique to the piconet, and is determined by the master device.

Bluetooth utilizes a slow hopping scheme, hopping in a pseudorandom fashion through 79 1-MHz channels. The frequency channels are located at $(2,402 + k)$ MHz, with $k = 0, 1, \ldots, 78$. A Bluetooth piconet hops through 1600 different frequencies per second. Each frequency hop corresponds to one slot, with each slot lasting $1/1600 =$ 625 μs. Each packet may be one, three, or five slots long. A frame consists of two packets, one packet being a transmit packet and the other being a receive packet.

A packet consists of an access code, a header, and a payload. The access code is 72 bits long and is used for clock synchronization, DC offset compensation, identification, and signaling. The header is 54 bits long and is used for addressing, identifying the packet type, controlling flow, sequencing to filter retransmitted packets, and verifying header integrity (ensuring that the header was not altered by another source). The payload is between zero and 2744 bits, depending on the type of packet. In packets that are one slot long, the payload is 240 bits long. In three-slots-long packets, the payload is 1500 bits long. In packets that are five slots long, the payload is 2744 bits long.

Each Bluetooth device includes a unique IEEE-type 48-bit address, called a Bluetooth device address, which is assigned to each Bluetooth device at the manufacture, and a 28-bit clock. The clock ticks once every 312.5 μs, which corresponds to half the residence time in a frequency band when the radio hops at the rate of 1600 hops per second.

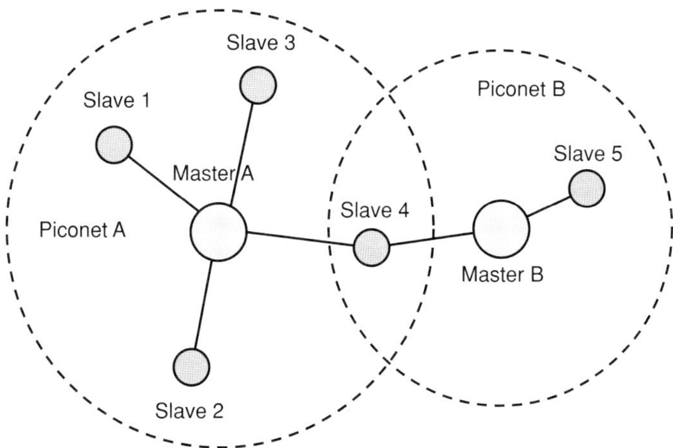

Figure 2.10 Piconet.

Bluetooth devices that are in communication with one another are organized into groups of two to eight devices called piconets as shown in Figure 2.10. A piconet consists of a single master device and between one and seven slave devices. A device may belong to more than one piconet, but it can be the master in no more than one piconet; thus, a device may be a slave in two piconets or a master in one piconet and a slave in another piconet.

The slaves utilize the Bluetooth clock of the master to maintain time synchronization. The pseudorandom hopping sequence is determined by the 48-bit Bluetooth device address of the master. The Bluetooth clock of the master clock determines the phase in the hopping pattern, thereby determining the particular frequency to be used in a particular time-slot. Thus, the communications channel in a particular piconet is fully identified by the master, and this communications channel serves to distinguish one piconet from another.

The master and the slaves alternate transmit opportunities for transmitting according to a time-division duplexing scheme. According to this scheme, the master transmits on even-numbered time-slots, as defined by the master's Bluetooth clock, while the slaves transmit on odd-numbered slots. A given slave may transmit only if the master has just transmitted to this slave.

In order to determine the presence and identities of other Bluetooth devices, Bluetooth devices engage in inquiry and page processes. The inquiry process is performed without knowledge of the identity or presence of other Bluetooth devices, whereas the paging process is performed with knowledge of the identity and presence of other Bluetooth devices.

During an inquiry process, a prospective master device makes its presence known by transmitting inquiry messages. Devices that are searching for inquiry messages respond with inquiry messages that contain their Bluetooth device addresses. After the master has acquired knowledge of the Bluetooth device address and presence of other Bluetooth devices within range, the master explicitly pages the other Bluetooth devices to join its piconet. Devices responding to the page will provide additional information, such as their clock phases, to the master.

Bluetooth devices have three low-power modes in which they reside when they are not in active communication. In sniff mode, a slave agrees with its master to listen for master transmissions periodically. In hold mode, a device agrees with another device in the piconet to remain silent for a given amount of time. A device that has gone into hold mode does not relinquish its temporary member address within the piconet. In park mode, a slave agrees with its master to park until further notice. In park mode, the slave device relinquishes its temporary member address within the piconet, and periodically listens to transmissions from the master. The slave may be invited back to active communications by the master, or may send a request to the master to be unparked.

Bluetooth devices typically provide link layer security between any two Bluetooth radios. A challenge/response system, such as an E1 algorithm [13], is used for authentication. The authentication is based on a link key, which is a 128-bit shared secret between the two Bluetooth devices. The link key is generated by a challenge and response process between the two Bluetooth devices. Data sent between two Bluetooth devices may also be encrypted, and may be ciphered with an E0 algorithm [13]. An encryption key may be between 8 and 128 bits long, and may be derived from the link key. The Bluetooth devices may use a configuration encryption key 0 to 16 bytes in length for key management and usage. The authentication and encryption keys may be generated with E2-E3 algorithms [13].

The specifications were formalized by the Bluetooth Special Interest Group (SIG). The SIG was formally announced on May 20, 1998. It was established by Ericsson, Sony Ericsson, IBM, Intel, Toshiba, and Nokia, and later joined by many other companies as Associate or Adopter members. Bluetooth is also known as IEEE 802.15.1.

For WPAN, besides Bluetooth technology, ZigBee is the name of a specification for a suite of high-level communication protocols using small, low-power digital radios based on the IEEE 802.15.4 standard for wireless personal area networks (WPANs). ZigBee operates in the ISM radio bands: 868 MHz in Europe, 915 MHz in the USA, and 2.4 GHz in most jurisdictions worldwide. The technology is intended to be simpler and cheaper than other WPANs such as Bluetooth. The specification supports data transmission rates of up to 250 kbps at a range of up to 30 m. ZigBee's technology is slower than 802.11b (11 Mbps) and Bluetooth (1 Mbps), but it consumes significantly less power.

Ultra-wideband

Ultra-wideband (UWB) is a technology for transmitting information spread over a large bandwidth (>500 MHz) that is able to share spectrum with other users. In 2002, the FCC authorized the unlicensed use of UWB in 3.1–10.6 GHz. The intention is to provide an efficient use of scarce radio bandwidth while enabling high data rate personal-area network (PAN) wireless connectivity. Deliberations in the International Telecommunication Union Radiocommunication Sector (ITU-R) resulted in a Report and Recommendation on UWB in November, 2005.

The FCC power spectral density emission limit for UWB emitters operating in the UWB band is –41.3 dBm/MHz. This is the same limit that applies to unintentional

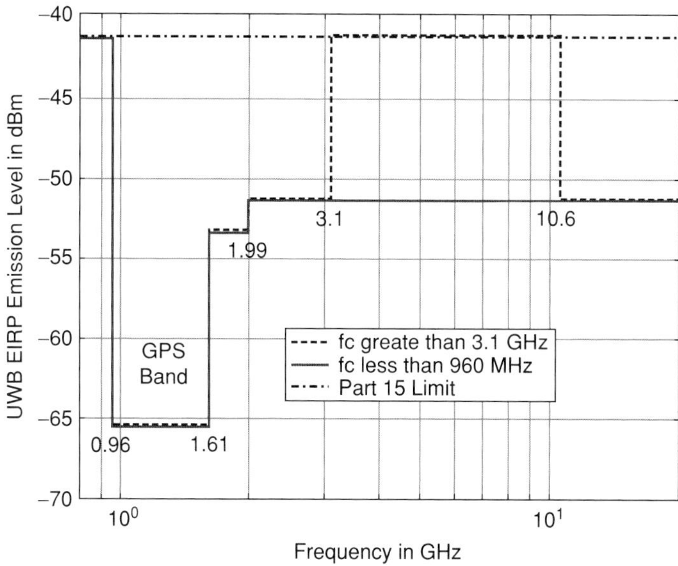

Figure 2.11 FCC UWB mask.

emitters in the UWB band, the so-called Part 15 limit [14]. However, the emission limit for UWB emitters can be significantly lower (as low as –75 dBm/MHz) in other segments of the spectrum to prevent interferences with other applications such as GPS. The FCC UWB spectrum mask is shown in Figure 2.11.

The ability of UWB technology to provide significantly high data rates within short ranges has made it an excellent alternative for Bluetooth for the physical layer of the IEEE 802.15.3a standard for wireless personal area networks (WPANs). However, like 802.11 standards, two opposing groups of UWB developers are competing over the IEEE standard. The two competing technologies are single-band UWB and multiband UWB. The single-band technique, backed by Motorola/XtremeSpectrum, supports the idea of impulse radio that occupies a wide spectrum. The multiband approach divides the available UWB frequency spectrum into multiple smaller and non-overlapping bands with bandwidths greater than 500 MHz to obey the FCC's definition of UWB signals. The multiband approach is supported by several companies, including Staccato Communications, Intel, Texas Instruments, General Atomics, and Time Domain Corporation.

For the single-band UWB, the most popular proposal, Direct Sequence (DS)-UWB, uses a combination of a single-carrier spread-spectrum design and wide coherent bandwidth. Unlike conventional wireless systems, which use narrowband modulated carrier waves to transmit information, DS-UWB transmits data by pulses of energy generated at very high rates: in excess of 1 billion pulses per second, providing support for data rates of 28 M, 55 M, 110 M, 220 M, 500 M, 660 M and 1320 M bit/s. A fixed UWB chip rate in conjunction with variable-length spreading code words enables this scalable support.

For the Multiband UWB, as shown in Figure 2.12, the available UWB spectrum, from 3.1 GHz to 10.6 GHz, is divided into $S = 14$ subbands. Each subband occupies a

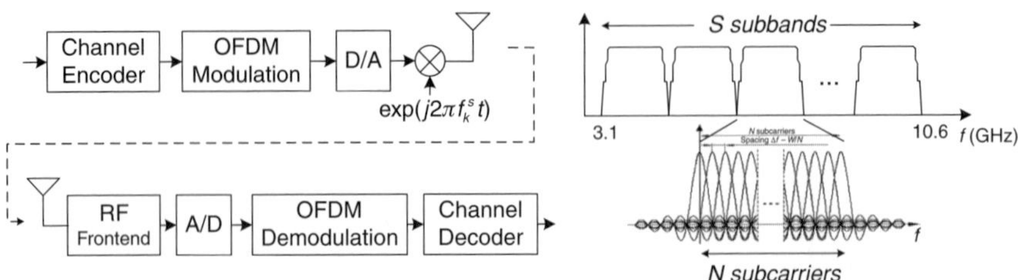

Figure 2.12 Multiband UWB system.

bandwidth of at least 500 MHz in compliance with the FCC regulations. The UWB system employs OFDM with $N = 128$ subcarriers, which are modulated using quadrature phase shift keying (QPSK). At each OFDM symbol period, the modulated symbol is transmitted over one of the S subbands. These symbols are time-interleaved across subbands. Different bit rates are achieved by using different channel-coding, frequency-spreading, or time-spreading rates. The frequency-domain spreading is obtained by choosing conjugate symmetric inputs to the IFFT, while the time-domain spreading is achieved by repeating the same information in an OFDM symbol on two different subbands [15]. The receiver combines the information transmitted via different times or frequencies to increase the signal-to-noise ratio (SNR) of received data.

Because of the extremely low emission levels currently allowed by regulatory agencies, UWB systems tend to be short range and high speed. High data rate UWB can enable wireless monitors, the efficient transfer of data from digital camcorders, wireless printing of digital pictures from a camera without the need for an intervening personal computer, and the transfer of files among cell phone handsets and other handheld devices like personal digital audio and video players. UWB is also used in "see-through-the-wall" precision radar imaging technology, precision positioning and tracking (using distance measurements between radios), and precision time-of-arrival-based localization approaches.

2.2.5 Wireless ad hoc networks

An ad hoc network is an autonomous collection of mobile users that communicate over bandwidth-constrained wireless links. The network is decentralized, so that all network activity that includes discovering the topology and delivering messages must be executed by the nodes themselves. Ad hoc networks need efficient distributed algorithms to determine network organization, link scheduling, and routing. For a special case of ad hoc network, Mobile Ad Hoc Networks (MANETs), since the nodes are mobile, the network topology may change rapidly and unpredictably over time.

The first generation of ad hoc networks was initiated in the early 1970s, when Packet Radio Networks (PRNET) were proposed from the Defense Advanced Research Projects Agency (DARPA) for multihop networks in a combat environment and Areal Locations of Hazardous Atmospheres (Aloha) was proposed in Hawai'i for distributed channel access management. The second generation of ad hoc networks emerged in the 1980s, when the ad hoc network systems were further enhanced and implemented as a part of the

Survivable Adaptive Radio Networks (SURAN) program. SURAN provided a packet-switched network to the mobile battlefield in an environment without infrastructure, so as to be beneficial in improving the performance of radios by making them smaller, cheaper, and resilient to electronic attacks. In the 1990s, the concept of commercial ad hoc networks arrived with notebook computers and other viable communications equipment. For example, the IEEE 802.11 subcommittee had adopted the term "ad hoc networks."

The advantages of ad hoc networks are the easiness and speed of deployment, which is an important requirement for military applications. For civil applications, ad hoc networks decrease dependence on expensive infrastructure. The set of applications for ad hoc networks is diverse, ranging from small, static networks that are constrained by power sources, to large-scale, mobile, highly dynamic networks. Some typical applications are personal area networks, emergency operations such as policing and firefighting, civilian environments such as taxi networks, and military use on the battlefields.

In contrast to the traditional wireless network with infrastructure, an ad hoc network needs its own design requirements so as to be functional. We list some important aspects as follows:

- Distributed operation and self-organization

  No node in the ad hoc network can depend on a network in the background to support the basic functions like routing. Instead, these functions must be implemented and operated efficiently in a distributed manner. Moreover, in the event topology changes due to mobility, the network can be self-organized to adapt to the changes.

  In addition, because the ad hoc nodes might belong to different authorities, they might not be willing to cooperate to fulfill the network functions. However, this non-cooperation can cause severe network breakdown. To motivate the distributed autonomous users is an important research and design topic. Traditionally, pricing anarchy is employed using the distributed control theory. Later in this book, we study how to explore the other methods such as game theory to motivate users' cooperative behaviors.
- Dynamic routing

  For MANET, the routing problem between any pair of nodes is challenging due to the mobility of nodes. The optimal source-to-destination route is time variant. Moreover, compared to the traditional network in which the routing protocols are proactive, the ad hoc dynamic routing protocols are reactive. The routes are determined only when the source requests the transmission to the destination. There are two types of ad hoc dynamic routing protocols: Table-driven routing protocols and source-initiated on-demand routing protocols.

  The table-driven routing protocols require each node to maintain one or more tables to store routing information. The protocols rely on an underlying routing table update mechanism that involves the constant propagation of routing information. Packets can be forwarded immediately since the routes are always available. However, this type of protocol causes substantial signaling traffic and power consumption problems. Some protocols existing in the literature are destination-sequenced distance-vector routing [16], clusterhead gateway switch routing [17], and wireless routing protocol [18].

Source-initiated on-demand routing creates routing only when desired by the source node. The disadvantage is that the packet at the source node must wait until a route can be discovered. But the advantage is that periodic route updates are not required. Some of the available routing protocols in the literature are ad hoc on-demand distance vector routing [19], dynamic source routing [20], temporally ordered routing algorithm [21], associativity-based routing [22], and signal stability-based adaptive routing protocol [23].

- Connectivity

  To achieve a connected ad hoc network, for any node there must be a multihop path to any other node. There are many types of connectivity definitions. In an undirected graph G, two vertices u and v are called connected if G contains a path from u to v. Otherwise, they are called disconnected. A graph is called connected if every pair of vertices in the graph is connected.

  One of the most adopted definitions is k-connectivity, which states that each node can at least connect to the rest of network if $k-1$ of its neighbor nodes are destroyed.

Definition 1. *A graph G with edge set $V(G)$ is said to be k-connected if $G\backslash Y$ is connected for all $Y \subseteq V(G)$ with $|Y| < k$. In other words, a graph is k-connected if the graph remains connected when fewer than k vertices are deleted from the graph.*

If a graph G is k-connected, and $k < |V(G)|$, then $k \leq \Delta(G)$, where $\Delta(G)$ is the minimum degree of any vertex $v \in V(G)$.

The following theorem due to Menger [24], a special case of the max-cut min flow theorem, states how to calculate k for k-connectivity:

Theorem 1. *Let G be a finite undirected graph and i and j be two nonadjacent vertices. Then the size of the minimum vertex cut for i and j (the minimum number of vertices whose removal disconnects i and j) is equal to the maximum number of pairwise vertex-independent paths from i to j.*

- Mobility

  To test a new protocol for MANET, it is important to use a mobility model that accurately represents the mobile users using the protocol, so that the performances of practical implementation match the simulation results. There are two types of mobility models, traces and synthetic models. For traces, the mobility patterns are observed in real-life systems, so that the accurate information is provided. However, traces are limited only for an existing environment. For an unknown environment, accurate synthetic models are necessary.

  A synthetic model tries to simulate the real movement of mobile users. In [25], several mobility models are discussed and explained. If the different mobile users are moving randomly relative to one another, some mobility mobiles are listed as follows:

  1. Random Walk Mobility Model: a simple model based on random directions and speeds.
  2. Random Waypoint Mobility Model: a model that includes pause times between changes in directions and speeds.

3. Random Direction Mobility Model: a model forcing mobile users to travel to the edge of the simulation area before changing directions and speeds.
4. Boundless Simulation Area Mobility Model: a model converting a 2D rectangular simulation area into a torus-shaped simulation area.
5. Gauss–Markov Mobility Model: a model using a set of parameters to change the degree of randomness in mobility patterns.
6. Probabilistic Version of Random Walk Mobility Model: a model with a probability to determine the next position of mobile users.
7. City Section Mobility Model: a model in which movement is on the streets of a city.
8. Random Trip Mobility Model [26]: a model that contains as special cases: the random waypoint on convex or non-convex domains, random walk, billiards, city section, space graph, and other models.

If the mobile users are moved in groups, some mobility models are listed as follows:

1. Exponential Correlated Random Mobility Model: a model using motion function to create movement.
2. Column Mobility Model: a model where the mobile users form a line and are uniformly moving forward in a certain direction.
3. Nomadic Community Mobility Model: a model in which mobile users move together from one position to another.
4. Pursue Mobility Model: a model in which mobile users follow a given target.
5. Reference Point Group Mobility Model: a model in which group movements are based on the path traveled by a logical center.

- Other issues such as lower power, security, and localization will be discussed in the next section together with sensor networks.

Finally, we discuss two ad hoc standards in 802.11 for wireless local networks and 802.15 for wireless personal area networks. One future design goal of ad hoc networks is to let mobile users be able to form connections and perform basic functions. These mobile users belong to different types of networks such as cellular, WiFi, and WPAN.

Most installed wireless LANs today utilize “infrastructure” mode that requires the use of one or more access points. With this configuration, the access point provides an interface to a distribution system (e.g., Ethernet), which enables wireless users to utilize corporate servers and internet applications. As an option, the 802.11 standard specifies “ad hoc” mode, which allows the radio network interface card (NIC) to operate in what the standard refers to as an independent basic service set (IBSS) network configuration. With an IBSS, there are no access points. User devices communicate directly with one another in a peer-to-peer manner.

For wireless personal area network applications such as Bluetooth, the ad hoc network is set up by forming piconets. Within each piconet, only one master device and possibly several slave devices form connections. A slave device can belong to different piconets and serve as connection between piconets. The major functionality of the piconets are piconet forming and maintenance, packet forwarding, and intra-piconet/inter-piconet scheduling.

2.2.6 Wireless sensor networks

A wireless sensor network (WSN) is a wireless network consisting of spatially distributed autonomous devices using sensors to cooperatively monitor physical or environmental conditions, such as temperature, sound, vibration, pressure, motion, or pollutants, at different locations. The goals and tasks of sensor networks are to determine the value of some parameter at a given location, to detect the occurrence of events of interest and estimate parameters of the detected events, to classify a detected object, or to track an object. The development of wireless sensor networks was originally motivated by military applications such as battlefield surveillance. However, wireless sensor networks are now used in many civilian application areas, including environment and habitat monitoring, healthcare applications, home automation, and traffic control. Some examples are listed as follows:

- Military sensor networks to detect and gain as much information as possible about enemy movements, explosions, and other phenomena of interest.
- Sensor networks to detect and monitor environmental changes such as in plains, forests, and oceans.
- Wireless traffic sensor networks to monitor vehicle traffic on highways or in congested parts of a city.
- Wireless surveillance sensor networks provide security in shopping malls, parking garages, and other civil facilities.
- Manufacturing sensors can facilitate the the monitoring and control process, which can reduce the cost, improve the flexibility, and enhance the accuracy.
- Sensors for supermarkets can speed products to shelves and provide customers with better-quality products.
- Medical sensors, especially those implanted sensors, need to constantly monitor the patients, have sufficient battery life, and be able to transmit the sensed information out of the body via wireless channels.

In Figure 2.13, we show the typical structure of one unit of sensor networks. In addition to one or more sensors, each node in a sensor network is typically equipped with a radio transceiver or other wireless communications device, a small microprocessor, some memory, and an energy source, usually a battery. For the sensors, different sensing applications can include temperature, light, humidity, pressure, accelerometers, magnetometers, chemical, acoustics, and image/video. The microprocessor has a significant constraint in computational power. Currently, the devices typically have the component-based embedded operating system. The available memory is also very limited. The current radio transceivers for sensor networks are low rate and short range. Some sensors can be powered by a wired power source, while most of the widely deployed sensors are powered by battery. Exchanging the energy-depleting sensors is a challenging job, so power saving is critical in the design of such wireless sensor networks.

The size of a single sensor node can vary from shoe-box-sized nodes down to devices the size of a grain of dust. The cost of sensor nodes is similarly variable, ranging

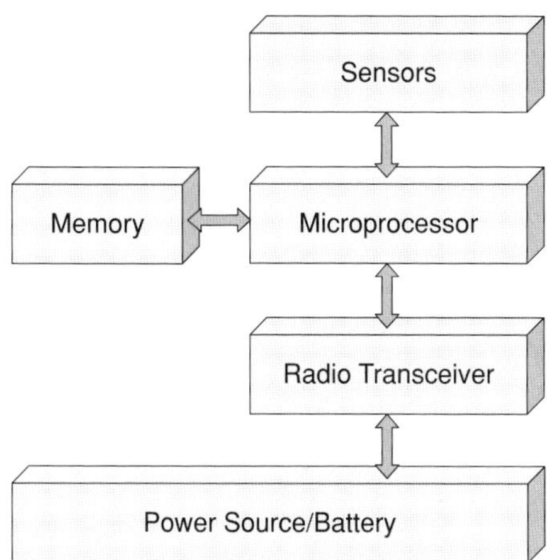

Figure 2.13 Sensor network structure.

from hundreds of dollars to a few cents, depending on the size of the sensor network and the complexity required of individual sensor nodes. Size and cost constraints on sensor nodes result in corresponding constraints on resources such as energy, memory, computational speed, and bandwidth. Basically, to design a wireless sensor network, the following requirements should be considered [27]:

1. large number of (mostly stationary) sensors;
2. low energy use to extend sensor network lifetime;
3. network self-organization;
4. collaborative signal processing;
5. querying ability.

In practice, there are some other design issues [28] such as how to deploy the sensor networks, how to locate a specific sensor, how to shut down and reactivate a sensor to save energy, how to route the information back to data-collecting points, how to reduce the packet forwarding loads by data fusion, and finally how to have secure sensor networks.

- Sensor network deployment

  The problem is to select the locations to place the sensor networks, given a particular application context, an operational region, and a set of wireless sensor devices. The sensors can be deployed in a structured sense or in a randomly scattering manner. The density of sensors is judged by the robustness and cost of the networks.

  Usually, the sensor networks have two types of sensors. The first type is the low-capability sensor that is in charge of collecting data, while the other type is the cluster head or data sink that is more powerful in computation and data transmission. The network topology between these two types of sensors can be star-connected single hop, multihop mesh/grid, or a multiple-tier hierarchical cluster.

Because the transmit power is bounded, the sensor can reach the other sensor with limited distance. This fact causes connectivity problems for sensor networks. Moreover, the sensing range is also restrained. The coverage of the sensing area is a function of the density of the sensors. There are different types of coverage metrics such as k-coverage, minimum coverage, and maximal breach distance [29].

- Localization

Localization means determining the location of a certain event sensed by the sensor. This location information can provide a location stamp over the event, track the monitored object, determine the coverage, form the cluster, facilitate routing, and perform efficient querying. Even though the information can be obtained by GPS, the cost and indoor environments prohibit the sensors from being equipped with GPS.

Nevertheless, the task of localization captures multiple aspects of sensor networks: The physical layer imposes measurement challenges, due to multipath, shadowing, sensor imperfections, and changes in propagation properties. Extensive computation is necessary for many formulations of localization problems. Moreover, the problems sometimes should be solved in a distributed manner or on a memory-constrained processor. Next, for networking and coordination issues, sensor nodes have to collaborate and communicate with each other to know the topology of the whole network. Finally, for system integration issues, it is challenging to integrate location services with other applications.

There are several types of localization mechanisms. First, an active localization system sends signals to localize a target. The examples are RADAR or LIDAR (LADAR). Secondly, in cooperative localization, the target cooperates with the system. For example, the target emits a signal with known characteristics, and then the system deduces location by detecting a signal. Thirdly, a passive localization system deduces location from observation of signals that are "already present." The example technique is to use the geometric methods to calculate the location by measuring the signal strength over the receivers in different locations. Finally, a blind localization system deduces the location of target without a priori knowledge of its characteristics.

- Time synchronization

Localization provides the sensor networks with spatial information, while accurate time synchronization is also essential, as shown in the following examples: First, since the delay for the information to the sink is unpredictable, each sensor needs to have a consistent time stamp for the message. This is more important for some types of data such as tsunami alarming, since the time information provides many scientific clues. For localization, the transmitter and receiver need to have synchronized time so that the time-of-flight can be calculated. For multiple access such as TDMA-based schemes, each sensor needs to transmit at the exact time-slots. For sleeping scheduling, the energy is saved by turning on and turning off of sensors at certain times.

The accurate time can be obtained by a GPS signal. But this approach is very expensive. Quartz crystal oscillators can provide accuracy of about several μs. For better accuracy, some techniques such as phase-locked loops need to be implemented to synchronize the clocks.

- Sleeping mechanism
  In most sensors, the primary source of power consumption is the radio for transmitting, receiving, and listening. If the sensors only wake up during the time when the radio is active and sleep in the remaining time, the limited energy can be conserved and the lifetime of sensor networks can be prolonged.
  However, the sleep-and-wake-up mechanism causes other design problems. First, there is a tradeoff between the delay of information and energy consumption. Moreover, the design of the MAC layer multiple access protocol needs to consider the wake-up time. Further, the transmitter and receiver should be synchronized to wake up at the same time. Finally, the fairness issue needs to be considered so that some sensors will not be overloaded to be energy depleted too early.
- Energy-efficient routing
  Since the energy is a major concern for the design of wireless sensor networks, to select energy-efficient routes from the data-collecting sensors to the data sink can significantly improve the network lifetime. In addition, when multiple routes are considered, individual optimal energy-efficient routes are not optimal, in the sense that some sensors on the critical paths might be depleted first. So joint optimization is necessary.
  In addition to the energy concern, the routing protocols also should take consideration of latency due to the sleeping mechanism of sensors. The routing protocols should also consider the data fusion/aggregation. Finally, for large sensor networks, scalability is an important issue. For the situations in which the sensors are mobile or can join/leave the network frequently, the adaptive ability is also a design challenge.
- Fusion/aggregation
  Fusion/aggregation is a process dealing with the association, correlation, and combination of data and information from single and multiple sources to achieve refined position and identity estimates for observed entities, and to achieve complete and timely assessments of situations and threats, and their significance. In wireless sensor networks especially large ones, it is impossible and energy inefficient to gather the information to make the decision. Instead, along the path to the data sink, a data fusion node collects the results from multiple nodes, it fuses the results with its own based on a decision criterion, and then sends the fused data to another node/base station. By doing this, the traffic load can be greatly reduced and energy can be conserved.
  There are two forms of data fusion/aggregation. In the first form, data from different node measurements are combined together to form larger packets. It is simple to implement, but requires higher computational burden, higher communication burden, and larger training data requirement. The second type is decision fusion, in which the decisions (hard or soft) based on node measurements are combined. The decision fusion solves the problems of the first form. The decision can be made by mechanisms such as voting. For example, a fusion node arrives at a consensus by a voting scheme: majority voting, complete agreement, and weighted voting. Other fusion decision algorithms include the probability-based Bayesian model [30] and stacked generalization [31].

For sensor networks, there are different fusion architectures from the data-collecting sensors to the data sink. In the centralized architecture, a central processor fuses the reports collected by all other sensing nodes. The centralized one has the advantages that an erroneous report can be easily detected, and it is simple to implement. On the other hand, it has the disadvantage that it is inflexible to sensor changes and the workload is concentrated at a single point. In the decentralized architecture, data fusion occurs locally at each node on the basis of local observations and the information obtained from neighboring nodes. There is no central processor node. The advantages are scalable and tolerant to the addition or loss of sensing nodes or dynamic changes in the network. In the hierarchical architecture, nodes are partitioned into hierarchical levels. The sensing nodes are at level 0 and the data sink at the highest level. Reports move from the lower levels to higher ones. This architecture has the advantage that workload is balanced among nodes.

- Security
  Because sensor networks may interact with sensitive data and/or operate in hostile unattended environments like military sensors, it is important for the security to be addressed from the system design. Moreover, because of inherent resource and computing constraints, security in sensor networks poses more challenges than traditional network/computer security. The possible security attacks include denial of service attack, Sybil attack, traffic analysis attacks, node replica attack, privacy attack, physical attack, and collusion attack. In the literature, there are some defensive mechanisms such as key cryptography and trust management. The security issue is beyond the scope of this book. A good survey of security issues in wireless sensor networks can be found in [32].

2.3 Advanced wireless technology

2.3.1 OFDM technology

Orthogonal Frequency-Division Multiplexing (OFDM) is a technique of transmitting multiple digital signals simultaneously over a large number of orthogonal subcarriers. Based on the fast Fourier transform algorithm to generate and detect the signal, data transmission can be performed over a large number of carriers that are spaced apart at precise frequencies. The frequencies (or tones) are orthogonal to each other. Therefore, the spacing between the subcarriers can be reduced and hence high spectral efficiency can be achieved. OFDM transmission is also resilient to interference and multipath distortion, which causes intersymbol interference (ISI).

OFDM transmitter and receiver block diagrams are shown in Figure 2.14 and Figure 2.15, respectively. $s[n]$ is a serial stream of binary digits to transmit. After the serial-to-parallel converter, the data are split into N streams. Each stream is then coded to $X_0, \ldots, X_{N-1}$ with possible different modulation methods (such as PSK and QAM), depending on the subchannel condition. An inverse FFT is computed on each set of symbols, giving a set of complex time-domain samples. These samples are then

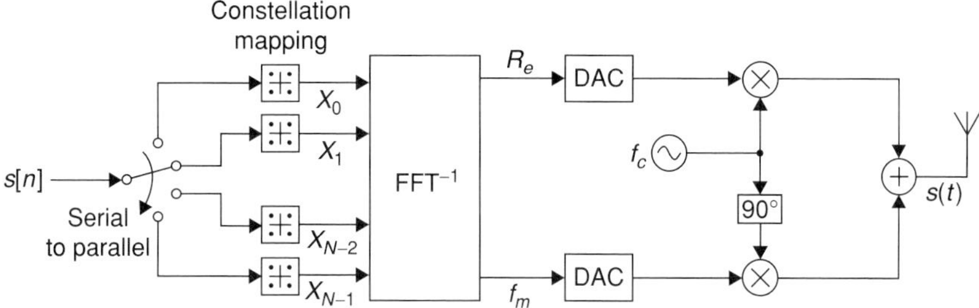

Figure 2.14 Illustration of OFDM transmitter.

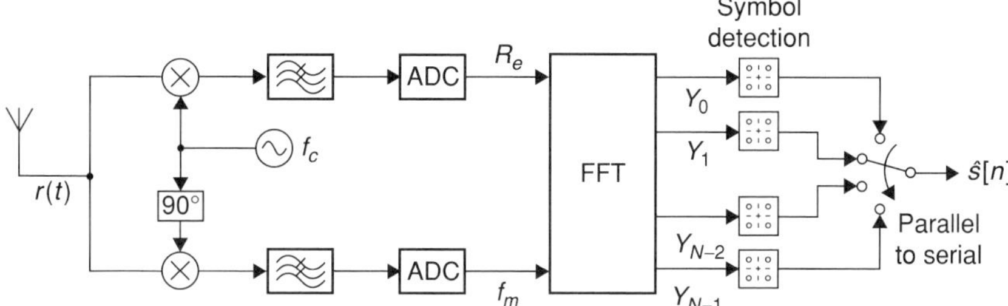

Figure 2.15 Illustration of OFDM receiver.

quadrature-mixed to passband in the standard way: The real and imaginary components are first converted to the analog domain using digital-to-analog converters (DACs); the analog signals are then used to modulate cosine and sine waves at the carrier frequency, f_c, respectively. These signals are then summed to yield the transmission signal, $s(t)$. The receiver picks up the signal $r(t)$, which is $s(t)$ transmitted through radio channels and contaminated by noise. Then $r(t)$ is quadrature-mixed down to baseband using cosine and sine waves at the carrier frequency. The baseband signals are then sampled and digitized using analog-to-digital converters (ADCs), and a forward FFT is used to convert back to the frequency domain. This returns N parallel streams, each of which is converted to a binary stream using an appropriate symbol detector. These streams are then recombined into a serial stream, $\hat{s}[n]$, which is an estimate of the original binary stream at the transmitter.

For a mathematical description, the low-pass equivalent OFDM signal is expressed as:

$$V(t) = \sum_{k=0}^{N-1} X_k e^{j2\pi kt/T}, \quad 0 \le t < T, \tag{2.10}$$

where X_k are the modulated data symbol for the kth data stream, and T is the OFDM symbol time.

To avoid intersymbol interference in multipath fading channels, a guard interval of length T_g is inserted prior to the OFDM block. During this interval, a cyclic prefix is transmitted such that the signal in the interval $-T_g \le t < 0$ equals the signal in the interval $(T - T_g) \leq t < T$. Cyclic prefix is often used in conjunction with modulation in order to retain the sinusoids' properties in multipath channels. It is well known that sinusoidal signals are eigenfunctions of linear, and time-invariant, systems. Therefore, if the channel is assumed to be linear and time invariant, then a sinusoid of infinite duration would be an eigenfunction. However, in practice, this cannot be achieved, as real signals are always time limited. So, to mimic the infinite behavior, prefixing the end of the symbol to the beginning makes the linear convolution of the channel appear as though it were circular convolution, and thus, preserves this property in the part of the symbol after the cyclic prefix.

Orthogonal Frequency Division Multiple Access (OFDMA) is a multiple access scheme implemented based on OFDM. In OFDMA, different users are allocated with different subcarriers, and hence multiple users can transmit their data simultaneously. QoS can be achieved in OFDMA by allocating different numbers of subcarriers to the users with different QoS requirements. In OFDMA, different users can occupy different time-frequency slots so as to fully utilize the diversity. OFDM can be combined with the CDMA scheme (i.e., multicarrier code division multiple access (MC-CDMA) or OFDM-CDMA). In this case, different codes are assigned to different users for concurrent transmissions.

Dynamic spectral management for OFDMA can be categorized into two groups based on the network control architecture.

1. Centralized control – A Spectral Management Center (SMC) monitors and controls the transmit spectra of all the users in the system. There must be a considerable amount of coordination and communication between the users and the SMC. This greatly increases the control signaling overhead but leads to a better performance when compared to the distributed control. Centralized control can be further categorized as follows [7, 33]:
   - Level 1 – The data rate and transmit power of the user are reported to a central controller and corresponding control signals are generated to control the rate and transmit power.
   - Level 2 – The noise spectra and the received signal spectra are monitored, and the transmit power is controlled by the central controller.
   - Level 3 – It allows complete coordination in real-time control of transmit power while monitoring the signals and noise spectra.
2. Distributed control – The users have the capability of sensing the channel conditions and adjust their transmit spectra accordingly. This may be efficient from the perspective of signaling and coordination but has the drawback of converging to a suboptimal point. It is also referred to as Level 0 [7, 33], where there is no Dynamic Spectrum Management (DSM) involved and the control is fully distributed.

The control exhibited by the SMC over the users determines the computational complexity of the overall system. A centralized control is efficient but consumes a lot of bandwidth in control messaging between the users and the base station. On the other hand, distributed algorithms increase the complexity of the receiver of the user. The resource allocation of OFDMA can be broadly categorized into three areas:

- Subcarrier assignment – The subcarriers with the best channel gains as seen by the user are allocated to the particular user.
- Rate allocation – The data rate is allocated depending on the user application requirements.
- Power control – Optimal transmit power is to be allocated to the user, in order to meet its rate requirements while not interfering with the other users.

OFDM and OFDMA are used in the emerging standards that include IEEE 802.11a/WiFi, IEEE 802.15/WiPAN, IEEE 802.16/WiMAX, IEEE 802.20/MobileFi, IEEE 802.22/WiRAN, digital audio broadcasting (DAB), terrestrial broadcasting of digital television (DVB-T, DVB-H), Flash-OFDM (Fast Low-latency Access with Seamless Handoff Orthogonal Frequency Division Multiplexing), SDARS for satellite radio, G.DMT (ITU G.992.1) for ADSL, and ITU-T G.hn for Power Line communication.

2.3.2 Multiple antenna system

For spatial diversity, transceivers employ antenna arrays and adjust their beam patterns such that they have good channel gain toward the desired directions, while the aggregate interference power is minimized at their output. Antenna array processing techniques such as beamforming, MIMO technology, and space-time coding can be applied to receive and transmit multiple signals that are separated in space. Hence, multiple co-channel users can be supported in each cell to increase the capacity by exploring the spatial diversity. In this subsection, we briefly discuss various multiple antenna technologies.

Beamforming technique

Beamforming is a technique in signal processing used for directional data transmission and reception. Using beamforming, the signal is received/transmitted in a particular direction to improve the receive/transmit gain. When transmitting, a beamformer controls the phase and amplitude of the signal to create a pattern of constructive and destructive interference in the wave front. When receiving, these signals are combined in such a way that the expected pattern of radiation is preferentially observed. Different weights can be assigned to the different signals so that the decoded information has the smallest error probability. In a conventional beamforming system, these weights are fixed and can be obtained according to the location of the receive antenna and the direction of the signal. Alternatively, the weights can be adaptively adjusted by considering the characteristics of the received signal to mitigate the interference from unwanted sources. In Figure 2.16, we show an illustration of beamforming technology.

MIMO technology and space-time coding

To enhance the performance of wireless transmission, Multiple-Input Multiple-Output (MIMO), or multiple antennas, can be used to transmit and receive the radio signals. Data transmitted from multiple antennas will experience different multipath fading, and at the receiver these different multipath signals are received by multiple antennas. By using advanced signal-processing techniques, multipath signals at the receiver can be

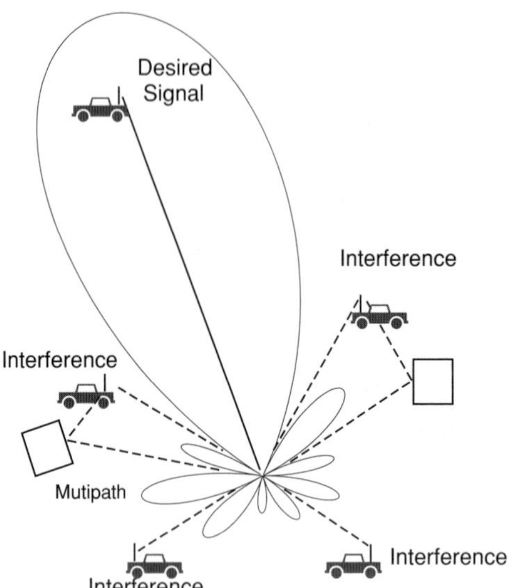

Figure 2.16 Illustration of beamforming.

combined to reconstruct original data. MIMO systems take advantage of this spatial diversity to achieve higher data rates or lower bit error rates. There are two basic types of MIMO systems, namely, the space-time coding MIMO system (for diversity maximization) and spatial multiplexing MIMO system (for data-rate maximization).

In a space-time coding MIMO system, a single data stream is redundantly transmitted over multiple antennas by using suitably designed transmit signals. At the receiver, multiple copies of the signal are received, which are used to construct the original data. Space-time coding can be categorized into space-time trellis coding (STTC), in which trellis code is transmitted over multiple antennas and multiple time-slots, and space-time block-coding (STBC) in which a block of data is transmitted over multiple antennas. Both STTC and STBC can achieve diversity gain, which improves the error performance. While STTC can also achieve coding gain (i.e., results in lower error rate), STBC can be implemented with less complexity.

Instead of transmitting the same data over multiple antennas, a spatial multiplexing MIMO system transmits different data streams over multiple antennas. In this case, the number of transmit antennas is equal to or larger than that of receive antennas and the data rate can increase by a factor of the number of transmit antennas [34]. There is a fundamental tradeoff between diversity gain and multiplexing gain. In simple words, the diversity gain improves the BER performance, while multiplexing gain will increase the rate, but on the other hand, the BER performance might be reduced. Detailed analysis can be found in the literature [35].

One simple example of an STBC is the Alamouti code [36], which was designed for a two-transmit antenna system and has the coding matrix:

$$C = \begin{bmatrix} s_1 & s_2 \\ -s_2^* & s_1^* \end{bmatrix}, \tag{2.11}$$

where s_1 and s_2 are two symbols, $*$ denotes complex conjugate, and components of the matrix are sent on two antennas and two time-slots, respectively. The coding rate is 1, and using the optimal linear decoding scheme the BER of this STBC is equivalent to maximal ratio combining (MRC).

Multiuser MIMO (MU-MIMO) was proposed to support data transmission from multiple users simultaneously. In this case, data from different users can be transmitted over different antennas. In this MU-MIMO, MIMO broadcast channels and MIMO multiple access channels are used for downlink and uplink transmissions, respectively. Alternatively, Space-Division Multiple-Access (SDMA) can exploit the information of users' locations to adjust the transmission and reception parameters to achieve the best path gain in the direction of each user. Phased array antenna techniques are generally used for SDMA. MIMO is an optional feature in the IEEE 802.16/WiMAX standard, while it is being used in the IEEE 802.11n standard.

2.3.3 Cognitive radios

The Federal Radio Act under FCC allows predetermined users the right to transmit at a given frequency. Unlicensed users are regarded as "harmful interference," and in most cases side bands were implemented to ensure that interference is not an issue. As technology advanced, higher-frequency bands were sold at auction bringing considerable revenue to the government. For example, since the 1994 PCS auction, more than $30 billion has been generated. As the demands for wireless communication become more and more pervasive, the wireless devices must find a way for the right to transmit at frequencies in the limited radio band. However, there exist a large number of frequency bands that have considerable, and sometimes periodic, dormant time intervals. For example, TV stations often do not work at night. There exist some spectrum holes at a given time over different spectrum bands. So there is a dilemma that on the one hand the mobile users have no spectrum to transmit, while on the other hand some spectrums are not fully utilized.

In order to cope with the dilemma, cognitive radio is a paradigm for wireless communication in which either a network or a wireless node changes its transmission or reception parameters to communicate efficiently without interfering with the licensed users. This alteration of parameters is based on the active monitoring of several factors in the external and internal radio environment, such as radio-frequency spectrums, user behaviors, and network states.

Depending on the set of parameters taken into account in deciding on transmission and reception changes, and for historical reasons, we can distinguish certain types of cognitive radio as

- Full Cognitive Radio ("Mitola radio"), in which every possible parameter observable by a wireless node or network is taken into account.
- Spectrum Sensing Cognitive Radio, in which only the radio-frequency spectrum is considered.
- Licensed Band Cognitive Radio, in which cognitive radio is capable of using bands assigned to licensed users.

- Unlicensed Band Cognitive Radio, which can only utilize unlicensed parts of the radio-frequency spectrum.

Cognitive radio can be designed as an enhancement layer on top of the Software-Defined Radio (SDR) concept. A SDR system is a radio communication system that can tune to any frequency band and receive any modulation across a large frequency spectrum by means of a programmable hardware that is controlled by software.

An SDR performs significant amounts of signal processing in a general-purpose computer, or a reconfigurable piece of digital electronics. The goal of this design is to produce a radio that can receive and transmit a new form of radio protocol just by running new software. The hardware of a software-defined radio typically consists of a superheterodyne RF front end that converts RF signals from (and to) analog IF signals, and analog-to-digital converters and digital-to-analog converters that are used to convert a digitized IF signal from and to analog form, respectively.

Software radios have significant utility for the military and cell phone services, both of which must serve a wide variety of changing radio protocols in real time. Software-defined radio can currently be used to implement simple radio modem technologies. In the long run, software-defined radio is expected by its proponents to become the dominant technology in radio communications. It is the enabler of the cognitive radio.

The cognitive transmitter and receiver can adapt to 3G, WIFI, and WPAN networks. By sensing the available spectrum, the cognitive radio can adapt to the most suitable available communication links. For example, if the user is at home, then he/she can communicate via Bluetooth; if the user travels to the airport, WIFI communication can be available; if the user drives on the highway, the cellular phone system can provide reliable communication links. Another analogy interpretation for cognitive radio is as follows: The licensed users are legally used to the spectrum, or like high-occupancy vehicles (HOV) can drive on the HOV lanes. However, the HOV lanes are not always occupied. So besides rush hours, the other vehicles can also drive on the HOV lanes. In this sense, the cognitive radios are similar to the other non-HOV vehicles.

Two major objectives of cognitive radio are to reliably communicate whenever and wherever and to efficiently utilize the radio spectrum. To achieve these objectives, three fundamental cognitive tasks [37] for cognitive radio must be fulfilled as follows:

1. Radio-Scene sensing analyzes the interferences and detects the spectrum holes.
2. Spectrum analysis such as channel-state estimation and predictions of channel capacity.
3. Transmit power control and dynamic spectrum management.

In Figure 2.17, we show the cognitive cycle, in which three tasks interact with each other to handle the outside worlds so that the best strategy can be calculated and implemented.

The IEEE 802.22 working group on Wireless Regional Area Networks (WRAN) is a group of the IEEE 802 LAN/MAN standards committee. The standard for WRAN Part 22 (Cognitive Wireless RAN Medium Access Control (MAC) and Physical Layer (PHY) Specifications) regulate policies and procedures for operation in the TV Bands.

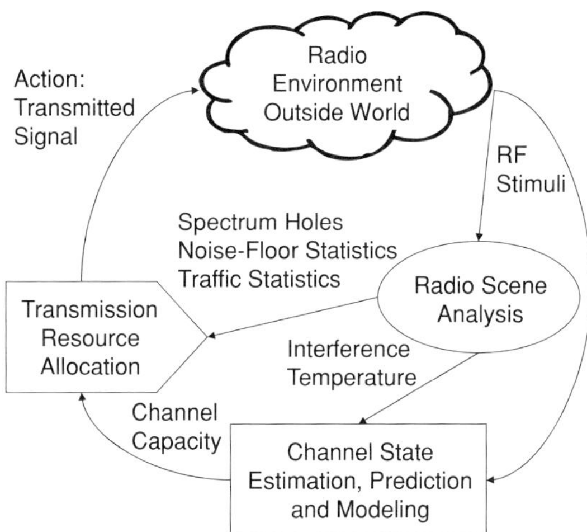

Figure 2.17 Cognitive cycle.

The standard focuses on constructing a consistent, fixed point-to-multipoint WRAN that will utilize UHF/VHF TV bands between 54 and 862 MHz. Specific TV channels as well as the guard bands of these channels are planned to be used for communication in IEEE 802.22.

The IEEE, together with the FCC, is pursuing a centralized approach for available spectrum discovery. Specifically each Access Point (AP) would be armed with a GPS receiver that would allow its position to be reported. This information would be sent back to centralized servers (in the USA these would be managed by the FCC), which would respond with the information about available free TV channels and guard bands in the area of the AP. Other proposals would allow local spectrum sensing only, where the AP would decide by itself which channels are available for communication. A combination of these two approaches is also envisioned. Table 2.5 describes the coverage and capacity of IEEE 802.22 WRAN.

Overall, cognitive radio can bring in a variety of benefits: For a regulator, it can lead to a significant increase in spectrum availability for new and existing applications. For a license holder, cognitive radio can reduce the complexity of frequency planning, facilitate the secondary spectrum market agreements, increase system capacity through access to more spectrum, and avoid interference. For equipment manufacturers, cognitive radio can increase demands for wireless devices. Finally, for a user, cognitive radio can provide a higher level of capacity per user, enhance interoperability and bandwidth-on-demand for Public Safety and Emergency Response operations, and provide ubiquitous mobility with a single-user device across disparate spectrum-access environments.

2.3.4 Scheduling and multiple access

The traffic scheduler in a wireless transmitter is responsible for selecting packets from the transmission queue based on a scheduling policy to transmit in the air. The scheduling

Table 2.5 WRAN coverage and capacity

RF channel bandwidth	6 MHz
Average spectrum efficiency	3 bit/s/Hz
Downlink user capacity	1.5 Mbit/s
Uplink user capacity	384 kbit/s
Oversubscription ratio	50
Number of users per downlink	600
Minimum no. of users	90
Assumed early take-up rate	3 bit/s/Hz
Potential no. of users	1800
No. of user per household	2.5
No. of user per coverage area	4500
WRAN base station power	98.3 W
Coverage	30.7 km
Min. population density	1.5/km^2

policy should consider the QoS requirements of user traffic as well as the wireless channel condition [38]. The traffic scheduling mechanism can be either work-conserving or non-work-conserving. In the case of work-conserving scheduling, the transmission channel is never idle as long as the transmission queues are non-empty. On the other hand, in the case of non-work-conserving scheduling, the transmission channel can be idle even though there are packets buffered in the transmission queues.

Some of the popular traffic-scheduling policies are round-robin, weighted round-robin, generalized processor sharing (GPS), weighted-fair queuing, and virtual clock scheduling schemes [39]. These packet scheduling policies were developed primarily for wired networks where transmission error is negligible compared to that in a wireless network. Similar scheduling schemes can be designed for wireless networks, considering the time-varying and location-dependent nature of the wireless channel errors. Because of the time- and location-dependent nature of channel variation, fairness (such as proportional fairness) is an important issue in wireless packet scheduling [40, 41]. A survey on wireless packet scheduling can be found in [39].

While traffic scheduling is used to choose among the queues (e.g., in a mobile terminal or in a base station/access point) or users for wireless transmission, a MAC protocol controls access of the users to the shared radio channel. Also, the MAC protocol must support service differentiation and QoS guarantee for different types of traffic/user in the network. A MAC protocol can be categorized as either a centralized (coordinate access) protocol or a distributed (random access) protocol. In the case of centralized MAC, a central controller (e.g., base station) controls channel access of the users (e.g., which channel, which user, and when to transmit). In this case, all users in the network must be able to communicate their QoS requirements to the central controller.

In the case of a distributed MAC protocol, all users can access the channel independently. Therefore, collisions can occur if more than one user transmits data on the same channel at the same time. Distributed MAC protocols can be designed based on Aloha, carrier sensing multiple access (CSMA) mechanisms [42]. In particular, before the user transmits data, it has to sense the channel. Since many users may sense the channel to be idle at the same time and transmit together, in case of collision each user must perform random backoff before data retransmission.

The problem of design and analysis of MAC protocols for different types of wireless networks has been extensively studied in the literature. A survey of MAC protocols for cellular, wireless ad hoc, sensor networks can be found in [43–45], respectively. MAC design issues for ultra-wideband wireless systems were addressed in [46]. MAC issues for multimedia traffic in wireless networks were surveyed in [47].

The IEEE 802.11 MAC protocol supports two kinds of access methods, namely, the distributed coordination function (DCF) and the point coordination function (PCF). In both mechanisms, only one user occupies all the bandwidth at each time-slot. PCF is based on polling, controlled by a point coordinator like Access Point, to communicate with a node listening and to see whether the airwaves are free. PCF seems to be implemented only in very few hardware devices as it is not part of the WiFi Alliance's interoperability standard.

In contrast, DCF is an access mechanism using carrier sense multiple access with collision avoidance (CSMA/CA). DCF mandates a station wishing to transmit to listen for the channel status for a DCF Interframe Space (DIFS) interval. If the channel is found to be busy during the DIFS interval, the station defers its transmission or proceeds otherwise. In a network that a number of stations contend for the multiaccess channel, if multiple stations sense the channel busy and defer their access, they will find that the channel is released virtually simultaneously and then try to seize the channel again at the same time. As a result, collisions may occur. In order to avoid such collisions, DCF also specifies random backoff, which forces a station to defer its access to the channel for an extra period. DCF also has an optional virtual carrier sense mechanism that exchanges short Request-to-send (RTS) and Clear-to-send (CTS) frames between the source and destination stations before the long data frame is transmitted. In RTS/CTS [48], a node wishing to send data initiates the process by sending a Request-to-send frame (RTS). The destination node replies with a CTS. Any other node receiving the RTS or CTS frame should refrain from sending data for a given time (solving the hidden node problem). The amount of time the node should wait before trying to get access to the medium is included in both the RTS and the CTS frame.

2.3.5 Wireless positioning and localization

The proliferation of wireless and mobile devices has fostered the demand for various location-aware applications, such as smart space [49], modern logistics [50], environmental monitoring, and object tracking [51]. Because of the high hardware cost and unsuitability for indoor environments, the traditional Global Positioning System (GPS) is often infeasible for large-scale deployments. These limitations have motivated great research efforts in designing novel schemes of wireless network localization. In such schemes, some special nodes called anchors, or beacons, are aware of their own positions and other nodes refer their locations through geographical relationships with the beacons. This subsection briefly reviews the state-of-the-art localization approaches

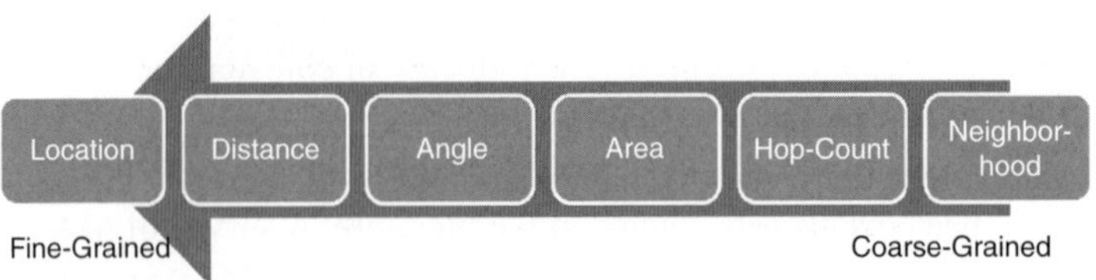

Figure 2.18 Physical measurements.

from two aspects: physical measurements for single-node localization and network-wide localization. Then we will discuss the techniques for controlling the localization errors caused by the measurement errors and algorithmic defects.

Single-node localization

Almost all existing localization algorithms need to measure the geographical information from the ground truth of the network deployment. The measuring techniques can be roughly classified into six categories: location, distance, angle, area, hop-count, and neighborhood, as shown in Figure 2.18. Directly obtaining the position without any further computation in a resource-constrained wireless networks is usually infeasible, and thus we will focus on the other five measurement methods.

The location of an unknown node can be inferred from its distance measurements to several reference nodes, which is the basic idea of multilateration. The Radio Signal Strength (RSS) based and time-based ranging are two of the most widely used distance measurement designs. RSS-based ranging techniques measure the distance according to the attenuation of the signal strength during propagation. The Time Difference of Arrival (TDOA) [52] technique estimates distances by determining the time difference of arrival of ultrasound/acoustic and radio signals. After the distance information is obtained, location of the unknown nodes can be solved by a numerical solution to an overdetermined linear system [53]. The Angle of Arrival (AoA) [54] localization methods estimate the location of a point using triangulation. According to the law of sines, the location of an unknown node can be determined by measuring angles to it from two known reference points. For area estimation [55, 56], the basic idea is to compute the intersection of all overlapping coverage regions that an unknown node is in and choose the centroid as the location estimate. The existing approaches can be categorized into two classes: single reference area estimation and multireference area estimation, where the former uses the coverage of a single to obtain the bounding region, whereas later defines the region according to any group of three reference nodes.

Hop count is a useful way to estimate the distance between two nodes in wireless ad hoc network. In isotropic network topologies, the Euclidean distance between a pair of nodes can be estimated by the product of the hop count and the node's radio range. Reference [57] adopts a better way to estimate the distance covered by one radio hop instead of directly using the radio range to achieve a higher accuracy. The neighborhood measurement approach is simple and economical. In k-neighbor proximity, we use the centroid of the polygon constructed by the k reference nodes with equal weights as the

Table 2.6 Comparison of physical measurement

Physical Measurement		Accuracy	Hardware Cost	Computation Cost
Distance	RSS	Median	Low	Low
	TDoA	High	High	Low
Angle	AoA	High	High	Low
Area	Single reference	Median*	Median*	Median
	Multireference	Median*	Median*	High
Hop-count	Per-hop distance	Median	low	Median
Neighborhood	Single neighbor	Low	Low	Low
	Multineighbor	Low	Low	Low

*: Depends on the diverse geometric constrains

estimated position of the unknown node. This approach benefits from the simplicity of computation and performs well when the the density of reference nodes is sufficiently high [58].

The overview of single-node localization is provided in Table 2.6 [59]. Note that no particular algorithm is an absolute favorite over the spectrum. Different approaches have their own advantages and drawbacks, which are suitable for different application scenarios.

Network-wide localization

According to the computational organization, the taxonomy of network-wide localization algorithm can be defined as centralized algorithms and distributed algorithms. In centralized algorithm, network nodes collect the physical information and send it back to a fusion center. Then the fusion center will distribute the computed positions back to the network. In contrast, the distributed approach processes data in a network. There are two important categories of distributed localization algorithms. The first group is a top-down approach, in which the location information diffuses from the beacon nodes around all the network. The second approach, a bottom-up approach, achieves the global coordinates by gradually merging local maps with nodes in relative coordinates.

Two major centralized localization approaches are Multidimensional Scaling (MDS) [60] and Semi-definite Programming (SDP) [55]. By the law of cosines and linear algebra, the MDS calculates the relative positions based on the pairwise distance. Then the relative positions are transformed into global coordinates based on some reference nodes. In SDP approach, the geometric constraints between the nodes in the network are reformulated into linear matrix inequalities (LMIs). Then the bounding region for each node can be obtained by solving a semi-definite program formed by the LMIs.

Beacon-based localization methods are top-down approaches that utilize distance measurements to reference nodes [61]. The location information is progressive propagated from beacons to an entire network using a Distance-Vector (DV) technique, like DV-hop and DV distance, or in an iterative way, where the newly localized nodes can be used as reference nodes to help locate other nodes. Coordinate system stitching [62] addressed the same problem in a bottom-up manner. Location information is originated

Table 2.7 Comparison of localization algorithms

Localization Algorithm		Accuracy	Node Density	Computation Cost	Accumulative Error
Centralized	MDS	High	Low	High	Low
	SDP	High	High	High	Low
Distributed	Beacon based	Low	High	Low	High*
	Coordinate stitching	Low	High	Median	High

*: In the case of iterative localization

obtained in a local group of nodes in relative coordinates, and entire network localization in global coordinates can be achieved by gradually merging these local maps.

The overview of network-wide localization is provided in Table 2.7 [59]. Since all approaches have their own merits and drawbacks, the design of localization algorithms is quite application specific. A good design should sufficiently investigate the application properties, as well as take algorithm generality and flexibility into consideration.

Error control for localization

Physical ranging measurement in most localization algorithms will incur extrinsic and intrinsic errors inevitably. The extrinsic error is due to the uncertainty of the physical environment, and is caused by the multipath, fading, and shadowing effects of the channel. On the other hand, the intrinsic error is attributed to the hardware and software limitations. These errors not only cause inaccuracy but also result in non-consistency in localization results. In order to achieve a higher accuracy, many methods have been proposed to reduce the localization errors. This part introduces the characteristics of localization errors and discusses two mechanisms of error control for localization.

Understanding the error characteristics of localization is an essential step for error control. For any unbiased localization estimate, the lower bound on its covariance can be computed by the Cramer–Rao Lower Bound (CRLB) [63]. CRLB serves as a benchmark for localization algorithms, it dictates how much gain can be achieved by improving current algorithm. Moreover, the dependence of CRLB on network parameters, such as node density and network topology, can provide us guidelines to design localization algorithms.

Location refinement algorithms [64] are proposed to deal with the error propagation in iterative localization methods. Every node in the network maintains a registry consists of location estimate and corresponding estimate confidence. The unknown node chooses the reference node through an algorithm-specific strategy [65, 66] to achieve the highest confidence, and then updates its own registry. Different strategies will result in different location refinement algorithms. Robust localization is used to handle the outliers of measurement that significantly deteriorate localization accuracy.

Part I

Compressive Sensing Technique

3 Compressive sensing framework

Despite the relatively short history of CS theory pioneered by the work by Candès, Romberg, and Tao [67–69] and Donoho [70], the numbers of studies and publications in this area have become amazingly large. On the other hand, the applications of CS are just beginning to appear. The inborn nature that many signals can be represented by sparse vectors has been recognized in many areas of applications. Examples in wireless communication include the sparse channel impulse response in the time domain, the sparse unitization of the spectrum, and the time and spatial sparsity in the wireless sensor networks. For each of these sparse signals, there are innovative signal acquisition schemes that not only satisfy the requirements by the CS theory, but are also easily realizable on hardware. Efficient signal-recovery algorithms for each system are also available. They guarantee stable signal recovery with high probability.

In this chapter, we provide a concise overview of CS basics and some of its extensions. In subsequent chapters, we focus on CS algorithms and specific areas of CS research in wireless communication.

This chapter begins with Section 3.1, which gives the motivation of CS, illustrates the typical steps of CS by an example, summarizes the key components CS, and discusses how nearly sparse signals and measurement noise are treated in robust CS. Following these discussions, Section 3.2 compares CS with traditional sensing and examines their advantages and disadvantages. The foundation of CS – sparse representation – is studied in Section 3.3, which briefly covers sparsifying transforms, analytic dictionaries, learned dictionaries, as well as a few extensions of sparse modeling. Next, Section 3.4 discusses a variety of conditions under which one can trust CS for encoding signals during sensing and recovering them faithfully after sensing. Finally, two synthetic examples are given in Section 3.5 to close this chapter.

3.1 Background

The Nyquist/Shannon sampling theory has been accepted as the doctrine for signal acquisition and processing ever since it was implied by the work of Nyquist in 1928 [71] and proved by Shannon in 1949 [72]. The theorem says that to exactly reconstruct an *arbitrary* band-limited signal from its samples, the sampling rate needs to be at least twice the bandwidth.

In many applications nowadays, including digital image and video cameras, the Nyquist rate is so high that the excessive number of samples make compression a necessity prior to storage or transmission. In other applications, including medical scanners, radars, and high-speed analog-to-digital converters, increasing the sampling rate is very expensive, either due to the sensor itself being expensive or the measurement process being costly. Often, we are faced with the situation of not being able to collect enough measurements to meet the Nyquist rate.

On the other hand, the chance is that we are often not dealing with arbitrary signals. Most signals we are interested in are highly compressible; namely, they can be represented by a set of sparse or nearly sparse coefficients. CS exploits this feature of signals and thus allows a sampling rate significantly lower than the Nyquist rate. Although this book does not cover CS imaging, we use images as an example to illustrate this point. Let us ignore the three channels of the RGB (red, green, blue) colors and imagine a gray-scale image of n pixels represented by a vector $\mathbf{u}^o \in \mathbb{R}^n$ (though when displayed on screens or printed on paper, it has a typical rectangle shape). A computer program can compress $\mathbf{u}^o$ (superscript "o" stands for original) into a much smaller vector because only a small amount of information such as objects and their boundaries, certain repeated patterns, textures, and hues in the image are visually important; thence, the rest is "thrown away." In a nutshell, an image compression algorithm compresses $\mathbf{u}^o$ by looking up a dictionary Ψ (e.g., a wavelet basis) for the useful pieces of information in $\mathbf{u}^o$. Mathematically, this is to represent $\mathbf{u}^o$ by a certain *sparse* vector $\mathbf{x}^o$ such that $\Psi\mathbf{x}^o$ contains nearly all the useful information in $\mathbf{u}^o$ and $\|\mathbf{u}^o - \Psi\mathbf{x}^o\|$ is small. The well-known JPEG2000 can compress a megapixel image ($n \approx 1\,000\,000$) taken from a camera into just a few thousand wavelet coefficients, reducing its size from a few megabytes to a few dozen kilobytes, leading to economical storage and fast transfer. This fact triggers a question: when only a few thousand coefficients are saved, why are the excessive measurements of a million pixels taken in the first place? A straightforward answer is that because we do not know which are the ones to measure (perhaps also because traditional image sensing cannot directly record wavelet coefficients, or it is very expensive to do so), we must record a sufficient number of pixels so that the subsequent image compression step can compute the wavelet transform at a sufficient resolution and keep the largest wavelet coefficients. In a new sensing paradigm called CS, however, only a small number of measurements of an image (say, $1/4$ of the number of pixels or fewer) need to be collected; and based upon the compressibility of the image, a sophisticated yet fast algorithm can recover the useful part of it. How is this done?

Steps of CS. The entire process of CS consists of three parts, which are depicted in Figure 3.1:

1. signal sparse representation;
2. linear encoding and measurement collection;
3. non-linear decoding (sparse recovery).

Part 1 is the foundation of CS, but it can be done after part 2. The importance of a *sparsifying* basis Ψ is clear from the discussions in the last paragraph, and we discuss

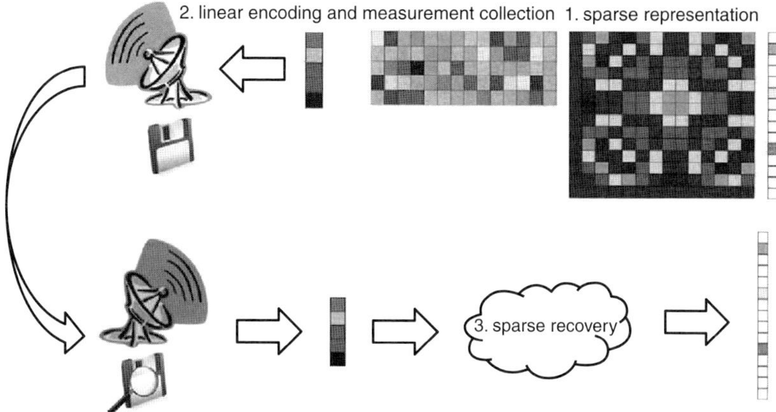

Figure 3.1 Scheme of CS.

it in more detail in Section 3.3. So for simplicity, let us ignore Ψ for now and set it to the identity, $\Psi = I$, and thus assume that $\mathbf{u}^o = \mathbf{x}^o$ is itself a k-sparse vector; namely, the number of non-zero entries of $\mathbf{x}^o$ is no more than k. A common notation for this is

$$\|\mathbf{x}^o\|_0 \leq k.$$

First, $\mathbf{x}^o$ is encoded into a smaller vector $\mathbf{b}$ by a matrix $\mathbf{A} \in \mathbb{R}^{m \times n}$ with $m < n$, which is chosen independently (non-adaptively) of $\mathbf{x}^o$ since $\mathbf{x}^o$ is unknown. This step is written as

$$\mathbf{b} := \mathbf{A}\mathbf{x}^o.$$

In CS applications, this encoding is typically not calculated by a computer but obtained by certain physical, optical, electrical, or electromagnetic means depending on the application. Since $m < n$, $\mathbf{b}$ is a compression of $\mathbf{x}^o$. Next, $\mathbf{b}$ is recorded by a sensor (while $\mathbf{x}^o$ remains unknown) and becomes digitally available. CS theory studies under what condition $\mathbf{x}^o$ can be recovered from $\mathbf{b}$, and when $\mathbf{b}$ is contaminated by noise, how to guarantee the recovery is close to $\mathbf{x}^o$. Clearly, $\mathbf{A}\mathbf{x} = \mathbf{b}$ is an underdetermined equation system and has an infinite number of solutions. This is where the sparsity of $\mathbf{x}^o$ plays its role. Since $\mathbf{x}^o$ is sparse, a natural approach is to find the sparsest solution of $\mathbf{A}\mathbf{x} = \mathbf{b}$, namely, solving

$$\min_{\mathbf{x}} \|\mathbf{x}\|_0 \quad \text{subject to } \mathbf{A}\mathbf{x} = \mathbf{b}. \tag{3.1}$$

However, this combinatorial problem is generally NP-hard [73], and the simple approach of trying all possible supports of cardinality k is computationally intractable. A tractable decoder is needed! Replacing the ℓ_0 "norm" by the ℓ_1 norm yields the problem

$$\min_{\mathbf{x}} \|\mathbf{x}\|_1 \quad \text{subject to } \mathbf{A}\mathbf{x} = \mathbf{b}, \tag{3.2}$$

which is a convex program (in particular, it can be transformed to a linear program) and has several fast solvers (see the next chapter). Ideally, we would like (3.2) to recover

$\mathbf{x}^o$ when m, the number of measurements, is equal to $2k$, which is the "amount of information" that $\mathbf{x}^o$ carries; after all, $\mathbf{x}^o$ is uniquely determined by the k indices and the k values of its non-zero entries. However, we must pay a reasonable price for not knowing the support of $\mathbf{x}^o$ (there are totally $\binom{n}{k}$ possibilities!), yet making sure that a single non-adaptive projection $\mathbf{A}$ good for the reconstruction of all, or at least most, k-sparse vectors $\mathbf{x}^o$ by solving a computationally tractable problem (3.2). As we shall see below, there are random matrices $\mathbf{A}$ with $m = O(k \log(n/k))$ rows that, with overwhelming probability, allow any k-sparse vector $\mathbf{x}^o$ to be recovered by problem (3.2). Besides solving (3.2) by linear programming algorithms, there are very efficient solvers for (3.2), which run almost as fast as the least-squares solvers; we review several such algorithms in the next chapter. Once the solution $\mathbf{x}^o$ is recovered in a computer, we obtain $\mathbf{u}^o$ and accomplish CS.

Robust CS and signal recovery. For CS to be useful in real-world sensing applications, it must deal with imperfect sparsity and sensors. Imperfect sparsity occurs when it is difficult to represent the underlying signal by an exactly spare vector (natural images are very good examples), even using a carefully chosen dictionary or transform in Section 3.3. However, it can be much easier to present it using vectors that are *close to being sparse*. To measure how sparse $\mathbf{x}$ is, we have used its number of non-zeros $\|\mathbf{x}\|_0$. We say that $\mathbf{x}$ is (exactly) sparse if $\|\mathbf{x}\|_0 \ll \dim(\mathbf{x})$ – the total number of entries of $\mathbf{x}$. If $\|\mathbf{x}\|_0$ is *not* so small compared to $\dim(\mathbf{x})$ yet $\mathbf{x}$ can be closely approximated by an exactly sparse vector, then $\mathbf{x}$ is called *nearly sparse*. A nearly sparse vector is nearly as useful and powerful as an exactly sparse vector during reconstruction and processing. The quantity below can be used to measure how nearly sparse a signal is.

Definition 2. *Let $\mathbf{x}_{[k]}$ – the best k-term approximation of $\mathbf{x}$ – be the vector obtained by setting all but the k largest (in magnitude) entries of $\mathbf{x}$ to zero (ties are broken arbitrarily). The approximate error is*

$$\sigma_{[k]}(\mathbf{x}) := \|\mathbf{x} - \mathbf{x}_{[k]}\|_p, \tag{3.3}$$

where common choices of p are 1 and 2, and $\|\mathbf{y}\|_p$ is defined as $(\sum_i |y_i|^p)^{1/p}$.

For k much smaller than $\dim(\mathbf{x})$, $\sigma_{[k]}$ measures the approximate sparsity of $\mathbf{x}$. A nearly sparse vector $\mathbf{x}$ is one with a small $\sigma_{[k]}(\mathbf{x})$ for some $k \ll \dim(\mathbf{x})$. The so-called *fast-decaying* signals have their $\sigma_{[k]}$ that decay very quickly in k, for example, in $O(1/k)$ or even $O(1/c^k)$ with some $c > 1$. As long as the coefficients are fast decaying, a large number of the coefficients can be discarded often without causing too much loss of the information. Figure 3.2 depicts a signal with entries that decay exponentially fast, as well as its best k-term approximation error for $k = 1, \ldots, 128$.

Imperfect sensors exist in many real-world applications, and they are subject to a variety of errors due to, for examples, design/manufacture imperfectness, quantization loss, thermal noise, limited response range, material aging, etc. As a result, the recorded data $\mathbf{b}$ may be contaminated by a certain amount of unknown error. As it is easy to see noise in an image, we add three different types of noise to image Cameraman and demonstrate them in Figure 3.3.

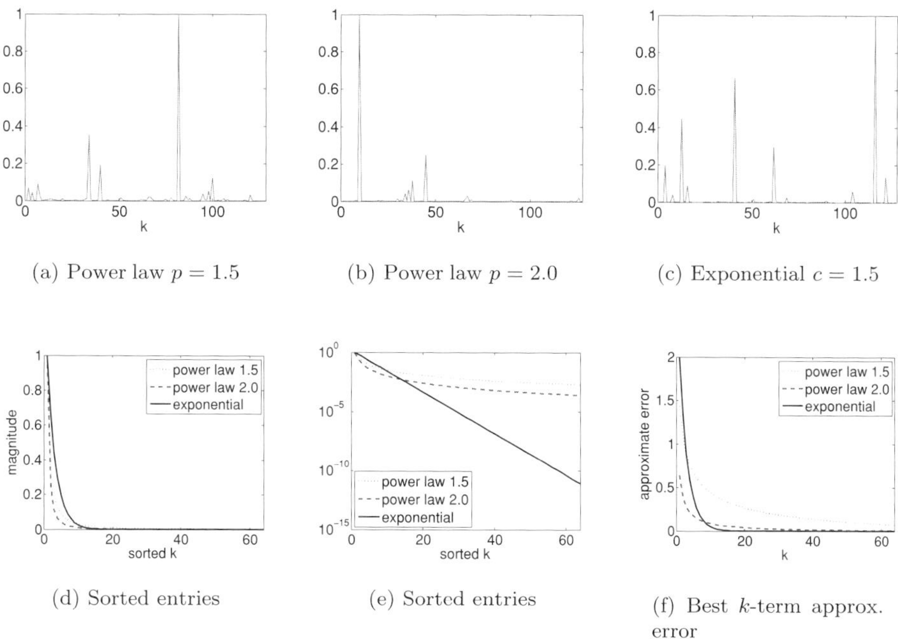

(a) Power law $p = 1.5$ (b) Power law $p = 2.0$ (c) Exponential $c = 1.5$

(d) Sorted entries (e) Sorted entries (f) Best k-term approx. error

Figure 3.2 Two power-law and one exponentially decaying signals.

(a) Original (b) Gaussian noise

(c) Random-valued impulsive (d) Salt-and-pepper

Figure 3.3 Image Cameraman with three different types of noise.

To deal with imperfect sparsity and sensors, we consider the abstract sensing model

$$\mathbf{b} = \mathbf{A}\mathbf{x}^o + \mathbf{w}, \tag{3.4}$$

where $\mathbf{A}$ is an m-by-n matrix and $\mathbf{w}$ is an n-dimensional vector of noise or measurement error. The entries of $\mathbf{b}$ are noisy linear measurements of $\mathbf{x}^o$. If $\mathbf{w} \neq \mathbf{0}$, model (3.2) no longer holds. One can consider the ℓ_1 minimization problem that constraints or penalizes the violation of $\mathbf{b} = \mathbf{A}\mathbf{x}^o$ as follows:

$$\min_{\mathbf{x}} \|\mathbf{x}\|_1 \quad \text{subject to } \|\mathbf{A}\mathbf{x} - \mathbf{b}\|_2 \leq \epsilon, \tag{3.5}$$

$$\min_{\mathbf{x}} \|\mathbf{x}\|_1 + \frac{\mu}{2}\|\mathbf{A}\mathbf{x} - \mathbf{b}\|_2^2. \tag{3.6}$$

The two programs are equivalent in the sense that the solution to one problem is also the solution to the other as long as the parameters ϵ and μ are properly set. However, the correspondence between ϵ and μ is not generally known a priori. Depending on the application and the available problem data, one of ϵ and μ might be easier to obtain than the other, so one of (3.5) and (3.6) is preferred to the other. How to choose ϵ or μ is an important practical issue. General principles include (i) making statistical assumptions on $\mathbf{w}$ and $\mathbf{x}^o$ and interpreting (3.5) or (3.6) as, for example, MAP estimations, (ii) cross-validation (performing reconstruction on one subset of measurements and validating the recovery on the other subset of measurements), and (iii) finding the best parameters on a test data set and using these parameters on the actual data with proper adjustments to address the differences in scale, dynamic range, sparsity, and noise level.

3.2 Traditional sensing versus compressive sensing

In traditional sensing, a signal is first acquired then compressed. Think of taking an image by a digital camera as an example. The CCD or CMOS chip in the camera turns photons into an 2D array of digital signals, which is then compressed and stored in an image file.

Compared to traditional sensing as shown in Figure 3.4, CS *senses less* and *computes more*. CS "compresses" a signal before it is recorded, whereas traditional sensing compresses a signal after sensing. Since CS senses less, in many of its applications, the save leads to much faster sensing and/or much smaller/cheaper sensors. For instance, in medical applications such as MR imaging, CS is reducing the acquisition time and thus making the imaging experience more comfortable, affordable, and in CT imaging, CS is reducing the amount of X-ray radiation used to image the subject body and thus making it safer. On the other hand, because traditional sensing senses more and senses before compression, the digital form of the signal is already available during compression; hence, the compression is more flexible, can be adaptive to each signal, and achieves much better compression ratios. Therefore, traditional sensing is more effective at reducing the signal size. The two sensing approaches have their own strengths; neither one will replace the other. They address different bottlenecks, fit different needs, and achieve different performance. They complement each other. However, since CS is a

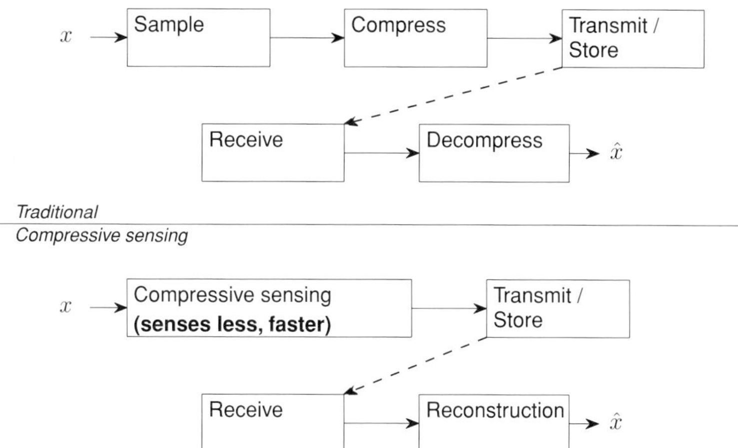

Figure 3.4 Traditional sensing versus CS.

rather new sensing paradigm that senses the information of a signal instead of the signal itself, its theory and computation are less familiar to the communities of researchers and engineers. On the other hand, the principle of CS that simple signals can be recovered from a small number of measurements has stimulated new research results far beyond sensing in areas such as optimization, random matrix theory, statistics, machine learning, medical imaging, etc.

The rest of this chapter provides the reader with brief yet fundamental results of CS in two aspects: sparse representation, and CS encoding and decoding.

3.3 Sparse representation

Sparse representation is the basis of CS. It is concerned with expressing the information of a signal by a small number of real or complex numbers. Mathematically, this is to express a signal $\mathbf{u}^o$ as

$$\mathbf{u}^o = \sum_{i=1}^{p} \psi_i x_i^o,$$

where all but a small number of entries x_i^o are zero (or small enough to safely neglect). $\Psi = [\psi_1 \; \psi_2 \; \cdots \; \psi_p]$ is called a *dictionary*. A dictionary is sometimes used to sparsely represent just one signal, but more often a set of similar signals. The dictionary atoms encode the most salient information of the signals. On the other hand, it is typical that unwanted artifacts and noise cannot be sparsely represented under the dictionary. Expressing artifacts and noise must involve a large number of ψ_is. So, combined with sparse optimization, denoising is often obtained. Besides using a dictionary, a signal can also become sparse under a certain transform Υ, namely, $\Upsilon(\mathbf{u})$ is a sparse vector. Examples include the gradient operator, curvelet transforms, etc. Figure 3.5 presents the DCT and wavelet coefficients, as well as the local variation, of image Cameraman. For

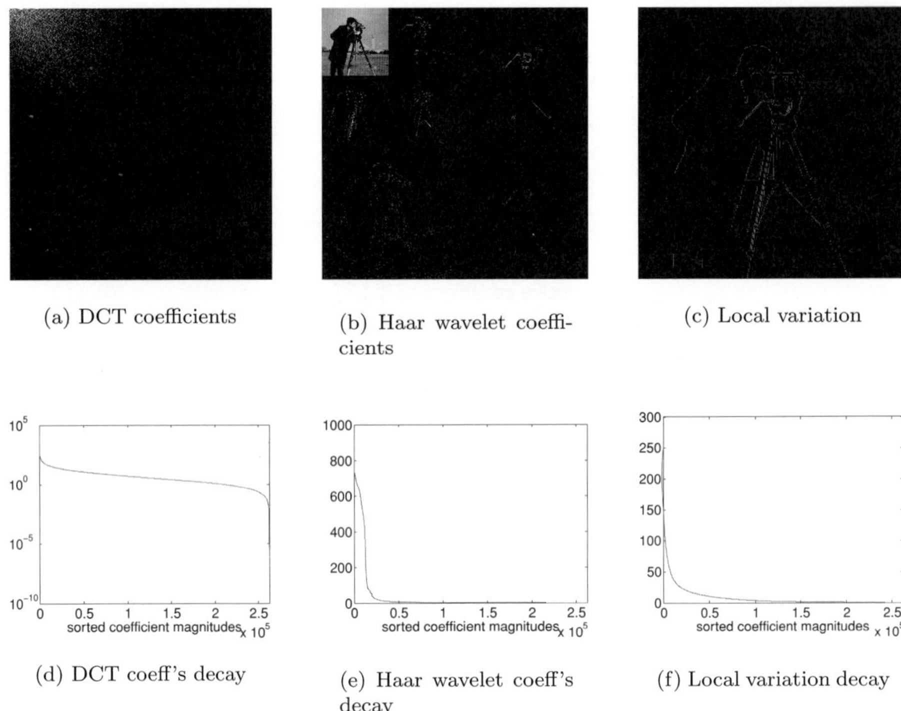

(a) DCT coefficients (b) Haar wavelet coefficients (c) Local variation

(d) DCT coeff's decay (e) Haar wavelet coeff's decay (f) Local variation decay

Figure 3.5 Sparsity of image Cameraman (the DCT and wavelet coefficients are scaled for better visibility).

a given two-dimensional image $\mathbf{u}$, its variation at pixel (i, j) is

$$\|Du_{i,j}\|_2 = \left((u_{i+1,j} - u_{i,j})^2 + (u_{i,j+1} - u_{i,j})^2\right)^{1/2}.$$

It is clear from Figure 3.5 that a non-sparse image has nearly sparse representations.

Signal reconstruction and analysis are based on how a signal is represented. The standard representation is discrete sample points in time or space. They are called pixels for images. Some signals such as a stellar map and a wireless channel power vector are sparse under the standard representation, but most signals are not. The standard representation is suitable for operations such as down/upsampling, display, scaling, etc., because these operations are applied to the sample points independently or locally; they do not involve relations between multiple distant sample points. However, for more complicated tasks such as reconstruction, segmentation, and recognition, the standard representation is often inefficient. We need a meaningful representation that describes the useful characteristics, structures, and features of the signal. For example, in signal denoising, the representation should be sparse only for the wanted part of the signal; for the unwanted parts such as noise and artifacts, the representation should be dense. To recognize the object in a signal, as another example, the representation should be sparse for the most salient features of the object. There are many more other tasks (cf. books [74, 75]) that all look for sparse representations.

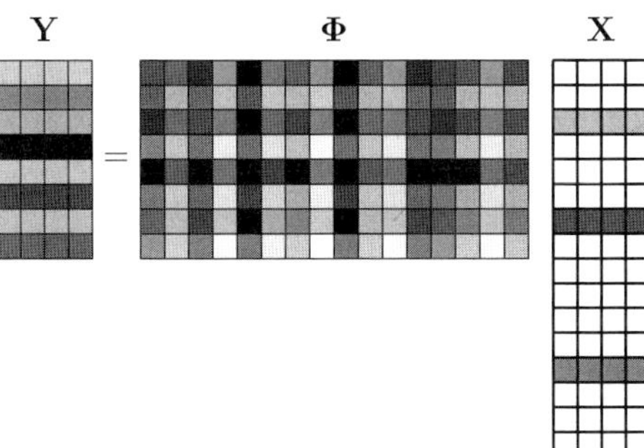

Figure 3.6 Joint sparse signal CS.

Sparse representation involves the choice of dictionary Ψ or sparsifying transform Υ. In the simplest cases, Ψ or Υ is orthogonal. Examples include the unitary discrete Fourier transform and various orthogonal wavelet bases. They often equip with fast computation at the complexity of $n \log n$ instead of n^2. Well-known non-orthogonal examples include total variation [76], the curvelet transform [77], and various tight frames. These dictionaries are analytic, often designed to handle specific kinds of signals. They have been extensively studied and widely considered in applications as they are easier to analyze and feature faster numerical implementations, compared to the learned (i.e., trained, as opposed to analytic) dictionaries, which we are going to discuss next.

The earliest major work in dictionary learning was due to Olshausen and Field [78], who trained an overcomplete dictionary for sparsely representing small image patches of a set of natural images. Remarkably, the atoms in the trained image were very similar to the simple cell receptive fields in early vision. This result suggests the potential of sparse representation in uncovering fundamental features in complex signals.

Modern training approaches such as MOD [79], kSVD [80], and many others can generate useful dictionaries out of various training sources, from just a single (incomplete) signal, to a few signals of the same kind, and to a database of signals. They can form dictionaries from scratch, with no structural restrictions whatsoever, or from certain functions or distributions with parameters learned from the training set. The training and usage of these dictionaries, especially those without any structures, are computationally expensive, but these dictionaries tend to do better on complex signals and complicated tasks for which no analytic dictionaries have been designed.

3.3.1 Extensions of sparse models

Joint sparse signals. A set of signals $\mathbf{u}^{(i)}, i = 1, \ldots, L$, of the same dimension are jointly sparse if each of them is sparse and their non-zero entries are colocated at (roughly) the same coordinates, or so under dictionaries or transforms. Figure 3.6 depicts joint sparse signal CS. The model is closely related to group LASSO [81]. Applications of the recovery of such signals [82] include multikernel machine learning [83], source

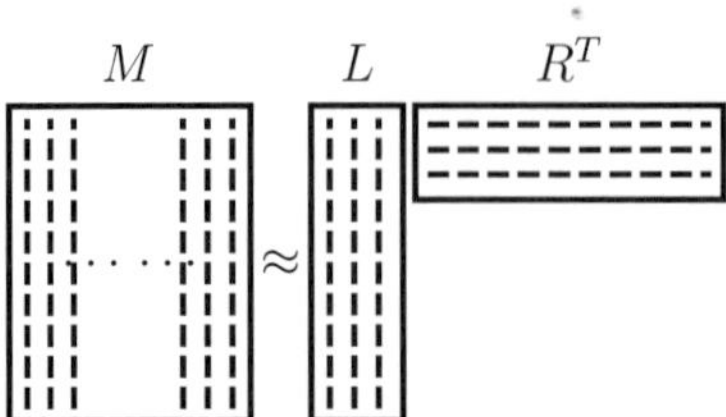

Figure 3.7 Approximation of a low-rank matrix.

localization [84], neuromagnetic imaging [85], wireless spectrum sensing [86], and many more.

Low-rank matrices. A matrix $\mathbf{M} \in \mathbb{R}^{m\times n}$ of rank $r \ll \min\{m, n\}$ has mn entries but only $r(m+n-r)$ degrees of freedom (consider the singular value decomposition $\mathbf{M} = \mathbf{U}\Sigma\mathbf{V}^T$; $\mathbf{U}$, Σ, and $\mathbf{V}$ have $\sum_{i=1}^{r}(m-i)$, r, and $\sum_{i=1}^{r}(n-i)$ degrees of freedom, respectively, which sum to $r(m+n-r)$). Figure 3.7 illustrates the approximation of a large low-rank matrix by a pair of skinny and fat matrices. Applications of low-rank matrix recovery [87–89] include model reduction [90], network Euclidean embedding [91], recovering shape and motion from image streams [92, 93], the Netflix recommendation problem [94], and more. In addition, there are models and applications of decomposing a matrix into the sum of low-rank and sparse parts [95].

Unions of subspaces and model-based CS. The support of a k-sparse vector is one of the $\binom{n}{k}$ possibilities, but for many signals in practice some or even most of these possibilities are not possible. For example, some transforms' coefficients follow a certain tree structure, some signals' non-zero entries tend to cluster, and some signals must lie in particular linear subspaces. These signals are easier to recover [96], and their structures can be generalized as the union of certain subspaces [97, 98].

3.4 CS encoding and decoding

In CS, the signal $\mathbf{u}^o = \Psi\mathbf{x}^o$ is encoded to $\mathbf{b} = \mathbf{A}\mathbf{u}^o$ with the hope of recovering $\mathbf{u}^o$ from $\mathbf{A}$ and $\mathbf{b}$. The recovery would be straightforward if $\mathbf{A}$ had full column rank; in this case, $\mathbf{u}^o$ would be the unique solution of

$$\min_{\mathbf{u}} \|\mathbf{b} - \mathbf{A}\mathbf{u}\|_2^2,$$

or $\mathbf{u}^o = (\mathbf{A}^T\mathbf{A})^{-1}\mathbf{A}^T\mathbf{b}$. However, CS uses the *downsampled* linear encoding, namely, $\mathbf{A}$ has fewer rows than columns. Such a matrix cannot have full column rank, and $\mathbf{b} = \mathbf{A}\mathbf{u}$ has multiple solutions. Recall that $\mathbf{x}^o$ is sparse or nearly sparse and, prior to CS signal reconstruction, the locations and values of its non-zero entries are unknown. So, the key question is: what kind of matrix $\mathbf{A}$ allows the recovery of $\mathbf{u}^o$ (or a good approximate of $\mathbf{u}^o$) from $\mathbf{b} = \mathbf{A}\mathbf{u}$ given merely that $\mathbf{x}^o$ is sparse? The answer is closely related to a concept called *coherence*.

Definition 3. *In $\mathbb{R}^n$ or $\mathbb{C}^n$, the coherence between the basis Φ, which has elements $\phi_1, \ldots, \phi_n$, and the basis Ψ, which has columns $\psi_1, \ldots, \psi_n$, is*

$$\mu(\Phi, \Psi) = \sqrt{n} \max_{1 \le i,j \le n} \frac{|\langle \phi_i, \psi_j \rangle|}{\|\phi_i\|_2 \|\psi_j\|_2}. \tag{3.7}$$

The quantity $\mu(\Phi, \Psi)$ measures how small the closest angle between any two elements of Φ and Ψ can be. If there exists a pair of elements ϕ_i and ψ_j obeying $\phi_i = \alpha\psi_j$ for some scalar α; i.e., they make a zero angle, then $\mu(\Phi, \Psi) = \sqrt{n}$, reaching its maximum, and we say that Φ and Ψ are coherent. On the other hand, no ψ_j can be orthogonal to all ϕ_is since $\Phi = \{\phi_i\}$ can represent any vector, including ψ_j. It turns out that the minimum of $\mu(\Phi, \Psi)$ is 1. Therefore, $1 \le \mu(\Phi, \Psi) \le \sqrt{n}$. When $\mu(\Phi, \Psi)$ is close to 1, we say that Φ and Ψ are incoherent. When Φ and Ψ are coherent, there exist signals that can be sparsely represented by either Φ or Ψ, and in particular, there exist ψ_i that is sparse under Φ and, conversely, ϕ_i that is sparse under Ψ. Such signals do not exist when they are incoherent.

How is this related to CS? One can associate the sensing matrix $\mathbf{A}$ with a certain orthogonal basis Φ in the way that the subspace spanned by the rows of $\mathbf{A}$ equals that by some m elements of Φ. Or, simply speaking, $\mathbf{Au}$ gives just m out of the n coefficients of $\Phi^T\mathbf{u}$. Then the question becomes what relation between two orthogonal bases Φ and Ψ allows the recovery of $\mathbf{x}^o$ (and thus $\mathbf{u}^o$) from incomplete coefficients of $\Phi^T\Psi\mathbf{x}^o$. The answer is *low coherence* between them, namely, $\mu(\Phi, \Psi)$ is much closer to 1 than $\sqrt{n}$. With low coherence, every element of Φ is "misaligned" with every element of Ψ, so $\langle \phi_i, \psi_j \rangle$s are roughly equal and stay uniformly away from zero. Hence, each coefficient i of $\Phi^T\Psi\mathbf{x}^o$, which equals $\sum_{1 \le j \le n} \langle \phi_i, \psi_j \rangle x_j^o$, encodes a guaranteed amount of information of each x_j^o. No x_j^o is left out, and no x_j^o dominates. In contrast, high coherence leads to uneven $\langle \phi_i, \psi_j \rangle$, and at least for some i, the coefficient i of $\Phi^T\Psi\mathbf{x}^o$ encodes just a few x_j^os while being nearly useless for the others. Since $\mathbf{x}^o$ is sparse, this coefficient i has a good chance to miss the non-zeros of $\mathbf{x}^o$ and thus become useless for recovery $\mathbf{x}^o$. Therefore, low coherence is a guarantee that every measurement does its job of carrying a useful amount of information of all the non-zeros of $\mathbf{x}^o$ no matter where they are. With high coherence, the quality of the measurements depends on the locations of the non-zeros of $\mathbf{x}^o$. The quality of measurements determines the number of measurements required for recovery, which is formally given in the following theorem.

Theorem 2 ([99]). *For a given $\mathbf{u}^o = \Psi\mathbf{x}^o$ where $\mathbf{x}^o$ has at most k non-zero entries, choose m entries of $\Phi^T\mathbf{u}^o$ uniformly at random, denoted as vector $\mathbf{b} = P_\Omega\Phi^T\mathbf{u}^o$, where P_Ω is the selection operator. As long as*

$$m \ge C \cdot \mu^2(\Phi, \Psi) \cdot (k \log n) \tag{3.8}$$

for some constant $C > 0$ independent of k and n, the solution to $\min\{\|\mathbf{x}\|_1 : \mathbf{b} = P_\Omega\Phi^T\Psi\mathbf{x}\}$ is $\mathbf{x}^o$ with overwhelming probability. (The result is shown for nearly all possible sign sequences of $\mathbf{x}^o$.)

The theorem claims that as long as Φ and Ψ are sufficiently incoherent and a signal is sparse when represented under basis Ψ, ℓ_1-minimization can exactly recover the signal from only $O(\mu^2(\Phi, \Psi) \cdot (k \log n))$ measurements, a number that can be much smaller than the dimension of the signal. Also, the recovery method does not need to know the number of non-zero entries in sparse representation or their locations. On the other hand, there are examples in [100, 101] on which no methods can achieve exact recovery with fewer than $k \log n$ measurements when $\mu(\Phi, \Psi) = 1$. Therefore, the required number of measurements in (3.8) cannot be significantly improved.

Besides low coherence, there are various results telling us when we can trust ℓ_1-minimization, namely, guarantee a successful CS recovery. To review these conditions, we would simply use the abstract model (3.4). When the measurements are $\mathbf{b} = P_\Omega \Phi^T \mathbf{u}^o$ and the signal is $\mathbf{u}^o = \Psi \mathbf{x}^o$, we can set $\mathbf{A} = P_\Omega \Phi^T \Psi$ in model (3.4). Among the widely used conditions are those based on the null-space property (NSP), the restricted isometry principle (RIP) [102], the spherical section property (SSP) [103], and a "RIPless" condition [104]. These properties have different strengths, and they together assert that a large number of sensing matrices such as those with entries sampled from sub-Gaussian distributions, the so-called Fourier and Walsh–Hadamard ensembles, and random Toeplitz and circulant ensembles are suitable for CS and have recovery guarantees. Next, we review these conditions and present the corresponding recovery guarantees.

The null-space property (NSP)

Definition 4. *Matrix* $\mathbf{A}$ *satisfies the NSP of order k if*

$$\|\mathbf{h}_S\|_1 < \|\mathbf{h}_{S^c}\|_1, \tag{3.9}$$

holds for all non-zero $\mathbf{h}$ *in the null space of* $\mathbf{A}$ *and all coordinate sets* $S \subset \{1, 2, \ldots, n\}$ *of cardinality* $|S| \leq k$.

In plain English, the condition examines every non-zero null-space vector of $\mathbf{A}$ and requires it not to concentrate its ℓ_1-energy on any set of k entries. This is a simple yet natural form of recovery conditions since all the feasible solutions of $\mathbf{Ax} = \mathbf{b}$ have the form $\mathbf{x} = \mathbf{x}^o + \mathbf{h}$, where $\mathbf{h}$ is a vector in the null space of $\mathbf{A}$. One of the very first steps toward recovery guarantees is to study the property of the null-space vectors of $\mathbf{A}$. The following theorem relates the null-space property to the exact recovery of k-sparse signals by ℓ_1 minimization.

Theorem 3 (NSP condition [105–107]). *Problem* (3.2) *uniquely recovers all k-sparse vectors* $\mathbf{x}^0$ *from measurements* $\mathbf{b} = \mathbf{Ax}^0$ *if and only if* $\mathbf{A}$ *satisfies the k-NSP.*

The proof is based on the following idea: whenever $\mathbf{x}^o$ has k or fewer non-zero entries, deviating from $\mathbf{x}^o$ to $\mathbf{x}^o + \mathbf{h}$ must cause an increase in ℓ_1-norm since in (3.9), $\|\mathbf{h}_S\|_1$ is the maximum possible decrease while $\|\mathbf{h}_{S^c}\|_1$ is the corresponding increase. Figure 3.8 illustrates a simple case where $\mathbf{x}^o$ is exactly recovered by ℓ_1 minimization since any point $\mathbf{x}^o + \mathbf{h}$ (which is on the gray plane) leads to a larger ℓ_1-norm.

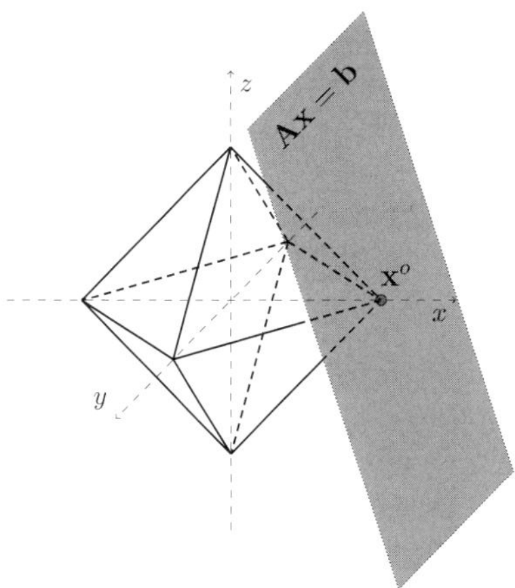

Figure 3.8 Exact recovery of $\mathbf{x}$ by ℓ_1 minimization.

Proof. Sufficiency. Pick any k-sparse vector $\mathbf{x}^o$. Let $\mathcal{S} := \text{supp}(\mathbf{x}^o)$ and $\mathcal{Z} = \mathcal{S}^c$. For any *non-zero* $\mathbf{h} \in \text{Null}(\mathbf{A})$, we have $\mathbf{A}(\mathbf{x}^o + \mathbf{h}) = \mathbf{A}\mathbf{x}^o = \mathbf{b}$ and

$$\begin{aligned}\|\mathbf{x}^0 + \mathbf{h}\|_1 &= \|\mathbf{x}^0_{\mathcal{S}} + \mathbf{h}_{\mathcal{S}}\|_1 + \|\mathbf{h}_{\mathcal{Z}}\|_1 \\ &\geq \|\mathbf{x}^0_{\mathcal{S}}\|_1 - \|\mathbf{h}_{\mathcal{S}}\|_1 + \|\mathbf{h}_{\mathcal{Z}}\|_1 \\ &= \|\mathbf{x}^0\|_1 + (\|\mathbf{h}_{\mathcal{Z}}\|_1 - \|\mathbf{h}_{\mathcal{S}}\|_1),\end{aligned} \tag{3.10}$$

where we have applied the triangle equality at the second line. Since the k-NSP property (3.9) of $\mathbf{A}$ guarantees $\|\mathbf{x}^0 + \mathbf{h}\|_1 > \|\mathbf{x}^0\|_1$, $\mathbf{x}^o$ is the unique solution of (3.2).

Necessity. The inequality (3.10) holds with equality if $\text{sign}(\mathbf{x}^o_{\mathcal{S}}) = -\text{sign}(\mathbf{h}_{\mathcal{S}})$ and $\mathbf{h}_{\mathcal{S}}$ has a sufficiently small scale. Therefore, for problem (3.2) to uniquely recover *all* k-sparse vectors $\mathbf{x}^0$, property (3.9) is also necessary. □

Note that although the NSP is necessary for exact recovery of all k-sparse vectors uniformly, it is no longer necessary if the uniformity is dropped, i.e., if one only considers certain k-sparse vector(s).

The NSP is widely used in the proofs of other conditions that guarantee the uniform recovery by (3.2) of sparse vectors. In addition, NSP of order $2k$ is necessary for stable uniform recovery. Consider an arbitrary decoder Δ, computationally tractable or not, which returns a vector from the input $\mathbf{b} = \mathbf{A}\mathbf{x}^o$. If one requires Δ to be stable in the sense

$$\|\mathbf{x}^o - \Delta(\mathbf{A}\mathbf{x}^o)\|_1 < C \cdot \sigma_{[k]}(\mathbf{x}^o) \tag{3.11}$$

for all vectors $\mathbf{x}^o$, where C is a universal constant and $\sigma_{[k]}$ is defined in (3.3) with $p = 1$ (using ℓ_1-norm to measure the best k-term approximation error), then $\mathbf{A}$ must obey

$$\|\mathbf{h}_{\mathcal{S}}\|_1 < C \cdot \|\mathbf{h}_{\mathcal{S}^c}\|_1, \tag{3.12}$$

for all non-zero $\mathbf{h}$ in the null space of $\mathbf{A}$ and all coordinate sets $\mathcal{S} \subset \{1, 2, \ldots, n\}$ of cardinality $|\mathcal{S}| \leq 2k$. One can show this by applying an argument in [108].

The restricted isometry principle (RIP). Next, we review a property regarding the "orthogonality" of $\mathbf{A}$. This property is more restrictive than the NSP, and several recovery results based on the RIP are proved through establishing the NSP. On the other hand, it gives robustness to measurement noise and has received more attention from researchers. Many recovery algorithms, including both optimization and greedy algorithms, have been analyzed under the RIP assumptions. Various random matrices satisfy the property with high probability. The property also has consequences beyond signal recovery such as stable embedding, dimension reduction, especially the relation to the Johnson–Lindenstrauss lemma.

Definition 5 ([102]). *Matrix* $\mathbf{A}$ *satisfies the RIP of order k if the smallest constant* δ_k *obeying*

$$(1-\delta_k)\|\mathbf{x}\|_2^2 \le \|\mathbf{A}\mathbf{x}\|_2^2 \le (1+\delta_k)\|\mathbf{x}\|_2^2 \tag{3.13}$$

for all k-sparse vectors $\mathbf{x} \in \mathbb{R}^n$ *satisfies* $\delta_k < 1$. δ_k *is called the restricted isometry constant of* $\mathbf{A}$.

Note that it is common practice to say that $\mathbf{A}$ satisfies the RIP as long as $\mathbf{A}$ can satisfy (3.13) after scaling. To apply (3.13) and its recovery results below, one might need to scale $\mathbf{A}$ first. Since CS is concerned with sensing matrices $\mathbf{A}$ with more columns than rows, such $\mathbf{A}$ cannot be orthogonal in the usual sense. On the other hand, the RIP requires $\mathbf{A}$ to be nearly orthogonal only restricted to the set of k-sparse vectors. It is sufficient for (3.2) to recover all k-sparse vector uniformly if the property holds for $2k$-sparse vectors with a certain maximum deviation from orthogonality in terms of δ_{2k}. Work [109] shows the sufficiency of $\delta_{2k} < 0.4142$, which is later improved to $\delta_{2k} < 0.4531$ [110], $\delta_{2k} < 0.4652$ [111], $\delta_{2k} < 0.4721$ [112], as well as $\delta_{2k} < 0.4931$ [113]. The last bound is up to date. We present a theorem from [113].

Theorem 4 (RIP condition for exact and stable recovery [113]). *Consider model* (3.4). *If* $\mathbf{A}$ *satisfies the RIP with* $\delta_{2k} \le 0.4931$, *then the solution* $\mathbf{x}^*$ *of* (3.5) *with* ϵ *set to* $\|\mathbf{w}\|_2$ *satisfies*

$$\|\mathbf{x}^* - \mathbf{x}^o\|_1 \le C_1 \cdot \sqrt{k}\|\mathbf{w}\|_2 + C_2 \cdot \sigma_{[k]}(\mathbf{x}^o), \tag{3.14}$$

$$\|\mathbf{x}^* - \mathbf{x}^o\|_2 \le \bar{C}_1 \cdot \|\mathbf{w}\|_2 + \bar{C}_2 \cdot \sigma_{[k]}(\mathbf{x}^o)/\sqrt{k}, \tag{3.15}$$

where C_1, C_2, $\bar{C}_1$, *and* $\bar{C}_2$ *are universal constants.*

The results claim that as long as the sensing matrix $\mathbf{A}$ has the RIP, the ℓ_1 recovery is stable against both the measurement noise and the signal noise, the latter of which is given in terms of the best k-term approximation error. When neither noise exists, namely, $\mathbf{w} = \mathbf{0}$ and $\mathbf{x}^o$ is k-sparse (thus $\sigma_{[k]}(\mathbf{x}^o) = 0$), inequalities (3.14) and (3.15) give the exact recovery of $\mathbf{x}^* = \mathbf{x}^o$. When there is measurement noise, i.e., $\mathbf{w} \ne \mathbf{0}$, the order of error $O(\|\mathbf{w}\|_2)$ in the recovered signal cannot be significantly improved since even if the locations of the largest k entries of $\mathbf{x}^o$ are given, the least-squares method will still give an error of $O(\|\mathbf{w}\|_2)$. When the signal is not exactly k-sparse, i.e., $\sigma_{[k]}(\mathbf{x}^o) \ne 0$, the error $O(\sigma_{[k]}(\mathbf{x}^o))$ is also tight since it is attained by power-law decaying signals. If the decay of

$\mathbf{x}^o$ follows the power law: the ith largest entry of $\mathbf{x}^o$ in magnitude obeys $|x_{(i)}| = C \cdot i^{-r}$, $r > 1$, then through basic integration, one gets $\sigma_{[k]}(\mathbf{x}^o)/\sqrt{k} = O(k^{-r+1/2})$. On the other hand, even if we let $\mathbf{x}^* = \mathbf{x}^o_{[k]}$, we still get $\|\mathbf{x}^* - \mathbf{x}^o\|_2 = O(k^{-r+1/2})$. Therefore, the bounds in (3.14) and (3.15) cannot be essentially improved. However, note that given additional properties of $\mathbf{w}$ and $\mathbf{x}^o$ such as their statistical distributions, one can do better. Bounds (3.14) and (3.15) are optimal up to constant factors when applied to all noise and signals uniformly.

Furthermore, the first inequality in the RIP condition (3.13) is necessary for stable recovery against *arbitrary* unknown noise $\mathbf{w}$. A constructive proof can be found in [114].

Recall that CS requires a sensing matrix that is non-adaptive to the underlying signal, namely, the design of the matrix should be independent of any particular signal or its existing measurements. (The design of adaptive and online sensing matrices is interesting and very useful but out of the scope of our discussions.) Somewhat surprisingly, several kinds of random matrices $\mathbf{A}$ satisfy the RIP with at least high probability up to proper scaling. They include, but are not limited to, sensing matrices $\mathbf{A}$ whose entries are i.i.d. samples from the standard normal distribution, symmetric Bernoulli distribution, as well as other sub-Gaussian distributions, those whose columns are sampled uniformly from the unit sphere $\mathbb{S}^{m-1}$, and those formed by randomly projecting (not randomly picking the rows of) an orthogonal basis. As long as their dimension $m \times n$ obeys $m \geq O(k \log(n/k))$, these matrices obey the RIP of order k with the fail probably decreasing to 0 exponentially fast in m; see [115, 116]. Various structured random matrices, which are easier to implement or naturally exist in applications, also satisfy the RIP though the best-known requirements on m are typically $O(k\,\mathrm{poly}(\log(n/k)))$; see [117]. Common examples of such matrices include those formed by randomly selected rows of a discrete Fourier matrix or matrix $\Phi\Psi$, where Φ and Ψ are incoherent, as well as random circulant or Toeplitz matrices. There are also works such as [118–120] that design deterministic matrices with RIPs but require m (sometimes, significantly) more than $O(k \log(n/k))$ yet still much less than $O(n)$.

The spherical section property. Next, we present recovery conditions based on the property SSP [103, 121] of $\mathbf{A}$. The advantage of this condition over the RIP is that the SSP is invariant to left-multiplying non-singular matrices to the sensing matrix $\mathbf{A}$ as pointed out in [103]. On the other hand, it does not have robustness against arbitrary unknown measurement noise $\mathbf{w}$.

Definition 6 (Δ-SSP [121]). *Let m and n be two integers such that $m > 0$, $n > 0$, and $m < n$. An $(n - m)$-dimensional subspace $\mathcal{V} \subset \mathbb{R}^n$ has the Δ spherical section property if*

$$\frac{\|\mathbf{h}\|_1}{\|\mathbf{h}\|_2} \geq \sqrt{\frac{m}{\Delta}} \tag{3.16}$$

holds for all non-zero $\mathbf{h} \in \mathcal{V}$.

To see the significance of (3.16), we note that (i) $\frac{\|\mathbf{h}\|_1}{\|\mathbf{h}\|_2} \geq 2\sqrt{k}$ for all $\mathbf{h} \in \mathrm{Null}(\mathbf{A})$ is a sufficient condition for the NSP inequality (3.9) and (ii) due to [122, 123], a uniformly random $(n-m)$-dimensional subspace $\mathcal{V} \subset \mathbb{R}^n$ has the SSP for

$$\Delta = C_0(\log(n/m) + 1)$$

with probability at least $1 - \exp(C_1(n-m))$, where C_0 and C_1 are universal constants. Hence, $m > 4k\Delta$ guarantees (3.9) to hold, and furthermore, if Null($\mathbf{A}$) is uniformly random, $m = O(k \log(n/m))$ is sufficient for (3.9) to hold with overwhelming probability (cf. [103, 121]).

Theorem 5 (SSP condition for exact recovery [103]). *Suppose* Null($\mathbf{A}$) *satisfies the* Δ*-SSP. If*

$$m \geq 4k\Delta, \tag{3.17}$$

then the null-space condition (3.9) *holds for all* $\mathbf{h} \in$ Null($\mathbf{A}$) *and coordinate sets* $\mathcal{S}$ *of cardinality* $|\mathcal{S}| \leq k$. *By Theorem 3,* (3.17) *guarantees that problem* (3.2) *recovers any* k*-sparse* $\mathbf{x}^o$ *from measurements* $\mathbf{b} = \mathbf{A}\mathbf{x}^o$.

Theorem 6 (SSP condition for stable recovery [103]). *Suppose* Null($\mathbf{A}$) *satisfies the* Δ*-SSP. Let* $\mathbf{x}^o \in \mathbb{R}^n$ *be an* arbitrary *vector. If*

$$m \geq 4k\Delta, \tag{3.18}$$

then the solution $\mathbf{x}^*$ *of* (3.2) *satisfies*

$$\|\mathbf{x}^* - \mathbf{x}^o\|_1 \leq 4\sigma_{[k]}(\mathbf{x}^o). \tag{3.19}$$

"RIPless" analysis. Unlike the NSP, RIP, and SSP, the "RIPless" analysis [104] gives *non-uniform* recovery guarantees in the sense that the recovery guarantee is not given for all (exactly or nearly) k-sparse signals uniformly but a single, arbitrary signal. However, it applies to a wide class of CS matrices such as those with i.i.d. sub-Gaussian entries, orthogonal transform ensembles satisfying an incoherence condition, random Teoplitz/circulant ensembles, as well as certain tight and continuous frame ensembles, at $O(k \log(n))$ measurements. This analysis is especially useful in situations where the RIP, as well as the NSP and the SSP, is difficult to check or does not hold.

Theorem 7 (RIPless for exact recovery [104]). *Let* $\mathbf{x}^o \in \mathbb{R}^n$ *be a fixed k-sparse vector. With probability at least* $1 - 5/n - e^{-\beta}$, $\mathbf{x}^0$ *is the unique solution to problem* (3.2) *with* $\mathbf{b} = \mathbf{A}\mathbf{x}^o$ *as long as the number of measurements*

$$m \geq C_0(1+\beta)\mu(\mathbf{A}) \cdot k \log n,$$

where C_0 *is a universal constant and* $\mu(\mathbf{A})$ *is an incoherence parameter of* $\mathbf{A}$ *(see [104] for its definition and values for various CS matrices).*

Non-ℓ_1 decoding methods

Besides ℓ_1 and ℓ_1-like minimization, there exist a large number of other optimization and non-optimization models and algorithms that can efficiently recover sparse or structured

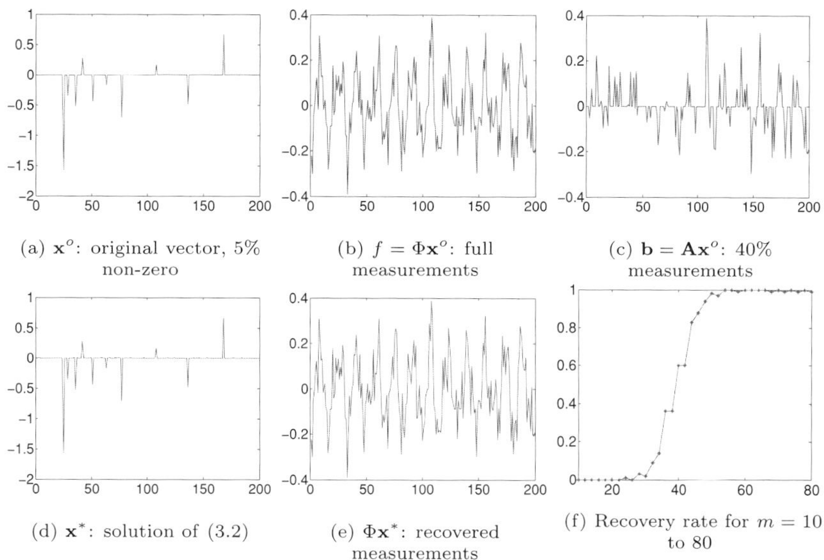

(a) $\mathbf{x}^o$: original vector, 5% non-zero

(b) $f = \Phi\mathbf{x}^o$: full measurements

(c) $\mathbf{b} = \mathbf{A}\mathbf{x}^o$: 40% measurements

(d) $\mathbf{x}^*$: solution of (3.2)

(e) $\Phi\mathbf{x}^*$: recovered measurements

(f) Recovery rate for $m = 10$ to 80

Figure 3.9 Sparse vector recovery from incomplete measurements. Φ is a discrete cosine transform.

signals from their CS measurements. Approaches such as greedy algorithms (e.g., [124–128]), iterative hard-thresholding algorithms (e.g., [129, 130]), combinatorial algorithms (e.g., [131–133]), and sublinear-complexity algorithms (e.g., [134, 135]) also enjoy recovery guarantees of different forms under certain assumptions on the sensing matrix and the signal. We review some of these methods in the next chapter but do not present the theorems of their recovery guarantees.

3.5 Examples

We close this chapter with two examples of CS encoding and decoding.

In the first example, we try to acquire a sparse vector $\mathbf{x}^o \in \mathbb{R}^n$ of length $n = 200$ with 10 non-zero entries; it is depicted in Figure 3.9(a). We let $\mathbf{A} \in \mathbb{R}^{m\times n}$ be formed from a random subset of $m = 80$ rows of the n-dimensional discrete cosine transform (DCT) Φ. Figure 3.9(b) depicts the full measurements $\Phi\mathbf{x}^o$, and Figure 3.9(c) depicts the CS measurements $\mathbf{b} = \mathbf{A}\mathbf{x}^o$, where missing measurements – those in $\Phi\mathbf{x}^o$ but not in $\mathbf{A}\mathbf{x}^o$ – are replaced by zeros. We solved model (3.2) to recover $\mathbf{x}^o$. The solution $\mathbf{x}^*$ is given in Figure 3.9(d), and it is equal to $\mathbf{x}^o$.

From a different perspective, this is also an example of missing data recovery [136]. Given $\mathbf{b}$, which is a portion of the full data $\mathbf{f} = \Phi\mathbf{x}^o$, one can recover the full data $\mathbf{f}$ by exploiting its sparsity under the DCT dictionary Φ. We plot the recovery $\Phi\mathbf{x}^*$ in Figure 3.9(e), and it is equal to $\Phi\mathbf{x}^o$.

Recall that one would like to take as few measurements as possible, that is, $m = k = 10$, but the use of non-adaptive sensing matrices in CS requires more measurements.

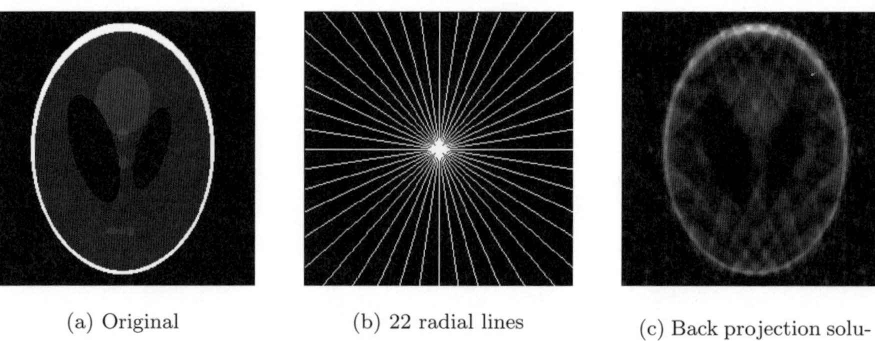

(a) Original (b) 22 radial lines (c) Back projection solution

Figure 3.10 Recovering the Shepp–Logan phantom from k-space samples on 22 radial lines.

To illustrate this point, we performed similar calculations for $m = 10, 11, \ldots, 80$, each with 100 repetitions of randomly chosen m measurements of $\Phi \mathbf{x}^o$. The percentages of successful recovery for all m are plotted in Figure 3.9(f), which shows that it is sufficient to have $m > 6k = 60$ to recover $\mathbf{x}^o$.

In the second example, we simulate ideal magnetic resonance imaging (MRI) of an ideal phantom depicted in Figure 3.10(a). The CS samples are a subset of its 2D Fourier coefficients, which correspond to the points on the 22 radial lines depicted in Figure 3.10(b). (One can further reduce the number of radial lines to as few as 8 by using more accurate algorithms and better models.) Roughly speaking (ignoring the various constraints in MRI sensing), the amount of sensing time required is proportional to the number of samples. CS can potentially reduce the acquisition time by recording only partial Fourier coefficients yet still return faithful reconstructions. Since the phantom image is piece-wise constant and the simulation does not introduce noise, we are able to *exactly* recover the original phantom by total variation minimization given the measurements. On the other hand, if the unsampled Fourier coefficients are replaced by 0, then the inverse Fourier transform gives the image in Figure 3.10(c).

4 Sparse optimization algorithms

This chapter reviews a collection of sparse optimization models and algorithms. The review focuses on introducing these algorithms along with their motivations and basic properties.

This chapter is organized as follows. Section 4.1 gives a short overview of convex optimization, including the definitions of convex sets and functions, the concepts of local and global optimality, as well as optimality conditions. A list of sparse optimization models is presented in Section 4.2; it deals with different types of signals and different kinds of noise and can also include additional features as objectives and constraints that arise in practical problems. Section 4.3 demonstrates that the convex sparse optimization problems can be transformed in equivalent cone programs and solved by off-the-shelf algorithms, yet it argues that these algorithms are usually inefficient or even infeasible for large-scale problems, which are typical in the CS applications. Sections 4.4–4.13 cover a large variety of (yet hardly complete) algorithms for sparse optimization. The list is not short because sparse optimization is a common ground where lots of optimization techniques, old or new, are found useful in varying senses. They have different strengths and fit different applications. One common reason that makes many of these algorithms efficient is their use of the shrinkage-like proximal operations, which can be computed very efficiently; so, we start with shrinkage in Section 4.4. Then, Section 4.5 presents a prox-linear framework and gives several algorithms under this framework that are based on gradient descent and take advantages of shrinkage-like operations. Duality is a very powerful tool in modern convex optimization; sparse optimization is not an exception. Section 4.6 derives a few dual models and algorithms for sparse optimization and discusses their properties. One class of dual algorithms is very efficient and extremely versatile, applicable to nearly all convex sparse optimization problems; it is based on variable/operator splitting and alternating minimization, which decomposes a difficult problem into a sequence of simple steps. The framework and applications of these algorithms are given in Section 4.7. Next, Section 4.8 briefly discusses the (block) coordinate descent algorithm, which has been a popular tool for solving many convex and non-convex problems in engineering practice. It can be integrated with prox-linear algorithms in Section 4.5 and becomes especially efficient on sparse optimization problems. Unlike other algorithms, the homotopy algorithms in Section 4.9 produces not just one solution but a path of solutions for the LASSO model (which is given in (4.13b) below); not only are they efficient at producing multiple solutions corresponding to different parameter values, they are especially fast if the solution is extremely sparse.

All the above algorithms can be (often significantly) accelerated by appropriately setting their step-size parameters that determine the amount of progress at each iteration. Such techniques are reviewed in Section 4.10. Since ℓ_q quasi-norms, $0 < q < 1$, are better approximations to the ℓ_0 "norm" than the ℓ_1 norm, solving non-convex ℓ_q minimization problems, when the global minimizers are successfully obtained, can give more faithful recoveries than ℓ_1 minimization. That's a big "if"; in Section 4.11, we explain why on problems with fast-decaying signals certain smoothing algorithms for ℓ_q minimization can indeed find global minimizers and perform better than ℓ_1 minimization. Unlike all previous algorithms, greedy algorithms do not necessarily correspond to an optimization model; instead of systematically searching for solutions that minimize objective functions, they build sparse solutions by constructing their supports step by step in a greedy manner. They have very good performance on fast-decaying signals. A few greedy algorithms are discussed in Section 4.12. Most of the algorithms and techniques in Sections 4.4–4.12 can be extended from sparse vector recovery to low-rank matrix recovery. However, for the latter problem, there is also a class of algorithms based on decomposing a low-rank matrix into the product of a skinny matrix and a fat one; algorithms in this class require relatively less memory and run very efficiently; they are briefly discussed in Section 4.13. Finally, Section 4.14 summarizes these algorithms and compares their advantages and suitable applications.

A large number of major algorithms for sparse optimization are covered in this chapter, but it would be impossible to cover all the sparse optimization algorithms in one chapter, and if we attempted to do so, the chapter would not fit the book and would be out of date in just a couple of months. Hence, we merely give a part of the big picture. Detailed analysis is omitted and referred to related papers.

4.1 A brief introduction to optimization

Before the discussions of the sparse optimization problems and algorithms, we first go over some necessary concepts of optimization.

A set $\mathcal{A}$ is called *affine* if the *line* through any two points in $\mathcal{A}$ lies entirely in $\mathcal{A}$; i.e., for any $\mathbf{x}_1, \mathbf{x}_2 \in \mathcal{A}$ and $\theta \in \mathbb{R}$, we have $\theta\mathbf{x}_1 + (1-\theta)\mathbf{x}_2 \in \mathcal{A}$. A function $f : \mathbb{R}^n \to \mathbb{R}$ is *affine* (also commonly called *linear*) if for any $\mathbf{x}_1, \mathbf{x}_2 \in \mathbb{R}^n$ and $\theta, \vartheta \in \mathbb{R}$, we have $f(\theta\mathbf{x}_1 + \vartheta\mathbf{x}_2) = \theta f(\mathbf{x}_1) + \vartheta f(\mathbf{x}_2)$.

A set $\mathcal{C}$ is called *convex* if the *line segment* between any two points in $\mathcal{C}$ lies entirely in $\mathcal{C}$; i.e., for any $\mathbf{x}_1, \mathbf{x}_2 \in \mathcal{C}$ and $\theta \in (0, 1)$, we have $\theta\mathbf{x}_1 + (1-\theta)\mathbf{x}_2 \in \mathcal{C}$. A function $f : \mathbb{R}^n \to \mathbb{R}$ is *convex* if $\mathrm{dom} f$ is convex and for any $\mathbf{x}_1, \mathbf{x}_2 \in \mathrm{dom} f$ and $\theta \in (0, 1)$, we have $f(\theta\mathbf{x}_1 + (1-\theta)\mathbf{x}_2) \le \theta f(\mathbf{x}_1) + (1-\theta) f(\mathbf{x}_2)$.

An optimization problem can be written in the form of

$$\min_{\mathbf{x}} \; f_0(\mathbf{x}) \quad \text{s.t. } f_i(\mathbf{x}) \le 0, \; i = 1, \ldots, m, \; h_j(\mathbf{x}) = 0, \; j = 1, \ldots, p, \tag{4.1}$$

where $\mathbf{x} = (x_1, \ldots, x_n)^T$ is the *decision variables*, $f_0 : \mathbb{R}^n \to \mathbb{R}$ is the *objective function*, $f_i : \mathbb{R}^n \to \mathbb{R}$, $i = 1, \ldots, m$, are the functions defining the *inequality constraints*, and

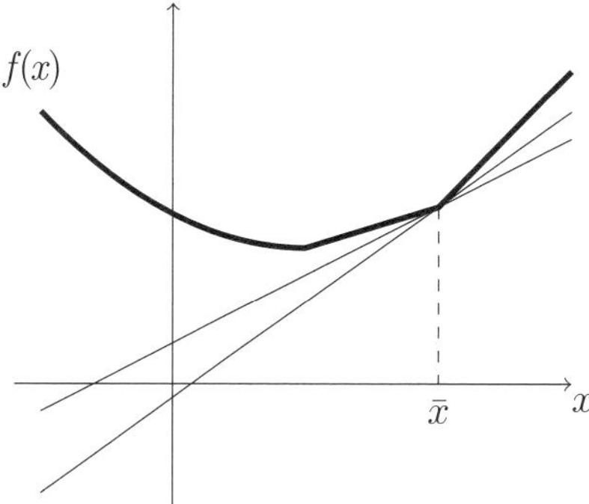

Figure 4.1 Two subgradients of a non-smooth function at $\bar{x}$.

$h_j : \mathbb{R}^n \to \mathbb{R}$, $j = 1, \ldots, p$ are the functions defining the *equality constraints*. If the functions $f_0, \ldots, f_m$ are convex and $h_1, \ldots, h_p$ are affine functions, problem (4.1) minimizes a convex function over a set of points lying in a convex set; such a problem is convex and called a convex optimization problem or a convex program.

All points $\mathbf{x}$ satisfying the constraints $f_i(\mathbf{x}) \leq 0$ $(i = 1, \ldots, m)$ and $h_j(\mathbf{x}) = 0$ $(j = 1, \ldots, p)$ are called *feasible solutions* or candidate solutions. The set of all feasible solutions is called the *feasible set*; we let it be denoted by $\mathcal{X}$. The feasible solution $\mathbf{x}^*$ that achieves the minimum objective value is called a *global (optimal) solution*. If $\mathbf{x}^*$ is feasible and achieves the minimal objective value among a neighborhood around $\mathbf{x}^*$, say, the set $\{\mathbf{x} : \|\mathbf{x} - \mathbf{x}^*\| \leq \delta\} \cap \mathcal{X}$ for some $\delta > 0$, then $\mathbf{x}^*$ is called a *local (optimal) solution*. In general, an optimization problem can have multiple local and global solutions. If the problem is convex, all local solutions are automatically global. If there are multiple global solutions, any point in their convex hull – the smallest convex set enclosing all of them – is also a global solution.

Next, we overview some optimality conditions used in developing algorithms, as well as to measure how far a current point is from optimality. First, we consider the unconstrained problem with $m = 0$ and $p = 0$, and the problem is to find $\mathbf{x}^*$ that minimizes $f_0(\mathbf{x})$. When f_0 is differentiable and convex, $\mathbf{x}^*$ is globally optimal if and only if $\nabla f_0(\mathbf{x}^*) = 0$. Without convexity, the condition is necessary but not necessarily sufficient.

When f_0 is convex but not differentiable, we need the concept of a subdifferential. The subdifferential of f at $\mathbf{x}$ is defined as the set

$$\partial f(\mathbf{x}) = \{\mathbf{y} : f(\mathbf{x}') \geq f(\mathbf{x}) + \langle \mathbf{y}, \mathbf{x}' - \mathbf{x} \rangle, \ \forall \mathbf{x}' \in \operatorname{dom} f\}. \tag{4.2}$$

If f is differentiable at $\mathbf{x}$, then $\partial f(\mathbf{x}) = \{\nabla f(\mathbf{x})\}$. A member of a subdifferential is called a subgradient. See Figure 4.1 for an example. At $\mathbf{x}^*$, $\mathbf{v}$ is a subgradient of f_0 if

$$f_0(\mathbf{x}) \geq f_0(\mathbf{x}^*) + \langle \mathbf{v}, \mathbf{x} - \mathbf{x}^* \rangle \tag{4.3}$$

for all $\mathbf{x}$. The optimality condition for minimizing a convex non-differentiable function f_0 is

$$\mathbf{0} \in \partial f_0(\mathbf{x}^*).$$

To deal with constraints, we need the Lagrangian $L : \mathbb{R}^n \times \mathbb{R}^m \times \mathbb{R}^p \to \mathbb{R}$ associated with problem (4.1), which is defined as

$$L(\mathbf{x}, \lambda, \nu) := f(\mathbf{x}) + \sum_{i=1}^{m} \lambda_i f_i(\mathbf{x}) + \sum_{j=1}^{p} \nu_j h_j(\mathbf{x}), \tag{4.4}$$

where λ_i and ν_j are the Lagrange multipliers associated with the ith inequality constraint $f_i(\mathbf{x}) \leq 0$ and the jth equality constraint $h_j(\mathbf{x}) = 0$, respectively.

The Lagrange dual function $g(\lambda, \nu) : \mathbb{R}^m \times \mathbb{R}^p \to \mathbb{R}$ is defined as

$$g(\lambda, \nu) = \inf_{\mathbf{x}} L(\mathbf{x}, \lambda, \nu), \tag{4.5}$$

and the dual problem is

$$\max_{\lambda, \nu} \; g(\lambda, \nu) \quad \text{s.t. } \lambda_i \geq 0, \; i = 1, \ldots, m. \tag{4.6}$$

Whether problem (4.1) is convex or not, the so-called *weak duality* relation holds, namely, for any feasible solutions $\mathbf{x}$ of (4.1) and λ obeying the constraints of (4.6), the objectives of (4.1) and (4.6) satisfy $g(\lambda, \nu) \leq f_0(\mathbf{x})$. Assuming that problem (4.1) is convex and has a finite optimal objective value, the *strong duality* relation holds if (4.6) has the same optimal objective value. For convex programs, strong duality usually (but not always) holds. Linear programs have strong duality unconditionally; Slater's condition, namely, the feasible set of (4.1) having a nonempty relative interior (there existing $\mathbf{x}$ feasible to (4.1) and obeying $f_i(\mathbf{x}) < 0, \; i = 1, \ldots, m$), is a sufficient condition of strong duality for general convex programs.

The Karush–Kuhn–Tucker (KKT) conditions are the first-order necessary optimality conditions when certain regularity conditions are satisfied (they are referred to as *constraint qualifications*, which are out of the scope of this book). Let $\mathbf{x}^*$ and (λ^*, ν^*) be any optimal solutions of problems (4.1) and (4.6) with $f_0(\mathbf{x}^*) = g(\lambda^*, \nu^*)$.

Definition 7. *Assume that in problem* (4.1)*, the objective function f and the constraint functions g_i and h_j are continuously differentiable at a point $\mathbf{x}^*$. The KKT conditions are*

$$\nabla f_0(\mathbf{x}^*) + \sum_{i=1}^{m} \lambda_i^* \nabla f_i(\mathbf{x}^*) + \sum_{j=1}^{p} \nu_j^* \nabla h_j(\mathbf{x}^*) = 0, \tag{4.7}$$

$$f_i(\mathbf{x}^*) \leq 0, \; i = 1, \ldots, m, \tag{4.8}$$

$$h_j(\mathbf{x}^*) = 0, \; i = 1, \ldots, p, \tag{4.9}$$

$$\lambda_i^* \geq 0, \; i = 1, \ldots, m, \tag{4.10}$$

$$\lambda_i f_i(\mathbf{x}^*) = 0, \; i = 1, \ldots, m. \tag{4.11}$$

When problem (4.1) is convex, the KKT conditions are also sufficient for global optimality. The points $\mathbf{x}^*$ and (λ^*, ν^*) are globally optimal to problems (4.1) and (4.6), respectively.

4.2 Sparse optimization models

In the next few sections, we overview a set of widely used algorithms for solving problems in the form of

$$\min_{\mathbf{x}}\{r(\mathbf{x}) : \mathbf{A}\mathbf{x} = \mathbf{b}\}, \tag{4.12a}$$

$$\min_{\mathbf{x}} r(\mathbf{x}) + \mu h(\mathbf{A}\mathbf{x} - \mathbf{b}), \tag{4.12b}$$

$$\min_{\mathbf{x}}\{r(\mathbf{x}) : \bar{h}(\mathbf{A}\mathbf{x} - \mathbf{b}) \leq \sigma\}, \tag{4.12c}$$

where $r(\mathbf{x})$, $h(\mathbf{x})$, and $\bar{h}(\mathbf{x})$ are convex functions.

Function $r(\mathbf{x})$ is a regularizer and is typically non-differentiable in sparse optimization problems; functions $h(\mathbf{x})$ and $\bar{h}(\mathbf{x})$ are data fidelity measures, which are often differentiable though there are exceptions such as ℓ_1 and ℓ_∞ fidelity measures (or loss functions), which are non-differentiable.

When introducing algorithms, our *first* focus is given to the ℓ_1 problems

$$\min_{\mathbf{x}}\{\|\mathbf{x}\|_1 : \mathbf{A}\mathbf{x} = \mathbf{b}\}, \tag{4.13a}$$

$$\min_{\mathbf{x}} \|\mathbf{x}\|_1 + \frac{\mu}{2}\|\mathbf{A}\mathbf{x} - \mathbf{b}\|_2^2, \tag{4.13b}$$

$$\min_{\mathbf{x}}\{\|\mathbf{x}\|_1 : \|\mathbf{A}\mathbf{x} - \mathbf{b}\|_2 \leq \sigma\}. \tag{4.13c}$$

We also consider the variants in which the ℓ_1 norm of $\mathbf{x}$ is replaced by regularization functions corresponding to transform sparsity, joint sparsity, low ranking, as well as those involving more than one regularization terms. These problems are convex and non-smooth. For example, as variants of problem (4.13a), most algorithms in this chapter can be extended to solving

$$\min_{\mathbf{s}}\{\|\mathbf{s}\|_1 : \mathbf{A}\Psi\mathbf{s} = \mathbf{b}\}, \tag{4.14a}$$

$$\min_{\mathbf{x}}\{\|\Upsilon\mathbf{x}\|_1 : \mathbf{A}\mathbf{x} = \mathbf{b}\}, \tag{4.14b}$$

$$\min_{\mathbf{u}}\{\mathrm{TV}(\mathbf{u}) : \mathcal{A}(\mathbf{u}) = \mathbf{b}\}, \tag{4.14c}$$

$$\min_{\mathbf{X}}\{\|\mathbf{X}\|_{2,1} : \mathbf{A}\mathbf{X} = \mathbf{b}\}, \tag{4.14d}$$

$$\min_{\mathbf{X}}\{\|\mathbf{X}\|_* : \mathcal{A}(\mathbf{X}) = \mathbf{b}\}, \tag{4.14e}$$

where $\mathcal{A}(\cdot)$ is a given linear operator, Υ is a linear transform, $\mathrm{TV}(\mathbf{u})$ is a certain discretization of total variation $\int |\nabla\mathbf{u}(x)|\mathrm{d}x$ (also, see [137] for a rigorous definition in the space of generalized functions of bounded variation), the $\ell_{2,1}$-norm is defined as

$$\|\mathbf{X}\|_{2,1} := \sum_{i=1}^{m} \left\|[x_{i1}\ x_{i,2} \cdots x_{in}]\right\|_2, \tag{4.15}$$

and the nuclear-norm $\|\mathbf{X}\|_*$ equals the sum of singular values of $\mathbf{X}$. Models (4.14a)–(4.14e) are widely used to obtain, respectively, a vector that is sparse under Ψ, a vector that becomes sparse by transform Υ, a piece-wise constant signal, a matrix consisted of joint-sparse vectors, and a low-rank matrix.

Decompose $\{1, \ldots, n\} = \mathcal{G}_1 \cup \mathcal{G}_2 \cup \cdots \cup \mathcal{G}_S$, where the $\mathcal{G}_i$s are certain sets of coordinates and $\mathcal{G}_i \cap \mathcal{G}_j = \emptyset$, $\forall i \neq j$. Define the (weighted) group $\ell_{2,1}$-norm as

$$\|\mathbf{x}\|_{\mathcal{G},2,1} = \sum_{s=1}^{S} w_s \|\mathbf{x}_{\mathcal{G}_s}\|_2,$$

where $w_s \geq 0$, $s = 1, \ldots, S$, are a given set of weights and $\mathbf{x}_{\mathcal{G}_s}$ is a subvector of $\mathbf{x}$ corresponding to the coordinates in $\mathcal{G}_s$. The model

$$\min_{\mathbf{x}} \{\|\mathbf{x}\|_{\mathcal{G},2,1} : \mathbf{A}\mathbf{x} = \mathbf{b}\} \tag{4.16}$$

tends to give solutions with all but a few non-zero subvectors $\mathbf{x}_{\mathcal{G}_s}$.

When the equality constraints of the problems in (4.14) and (4.16) do not hold for the target solutions, one can apply relaxation and penalty to the constraints like those in (4.13b) and (4.13c).

Depending on the type of noise, one may replace the ℓ_2-norm in (4.13b) and (4.13c) by other distance measures or loss functions such as the Kullback–Leibler divergence [138, 139] of two probability distributions p and q and the logistic loss function [140].

Practical problems sometimes have additional constraints such as the non-negativity constraints $\mathbf{x} \geq 0$ or bound constraints $\mathbf{l} \leq \mathbf{x} \leq \mathbf{u}$.

The $\ell_{2,1}$ norm in (4.15) is generalizable to the $\ell_{p,q}$-norm

$$\|\mathbf{X}\|_{p,q} = (\sum_i (\sum_j |x_{i,j}|^p)^{q/p})^{1/q},$$

which is convex if $1 \leq p, q \leq \infty$. For these problems and extensions, we study their algorithms extending from those for ℓ_1 minimization.

Furthermore, it is not rare to see problems with *complex-valued* variables and data. We discuss how to handle them in Section 4.6.6.

This chapter does not cover the model

$$\min_{\mathbf{x}} \{\|\mathbf{A}\mathbf{x} - \mathbf{b}\|_2 : \|\mathbf{x}\|_1 \leq k\},$$

as well as its variants. Solving such models may require projection to a polyhedral set such as the ℓ_1-ball $\{\mathbf{x} \in \mathbb{R}^n : \|\mathbf{x}\|_1 \leq 1\}$, which is more expensive than solving subproblems of minimizing ℓ_1-norm. However, there are efficient projection methods [141–143], as well as algorithms [144, 145] for ℓ_1-constrainted sparse optimization problems.

4.3 Classic solvers

Most non-smooth convex problems covered in this chapter can be transformed to optimization problems of standard forms, for which off-the-shelf algorithms and codes are available. In particular, using the identity

$$\|\mathbf{x}\|_1 = \min_{\mathbf{x}_1, \mathbf{x}_2} \{\mathbf{1}^T(\mathbf{x}_1 + \mathbf{x}_2) : \mathbf{x}_1 - \mathbf{x}_2 = \mathbf{x}, \mathbf{x}_1 \geq \mathbf{0}, \mathbf{x}_2 \geq \mathbf{0}\}, \tag{4.17}$$

where the minimizers $\mathbf{x}_1$ and $\mathbf{x}_2$ represent the positive and negative parts of $\mathbf{x}$, respectively, and $\mathbf{1} = [1, 1, \ldots, 1]^T$, one can transform problem (4.13a) to

$$\min_{\mathbf{x}_1, \mathbf{x}_2} \{\mathbf{1}^T(\mathbf{x}_1 + \mathbf{x}_2) : \mathbf{A}(\mathbf{x}_1 - \mathbf{x}_2) = \mathbf{b}, \mathbf{x}_1 \geq \mathbf{0}, \mathbf{x}_2 \geq \mathbf{0}\}. \tag{4.18}$$

From a solution $(\mathbf{x}_1^*, \mathbf{x}_2^*)$ to (4.18), one obtains a solution $\mathbf{x}^* = \mathbf{x}_1^* - \mathbf{x}_2^*$ to (4.13a).

Problem (4.18) is a standard-form linear program and can be solved by the simplex method, active-set method, as well as interior-point method, which have reliable academic and/or commercial solver packages.

Based on identity (4.17), problem (4.13b) can be transformed to a linearly constrained quadratic program (QP) and problem (4.13c) to a QP with quadratic and linear constraints, both of which can be further transformed to second-order cone programs (SOCPs) and solved by an interior-point method.

Likewise, we can model various nuclear-norm-based low-rank matrix recovery models as semi-definite programs (SDPs) based on the identity [146]

$$\|\mathbf{X}\|_* = \min_{\mathbf{Y},\mathbf{Z}} \left\{ \frac{1}{2}(\mathrm{tr}(\mathbf{Y}) + \mathrm{tr}(\mathbf{Z})) : \begin{bmatrix} \mathbf{Y} & \mathbf{X} \\ \mathbf{X}^T & \mathbf{Z} \end{bmatrix} \succeq \mathbf{0} \right\}, \tag{4.19}$$

where $\mathrm{tr}(\mathbf{M})$ is the trace of $\mathbf{M}$, and $\mathbf{M} \succeq \mathbf{0}$ means that $\mathbf{M}$ is symmetric and positive semi-definite. One can show (4.19) by considering the singular-value decomposition $\mathbf{X} = \mathbf{U}_{m\times m}\Sigma_{m\times n}\mathbf{V}_{n\times n}^T$ and noticing that due to the constraint in (4.19),

$$\begin{aligned} 0 &\leq \frac{1}{2}\mathrm{tr}\left(\begin{bmatrix} \mathbf{U}^T & -\mathbf{V}^T \end{bmatrix} \begin{bmatrix} \mathbf{Y} & \mathbf{X} \\ \mathbf{X}^T & \mathbf{Z} \end{bmatrix} \begin{bmatrix} \mathbf{U} \\ -\mathbf{V} \end{bmatrix} \right) \\ &= \frac{1}{2}\left(\mathrm{tr}(\mathbf{U}^T\mathbf{Y}\mathbf{U}) + \mathrm{tr}(\mathbf{V}^T\mathbf{Z}\mathbf{V})\right) - \mathrm{tr}(\Sigma) \\ &= \frac{1}{2}(\mathrm{tr}(\mathbf{Y}) + \mathrm{tr}(\mathbf{Z})) - \|\mathbf{X}\|_*, \end{aligned}$$

which equals 0 if we let $\mathbf{Y} = \mathbf{U}\Sigma_{m\times m}\mathbf{U}^T$ and $\mathbf{Z} = \mathbf{V}\Sigma_{n\times n}\mathbf{V}^T$.

While linear programming, SOCP, and SDP are well known and have very reliable solvers, many sparse optimization problems – especially those involving images and high-dimensional data or variables – are often too large in size for their solvers. For example, to recover an image of 10 million pixels, there will be at least 10 million variables along with, say, 1 million constraints in the problem, and even worse, in sparse optimization problems the constraints tend to have *dense* coefficient matrices. Such a data scale is far beyond the limit of the most standard solvers of linear programming, SOCP, and SDP. As a result, these solvers either run very slowly or report out of memory.

In the simplex and active-set methods, a matrix containing $\mathbf{A}$ must be inverted or factorized to compute the next point. This becomes very difficult when $\mathbf{A}$ is large and dense – which is often the case in sparse optimization applications. Each interior-point iteration approximately solves a Newton system and thus also factorizes a matrix involving $\mathbf{A}$.

Even when the factorization of a large and dense matrix can be done, it is still very disadvantageous to omit the facts that the solution will be sparse and that $\mathbf{A}$ may have structures permitting fast multiplications of $\mathbf{A}$ and $\mathbf{A}^T$ to vectors.

Nevertheless, simplex and interior-point methods are *reliable* and can return *highly accurate* solutions. They are good benchmark solvers for small-sized sparse optimization problems, and they can be integrated with fast but less-accurate solvers and in charge of returning accurate solutions to smaller subproblems.

For convex and non-smooth optimization problems, it is also natural to consider subgradient methods, namely, applying subgradient descent iterations to the unconstrained model (4.12b) and projected subgradient descent iterations to the constrained models (4.12a) and (4.12c). These methods are typically simple to implement, and the subgradient formulas for many non-smooth functions in this chapter are easy to derive. On the down side, the convergence guarantee and practical convergence speed of (projected) subgradient descents are somewhat weak. The convergence typically requires careful choice of step sizes, for example, those that diminish in the limit but diverge in sum (cf. [147]).

Unlike the above off-the-shelf methods, the algorithms we overview in this chapter do not necessarily invert or factorize large matrices or rely on the subgradients of non-smooth functions. Many of them are based on multiplying $\mathbf{A}$ and $\mathbf{A}^T$ to vectors, and exactly solving simpler subproblems involving the non-smooth functions. For problems with sparse solutions, such algorithms are more CPU and memory efficient than the off-the-shelf solutions above. Certain algorithms can even return highly accurate solutions.

4.4 Shrinkage operation

Shrinkage operation and its extensions are often used in sparse optimization algorithms. While ℓ_1-norm is non-differentiable, it has the excellent property of being component-wise separable (in the sense $\|\mathbf{x}\|_1 = \sum_{i=1}^n |x_i|$). It is very easy to minimize $\|\mathbf{x}\|_1$ together with other component-wise separable functions on $\mathbf{x}$ since it reduces to a sequence of independent subproblems. For example, the problem

$$\min_{\mathbf{x}} \|\mathbf{x}\|_1 + \frac{1}{2\tau}\|\mathbf{x} - \mathbf{z}\|_2^2, \tag{4.20}$$

where $\tau > 0$, is equivalent to solving $\min_{x_i} |x_i| + \frac{1}{2\tau}|x_i - z_i|^2$ over each i. Applying an analysis over the three cases of solution $(x_{\text{opt}})_i$ – positive, zero, and negative – one can obtain the closed-form solution

$$(x_{\text{opt}})_i = \begin{cases} z_i - \tau, & z_i > \tau, \\ 0, & -\tau \le z_i \le \tau, \\ z_i + \tau, & z_i < -\tau. \end{cases} \tag{4.21}$$

The solution is illustrated in Figure 4.2. Visually, the problem takes the input vector $\mathbf{z}$ and shrinks it to zero component-wise. Therefore, $(x_{\text{opt}})_i$ is called the *soft-thresholding*

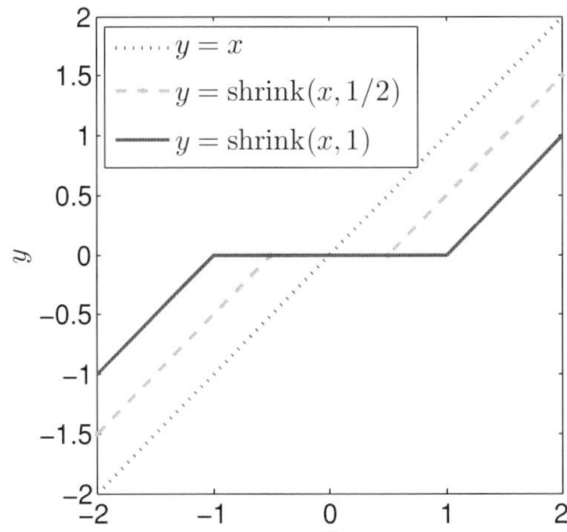

Figure 4.2 Illustration of shrink(x, τ).

or *shrinkage* of z_i, written as $(x_{\text{opt}})_i = \text{shrink}(z_i, \tau)$. The solution of (4.20) is given by

$$\mathbf{x}_{\text{opt}} = \text{shrink}(\mathbf{z}, \tau). \tag{4.22}$$

If we slightly abuse the notation and treat $0/0 = 0$, we get $\text{shrink}(z_i, \tau) = \max\{|z_i| - \tau, 0\} \cdot (z_i/|z_i|)$ and thus

$$\text{shrink}(\mathbf{z}, \tau) = \max\{|\mathbf{z}| - \tau, \mathbf{0}\} \cdot \text{sign}(\mathbf{z}), \quad \text{component-wise,} \tag{4.23}$$

where

$$\text{sign}(x) = \begin{cases} 1, & x > 0, \\ 0, & x = 0, \\ -1, & x < 0. \end{cases}$$

Operation (4.23) has been applied to image denoising [148] (applied to the wavelet coefficients of images) and has been found in earlier literature. It is now underlying a large number of sparse optimization algorithms and solvers.

If problem (4.20) is subject to constraints $\mathbf{x} \geq \mathbf{0}$, the solution becomes

$$\mathbf{x}_{\text{opt}} = \max\{\mathbf{z} - \tau, \mathbf{0}\}, \quad \text{component-wise.} \tag{4.24}$$

Unlike general non-smooth optimization problems, sparse optimization problems generally have rather simple non-smooth terms $r(\mathbf{x})$. One kind of simplicity is that $r(\mathbf{x})$ is easy to minimize together with simple functions such as $\|\mathbf{x} - \mathbf{z}\|_2^2$. As a side note, the general problem of minimizing $r(\mathbf{x}) + \frac{1}{2}\|\mathbf{x} - \mathbf{z}\|_2^2$ is called the Moreau–Yosida regularization of r, which has a rich literature.

4.4.1 Generalizations of shrinkage

As long as the second term in (4.20) is *convex* and *component-wise separable*, the solution of (4.20) is easy to obtain. We next study a few useful generalizations to (4.20), which are not component-wise separable but still easy to handle. They are used in algorithms for obtaining, for example, transform, group, or jointly sparse vectors, as well as low-rank matrices.

If ℓ_1 is replaced by ℓ_2 in (4.20), we obtain the problem

$$\min_{\mathbf{x}} \|\mathbf{x}\|_2 + \frac{1}{2\tau}\|\mathbf{x} - \mathbf{z}\|_2^2, \tag{4.25}$$

whose solution is given by

$$\max\{\|\mathbf{x}\|_2 - \tau, 0\} \cdot (\mathbf{x}/\|\mathbf{x}\|_2). \tag{4.26}$$

Unlike component-wise shrinkage (4.23), (4.26) shrinks a vector toward the origin along the radial line. Problem (4.26) may appear as subproblems in joint/group-sparse recovery algorithms.

When $\mathbf{x}$ and $\mathbf{z}$ are *complex-valued*, (4.23) and (4.26) remain correct under the definition $\|\mathbf{y}\|_2 = \sum_i |\mathrm{r}(y_i)^2 + \mathrm{i}(y_i)^2|^{1/2}$, where $\mathrm{r}(\cdot)$ and $\mathrm{i}(\cdot)$ are the real and imaginary parts of the input, respectively.

Let $\mathbf{Z} = \mathbf{U}\Sigma\mathbf{V}^T$ be the singular value decomposition of matrix $\mathbf{Z}$ and $\hat{\Sigma}$ be the diagonal matrix with diagonal entries

$$\mathrm{diag}(\hat{\Sigma}) = \mathrm{shrink}(\mathrm{diag}(\Sigma), \tau), \tag{4.27}$$

then we have from [149]:

$$\mathbf{U}\hat{\Sigma}\mathbf{V}^T = \arg\min_{\mathbf{X}} \|\mathbf{X}\|_* + \frac{1}{2\tau}\|\mathbf{X} - \mathbf{Z}\|_F^2, \tag{4.28}$$

where we recall that $\|\cdot\|_*$ and $\|\cdot\|_F^2$ equal the sum of the singular values and squared singular values of the input matrix, respectively. This result follows essentially from the unitary invariant properties: $\|\mathbf{X}\|_* \equiv \|\mathbf{AXB}\|_*$ and $\|\mathbf{X}\|_F \equiv \|\mathbf{AXB}\|_F$ for any matrix $\mathbf{X}$ and unitary matrices $\mathbf{A}$ and $\mathbf{B}$. In general, matrix problems with only unitary-invariant functions (e.g., $\|\cdot\|_*$, $\|\cdot\|_F$, spectral norm, trace, all Schatten (semi)-norms) and constraints (e.g., positive or negative (semi)-definiteness) typically reduce to vector problems regarding singular values. Computationally, singular value decomposition is expensive for large matrices, and, if it is needed every iteration, it can become the performance bottleneck. Therefore, algorithms that iteratively call (4.28) often use its approximate; see [150, 151], for example.

If we generalize (4.20) to

$$\min_{\mathbf{x}} \|\Psi\mathbf{x}\|_1 + \frac{1}{2\tau}\|\mathbf{x} - \mathbf{z}\|_2^2, \tag{4.29}$$

where Ψ is an orthogonal linear transform[1] then by introducing variable $\mathbf{y} = \Psi\mathbf{x}$, we have $\mathbf{x} = \Psi^T\mathbf{y}$ and $\|\mathbf{x} - \mathbf{z}\|_2 = \|\Psi(\mathbf{x} - \mathbf{z})\|_2 = \|\mathbf{y} - \Psi\mathbf{z}\|_2$ and obtain the new problem

$$\min_{\mathbf{y}} \|\mathbf{y}\|_1 + \frac{1}{2\tau}\|\mathbf{y} - \Psi\mathbf{z}\|_2^2. \tag{4.30}$$

Given the solution $\mathbf{y}_{\text{opt}}$ of (4.30), applying $\mathbf{x}_{\text{opt}} = \Psi^T\mathbf{y}_{\text{opt}}$, we recover the solution of (4.29):

$$\mathbf{x}_{\text{opt}} = \Psi^T \operatorname{shrink}(\Psi\mathbf{z}, \tau). \tag{4.31}$$

Another generalization involves total variation, which is often used in image-processing tasks [76] and models for piece-wise constant signals:

$$\min_{\mathbf{u}} \operatorname{TV}(\mathbf{u}) + \frac{1}{2\tau}\|\mathbf{u} - \mathbf{z}\|_2^2, \text{ where } \mathbf{u} \text{ is a matrix.} \tag{4.32}$$

There is no known closed-from solution for (4.32) except for $\mathbf{z}$ containing simple geometries such as disks. However, problem (4.32) has very efficient solvers that take advantage of the "local" structure of total variation such as graph-cut [152, 153], primal-dual algorithm [154], as well as [155–157]. The alternating direction method in Section 4.7 below is also efficient, and on total variation problems, it can be applied to models that are more complicated than (4.32); see [158, 159].

A challenge arises when the regularization function consists of more than one term, for example, $r(\mathbf{u}) = \|\Psi\mathbf{u}\|_1 + \operatorname{TV}(\mathbf{u})$. Another challenge is $r(\mathbf{u}) = \|\Psi\mathbf{u}\|_1$, where Ψ is *not* orthogonal. Even if Ψ is a tight frame, which satisfies $\Psi^T\Psi = \mathbf{I}$ and enjoys Parseval's identity, the same trick of changing variable no longer guarantees the correctness of (4.31). In these cases, the Moreau–Yosida regularization of r is not trivial. Typically, solutions are variable splitting and decomposition; see Section 4.7 for an alternative method based on variable splitting.

4.5 Prox-linear algorithms

The problems in Section 4.4 with simple solutions appear as subproblems in the so-called prox-linear method. To introduce this method, we consider a general form. Problems (4.12b) and (4.13b) can be written in the general form

$$\min_{\mathbf{x}} r(\mathbf{x}) + f(\mathbf{x}), \tag{4.33}$$

where r is the regularization function and f is the data fidelity function. The prox-linear algorithm has a long history in the literature; see [160–162] for example. It is based on *linearizing* the second term $f(\mathbf{x})$ at each $\mathbf{x}^k$ and adding a *proximal term* hence, getting the name prox-linear, giving rise to the iteration

$$\mathbf{x}^{k+1} = \arg\min_{\mathbf{x}} r(\mathbf{x}) + f(\mathbf{x}^k) + \langle \nabla f(\mathbf{x}^k), \mathbf{x} - \mathbf{x}^k \rangle + \frac{1}{2\delta_k}\|\mathbf{x} - \mathbf{x}^k\|_2^2. \tag{4.34}$$

[1] Ψ can be represented by a matrix with orthonormal columns.

The last term keeps $\mathbf{x}^{k+1}$ close to $\mathbf{x}^k$, and the parameter δ_k determines the step size. If r is ℓ_1-norm and the last term of (4.34) was not there, the solution would be either 0 or unbounded. Note that the second term is independent of the decision variable $\mathbf{x}$, and the last two terms can be combined to a complete square plus a constant as follows

$$\langle \nabla f(\mathbf{x}), \mathbf{x} - \mathbf{x}^k \rangle + \frac{1}{2\delta_k} \|\mathbf{x} - \mathbf{x}^k\|_2^2 = \frac{1}{2\delta_k} \left\|\mathbf{x} - \left(\mathbf{x}^k - \delta_k \nabla f(\mathbf{x}^k)\right)\right\|_2^2 + c,$$

where $c = -\delta_k \|\nabla f(\mathbf{x}^k)\|_2^2$ is independent of the decision variable $\mathbf{x}$. Therefore, iteration (4.34) is equivalent to

$$\mathbf{x}^{k+1} = \arg\min_{\mathbf{x}} \; r(\mathbf{x}) + \frac{1}{2\delta_k} \left\|\mathbf{x} - \left(\mathbf{x}^k - \delta_k \nabla f(\mathbf{x}^k)\right)\right\|_2^2. \tag{4.35}$$

It is now clear that the motivation of linearizing just function f is that iteration (4.35) is easy to compute for the aforementioned examples of regularization function r.

4.5.1 Forward-backward operator splitting

Besides the above prox-linear interpretation, there are other alternative approaches to obtain (4.34) such as forward-backward operator splitting, optimization transfers, and successive upper (first-order) approximation. We briefly review forward-backward operator splitting since it is general and very elegant. Various operator splitting schemes have been developed since the 1950s. Well-known ones are included in [163–167]; a good reference for forward-backward splitting and its applications in signal processing is [168]. It is based on manipulating subdifferential operators.

The first-order optimality condition of unconstrained, convex, non-smooth minimization is given based on the subdifferential of the objective function.

Lemma 1 ([169]). *Point $\mathbf{x}_{\text{opt}}$ is the minimizer of a convex function f if and only if*

$$\mathbf{0} \in \partial f(\mathbf{x}_{\text{opt}}), \tag{4.36}$$

where we have ignored the technical trivial to ensure the existence of ∂f.

The proof is trivial since $f(\mathbf{x}) \geq f(\mathbf{x}_{\text{opt}}) = f(\mathbf{x}_{\text{opt}}) + \mathbf{0}^T(\mathbf{x} - \mathbf{x}_{\text{opt}})$ if and only if (4.36) holds.

For (4.33), introduce operators T_1 and T_2:

$$T_1(\mathbf{x}) = \partial r(\mathbf{x}) \quad \text{and} \quad T_2(\mathbf{x}) = \partial f(\mathbf{x}).$$

The subdifferential obeys the rule

$$f(\mathbf{x}) = \alpha_1 f_1(\mathbf{x}) + \alpha_2 f_2(\mathbf{x}) \longrightarrow \partial f(\mathbf{x}) = \alpha_1 \partial f_1(\mathbf{x}) + \alpha_2 \partial f_2(\mathbf{x})$$

as long as $\alpha_1, \alpha_2 \geq 0$. Therefore, the optimality condition (4.36) for model (4.33) is $\mathbf{0} \in \partial(r + f)(\mathbf{x}_{\text{opt}}) = T_1(\mathbf{x}_{\text{opt}}) + T_2(\mathbf{x}_{\text{opt}})$, which also holds after multiplying a factor

$\tau = \delta_k^{-1} > 0$. Hence, we have

$$\begin{aligned}\mathbf{x}_{\text{opt}} \text{ solves (4.33)} &\Leftrightarrow \mathbf{0} \in \tau T_1(\mathbf{x}_{\text{opt}}) + \tau T_2(\mathbf{x}_{\text{opt}})\\ &\Leftrightarrow \mathbf{0} \in (\mathbf{x}_{\text{opt}} + \tau T_1(\mathbf{x}_{\text{opt}})) - (\mathbf{x}_{\text{opt}} - \tau T_2(\mathbf{x}_{\text{opt}}))\\ &\Leftrightarrow (I - \tau T_2)(\mathbf{x}_{\text{opt}}) \in (I + \tau T_1)(\mathbf{x}_{\text{opt}})\\ &\overset{(*)}{\Leftrightarrow} \mathbf{x}_{\text{opt}} = (I + \tau T_1)^{-1}(I - \tau T_2)(\mathbf{x}_{\text{opt}}), \end{aligned} \tag{4.37}$$

where for $(*)$ to hold, we have assumed that $\mathbf{z} \in (I + \tau T_1)(\mathbf{x})$ determines a unique $\mathbf{x}$ for a given $\mathbf{z}$. Following from the definition $T_1 = \partial r$ and Lemma 1, the solution $\mathbf{x}$ to $\mathbf{z} \in (I + \tau T_1)(\mathbf{x})$ is the minimizer of $\tau r(\mathbf{x}) + \frac{1}{2}\|\mathbf{x} - \mathbf{z}\|_2^2$. In other words,

$$(I + \tau T_1)^{-1}(\mathbf{z}) = \arg\min_{\mathbf{x}} \tau r(\mathbf{x}) + \frac{1}{2}\|\mathbf{x} - \mathbf{z}\|_2^2. \tag{4.38}$$

If we let $\mathbf{z} = (I - \tau T_2)(\mathbf{x}_{\text{opt}})$, we obtain (4.37); on the other hand, letting $\tau = \frac{1}{2\delta_k}$ and $\mathbf{z} = (I - \tau T_2)(\mathbf{x}^k) = \mathbf{x}^k - \frac{1}{2\delta_k}\nabla f(\mathbf{x}^k)$, we obtain

$$(I + \tau T_1)^{-1}(\mathbf{z}) = \arg\min_{\mathbf{x}} \tau r(\mathbf{x}) + \frac{1}{2}\|\mathbf{x} - (\mathbf{x}^k - \frac{1}{2\delta_k}\nabla f(\mathbf{x}^k))\|_2^2, \tag{4.39}$$

which only differs from the right-hand side of (4.34) by a quantity independent of $\mathbf{x}$. Hence, (4.34) and (4.39) yield the same solution, and thus (4.34) is precisely a fixed point iteration of (4.37):

$$\mathbf{x}^{k+1} = (I + \tau T_1)^{-1}(I - \tau T_2)(\mathbf{x}^k). \tag{4.40}$$

The operation $(I - \tau T_2)$ is called the forward step, or the gradient descent step. The operation $(I + \tau T_1)^{-1}$ is called the backward step, or the minimization step. Such splitting makes (4.37) and (4.40) easy to carry out for various interesting examples of r and f.

4.5.2 Examples

Since the forward step is straightforward for any differentiable function f, we focus on different choices of the function r.

Example 1 (Basis pursuit denoising). Let $r(\mathbf{x}) = \|\mathbf{x}\|_1$ and $f(\mathbf{x})$ be a differentiable function (e.g., $\frac{\mu}{2}\|\mathbf{A}\mathbf{x} - \mathbf{y}\|_2$ corresponding to (4.13b)). The backward step (4.38) is given by (4.22). Hence, (4.40) becomes

$$\mathbf{x}^{k+1} = \text{shrink}(\mathbf{x}^k - \delta_k \nabla f(\mathbf{x}^k), \delta_k). \tag{4.41}$$

The main computation at each iteration k is $\nabla f(\mathbf{x}^k)$. For problem (4.13b), it becomes $\mathbf{A}^T\mathbf{A}\mathbf{x}^k$.

If we generalize to $r(\mathbf{x}) = \|\Psi\mathbf{x}\|_1$ for an orthogonal linear transform, then following (4.31), (4.40) is given by

$$\mathbf{x}^{k+1} = \Psi^T \operatorname{shrink}(\Psi(\mathbf{x}^k - \delta_k \nabla f(\mathbf{x}^k)), \delta_k). \tag{4.42}$$

Example 2 (Total variation image reconstruction). Let $\mathbf{x}$ denote an image and $\mathrm{TV}(\mathbf{x})$ denotes its total variation. The backward step (4.38) solves the so-called Rudin–Osher–Fatemi model [76]

$$\min_{\mathbf{x}} \tau \mathrm{TV}(\mathbf{x}) + \frac{1}{2}\|\mathbf{x} - \mathbf{z}\|_2^2. \tag{4.43}$$

Since no closed-form solution is known, the iteration (4.40) is carried out in two steps:

$$\text{forward step:} \quad \mathbf{z}^k = \mathbf{x}^k - \delta_k \nabla f(\mathbf{x}^k),$$

$$\text{backward step:} \quad \mathbf{x}^{k+1} = \arg\min_{\mathbf{x}} \mathrm{TV}(\mathbf{x}) + \frac{1}{2\delta_k}\|\mathbf{x} - \mathbf{z}\|_2^2.$$

Example 3 (Joint sparse recovery for inexact measurements). Suppose we seek a solution $\mathbf{X} \in \mathbb{R}^{N \times L}$, with rows $\mathbf{x}_1, \ldots, \mathbf{x}_n$, that has all but a few rows containing non-zero entries. One approach is to set $r(\mathbf{X}) = \sum_{i=1}^N \|\mathbf{x}_i\|_p$ for some $p \geq 1$. The backward step (4.38) is separable across rows of $\mathbf{X}$ and, if $p > 1$, it is not further separable. The iteration (4.40) is carried out in two steps:

$$\text{forward step:} \quad \mathbf{Z}^k = \mathbf{X}^k - \delta_k \nabla f(\mathbf{X}^k),$$

$$\text{backward step:} \quad \mathbf{x}_i^{k+1} = \arg\min_{\mathbf{x}} \|\mathbf{x}\|_p + \frac{1}{2\delta_k}\|\mathbf{x} - \mathbf{z}_i^k\|_2^2, \; i = 1, \ldots, N,$$

where $\mathbf{z}_i^k$ denotes row i of $\mathbf{Z}^k$. For $p = 2$, the backward step is solved by (4.26).

Example 4 (Low-rank matrix recovery). In this case, we let $r(\mathbf{X}) = \|\mathbf{X}\|_*$ and consider the model

$$\min_{\mathbf{X}} r(\mathbf{X}) + f(\mathbf{X}), \tag{4.44}$$

where $f(\mathbf{X})$ is a convex function of $\mathbf{X}$. Problem (4.44) is a generalization of the unconstrained low-rank matrix recovery model

$$\min \|\mathbf{X}\|_* + \frac{\mu}{2}\|\mathcal{A}(\mathbf{X}) - \mathbf{b}\|_2^2, \tag{4.45}$$

where $\mathcal{A}$ is a linear operator and $\mu > 0$ is a penalty parameter. The iteration (4.40) gives

$$\mathbf{X}^{k+1} = \operatorname{shrink}_*(\mathbf{X}^k - \delta_k \nabla f(\mathbf{X}^k), \delta_k), \tag{4.46}$$

where shrink_* is given by (4.27) and (4.28). Algorithm FPCA [150] is based on (4.46).

Example 5 (Composite TV+ℓ_1 regularization). In this example, the model is

$$\min_{\mathbf{x}} r(\mathbf{x}) + \frac{\mu}{2}\|\mathbf{A}\mathbf{x} - \mathbf{b}\|_2^2,$$

where

$$r(\mathbf{x}) = \mathrm{TV}(\mathbf{x}) + \alpha\|\Psi\mathbf{x}\|_1.$$

The regularizer arises in image processing [170] to take advantage of both total variation and a sparsifying transform Ψ and also in signal processing and statistics [171] with $\Psi = \mathbf{I}$ to obtain a sparse solution with clustered non-zero entries. Sparsity and clustering are results of minimizing $\|\mathbf{x}\|_1$ and $\mathrm{TV}(\mathbf{x})$, respectively. An algorithm is proposed in [170] that uses forward-backward splitting twice; there are two forward steps and two backward steps, involving both (4.23) and (4.26).

The prox-linear iterations can be coupled with various techniques and achieve significant acceleration. Some of these techniques are reviewed in Section 4.10 below.

4.5.3 Convergence rates

If the objective function $F(\mathbf{x})$ is strongly convex, namely,

$$\alpha F(\mathbf{x}) + (1-\alpha)F(\mathbf{y}) \geq F(\alpha\mathbf{x} + (1-\alpha)\mathbf{y}) + \frac{\nu}{2}\alpha(1-\alpha)\|\mathbf{x} - \mathbf{y}\|_2^2$$

holds for any $\alpha \in [0, 1]$, then it is easy to show that a prox-linear algorithm with proper step sizes δ_k converges at a rate $O(1/c^k)$, where $0 < c < 1, c > 1$ is a certain number depending on ν and other properties of F and k is the number of iterations. However, problems such as (4.13b) are general not strongly convex, especially when $\mathbf{A}$ is a fat matrix, namely, the number of rows of $\mathbf{A}$ is less than that of its columns. With only convexity, the rate reduces to $O(1/k)$. For problems (4.12b) with $r(\mathbf{x}) = \|\mathbf{x}\|_1$ and smooth $\bar{h}(\mathbf{x})$ and (4.13b), linear convergence in the final stage of the algorithm can be obtained [172, 173], and [173] also shows that after a finite number of iterations, roughly speaking, $\mathbf{x}^k$ must have the optimal support $\mathrm{supp}(\mathbf{x}_{\mathrm{opt}})$. Nesterov's method [174] for minimizing smooth convex functions can be extended to the prox-linear algorithm in different ways for different settings [175–178]; they have a better global rate at $O(1/k^2)$ while only marginally increasing the computing cost per iteration. These are provable rates; on practical problems, researchers report that the advantage of $O(1/k^2)$-complexity methods is more apparent in the initial phase, and when the iterations reduce to a small set of coordinates, they are not necessarily faster and can even be slower.

4.6 Dual algorithms

While the algorithms in Section 4.5 solve the penalty model (4.13b), the algorithms in this section can generate a sequence of points that converges to a solution to the equality constrained model (4.13a). Interestingly, some of the points in this sequence, though not solving (4.13b) or (4.13c), have "better quality" in terms of sparsity and data fitness than the solutions of (4.13b) or (4.13c).

Coupled with splitting techniques, dual algorithms in this section can solve problems (4.13b) or (4.13c) as well.

4.6.1 Dual formulations

For a constrained optimization problem like (4.13a), duality gives rise to a companion problem. Between the original (primal) and the companion (dual) problems, there exists a weak, and sometimes also a strong, duality relation, which makes solving one of the problem somewhat equivalent to solving the other. This brings new choices of efficient algorithms to solve problem (4.12a). We leave the dual-formulation derivation of the general form (4.12a) to textbooks such as [146] and focus on sparse optimization examples.

Example 6 (Dual of basis pursuit). For ℓ_1 minimization problem (4.13a), its dual problem is

$$\max_{\mathbf{y}}\{\mathbf{b}^T\mathbf{y} : \|\mathbf{A}^T\mathbf{y}\|_\infty \leq 1\}. \tag{4.47}$$

To derive (4.47), we let the objective value of a minimization (maximization) problem be $+\infty$ ($-\infty$) at infeasible points[2] and establish

$$\begin{aligned}\min_{\mathbf{x}}\{\|\mathbf{x}\|_1 : \mathbf{A}\mathbf{x} = \mathbf{b}\} &= \min_{\mathbf{x}}\max_{\mathbf{y}} \|\mathbf{x}\|_1 - \mathbf{y}^T(\mathbf{A}\mathbf{x} - \mathbf{b}) \\ &= \max_{\mathbf{y}}\min_{\mathbf{x}} \|\mathbf{x}\|_1 - \mathbf{y}^T(\mathbf{A}\mathbf{x} - \mathbf{b}) \\ &= \max_{\mathbf{y}}\{\mathbf{b}^T\mathbf{y} : \|\mathbf{A}^T\mathbf{y}\|_\infty \leq 1\},\end{aligned}$$

where the first equality follows from that $\mathbf{A}\mathbf{x} = \mathbf{b}$ if and only if $\max_{\mathbf{y}} -\mathbf{y}^T(\mathbf{A}\mathbf{x} - \mathbf{b})$ is finite, the second equality holds for the mini-max principle, and the last equality follows from the fact that $\min_{\mathbf{x}}\{\|\mathbf{x}\|_1 - \mathbf{y}^T\mathbf{A}\mathbf{x}\}$ is finite if and only if $\|\mathbf{A}^T\mathbf{y}\|_\infty \leq 1$.

Example 7 (Group sparsity). The dual of (4.16) is

$$\max_{\mathbf{y}}\{\mathbf{b}^T\mathbf{y} : \|\mathbf{A}_{\mathcal{G}_s}^T\mathbf{y}\|_2 \leq w_i,\ s = 1, 2, \ldots, S\}, \tag{4.48}$$

which can be derived in a way similar to Example 6, or see [179].

---

[2] points that do not satisfy all the constraints.

Example 8 (Nuclear-norm). In (4.14e), we rewrite $\mathcal{A}(\mathbf{X}) = [\mathbf{A}_1 \bullet \mathbf{X}, \mathbf{A}_2 \bullet \mathbf{X}, \ldots, \mathbf{A}_M \bullet \mathbf{X}]$, where $\mathbf{Y} \bullet \mathbf{X} = \sum_{m=1}^{M} \sum_{n=1}^{N} y_{mn} x_{mn}$. Thus, $\mathcal{A}(\mathbf{X}) = \mathbf{b}$ can be written as $\mathbf{A}_m \bullet \mathbf{X} = b_m$, for $m = 1, 2, \ldots, M$. The dual of (4.14e) is obtained through

$$\begin{aligned}
\min_{\mathbf{X}}\{\|\mathbf{X}\|_* : \mathcal{A}(\mathbf{X}) = \mathbf{b}\} &= \min_{\mathbf{X}} \max_{\mathbf{y}} \|\mathbf{X}\|_* - \mathbf{y}^T(\mathcal{A}(\mathbf{X}) - \mathbf{b}) \\
&= \max_{\mathbf{y}} \min_{\mathbf{X}} \|\mathbf{X}\|_* - \Big(\sum_{m=1}^{M} y_m \mathbf{A}_m\Big) \bullet \mathbf{X} + \mathbf{b}^T\mathbf{y} \\
&\overset{(*)}{=} \max_{\mathbf{y}} \{\mathbf{b}^T\mathbf{y} : \|\sum_{m=1}^{M} y_m \mathbf{A}_m\|_2 \leq 1\},
\end{aligned}$$

where $\|\mathbf{X}\|_2 = \sigma_1(\mathbf{X})$ equals the maximum singular value of $\mathbf{X}$ (i.e., the operator norm of $\mathbf{X}$) and $(*)$ follows from the fact that $\|\cdot\|_*$ and $\|\cdot\|_2$ are dual to each other [89].

4.6.2 The augmented Lagrangian method

The augmented Lagrangian method (also called the method of multipliers) is one of the most well-known algorithms that returns a pair of primal–dual solutions by solving a sequence of unconstrained problems. The augmented Lagrangian iteration [180, 181] for problem (4.12a) is

$$\mathbf{x}^{k+1} = \arg\min_{\mathbf{x}} r(\mathbf{x}) - (\mathbf{y}^k)^T(\mathbf{A}\mathbf{x} - \mathbf{b}) + \frac{\delta}{2}\|\mathbf{A}\mathbf{x} - \mathbf{b}\|_2^2, \tag{4.49a}$$

$$\mathbf{y}^{k+1} = \mathbf{y}^k + \delta(\mathbf{b} - \mathbf{A}\mathbf{x}^{k+1}), \tag{4.49b}$$

starting from $k = 0$ and $\mathbf{y}^0 = \mathbf{0}$ or a fixed vector, where $\delta > 0$ is a penalty parameter for constraint violation. The objective function of (4.49a) is called the augmented Lagrangian, which is convex in $\mathbf{x}$ (if r is convex) and linear in $\mathbf{y}$. Besides being called the dual variables, $\mathbf{y}$ is often referred to as the Lagrange multipliers (of the constraints $\mathbf{A}\mathbf{x} = \mathbf{b}$).

The optimality conditions (KKT conditions) of (4.12a) are

$$\text{(primal feasibility)} \quad \mathbf{0} = \mathbf{A}\mathbf{x}_{\text{opt}} - \mathbf{b}, \tag{4.50a}$$

$$\text{(dual feasibility)} \quad \mathbf{0} \in \partial r(\mathbf{x}_{\text{opt}}) - \mathbf{A}^T\mathbf{y}_{\text{opt}}. \tag{4.50b}$$

At every iteration, $\mathbf{x}^{k+1}$ and $\mathbf{y}^{k+1}$ maintain condition (4.50b) since from (4.49a)

$$\mathbf{0} \in \partial r(\mathbf{x}^{k+1}) - \mathbf{A}^T(\mathbf{y}^k + \delta(\mathbf{b} - \mathbf{A}\mathbf{x}^{k+1})) = \partial r(\mathbf{x}^{k+1}) - \mathbf{A}^T\mathbf{y}^{k+1}. \tag{4.51}$$

Also, the iterations keep adding a penalty to the violation of $\mathbf{A}\mathbf{x} - \mathbf{b}$ in the sense that

$$-(\mathbf{y}^k)^T(\mathbf{A}\mathbf{x} - \mathbf{b}) + \frac{\delta}{2}\|\mathbf{A}\mathbf{x} - \mathbf{b}\|_2^2 = \frac{\delta}{2}\langle \mathbf{A}\mathbf{x} - \mathbf{b}, \sum_{i=1}^{k}(\mathbf{A}\mathbf{x}^i - \mathbf{b}) + (\mathbf{A}\mathbf{x} - \mathbf{b})\rangle.$$

So, the iteration maintains dual feasibility (4.50b) and works toward primal feasibility (4.50a). The conditions of convergence are quite mild; in particular, $r(\mathbf{x})$ is allowed to be non-differentiable. The reader is referred to textbook [182].

Here, we take the primal-dual problem pair (4.13a)–(4.47) with $r(\mathbf{x}) = \|\mathbf{x}\|_1$ as an example. Algorithms for other $r(\cdot)$ can be obtained similarly.

$$\mathbf{x}^{k+1} = \arg\min_{\mathbf{x}} \|\mathbf{x}\|_1 - (\mathbf{y}^k)^T(\mathbf{A}\mathbf{x} - \mathbf{b}) + \frac{\delta}{2}\|\mathbf{A}\mathbf{x} - \mathbf{b}\|_2^2, \tag{4.52a}$$

$$\mathbf{y}^{k+1} = \mathbf{y}^k + \delta(\mathbf{b} - \mathbf{A}\mathbf{x}^{k+1}). \tag{4.52b}$$

It is shown in [183] that the update (4.52b) is equivalent to the proximal iterations of (4.47):

$$\mathbf{y}^{k+1} = \arg\max_{\mathbf{y}}\{\mathbf{b}^T\mathbf{y} - \frac{1}{2\delta}\|\mathbf{y} - \mathbf{y}^k\|_2^2 : \|\mathbf{A}^T\mathbf{y}\|_\infty \leq 1\}. \tag{4.53}$$

For non-linear programs, one may have to choose a sufficiently large δ (possibly not existing) in order to guarantee convergence. For convex problems, however, any fixed $\delta > 0$ leads to the convergence $\mathbf{x}^k \to \mathbf{x}_{\text{opt}}$, a solution of (4.13a). In light of (4.49a), δ balances objective descent and constraint satisfaction, and thus properly updating δ can noticeably accelerate the convergence. To terminate the algorithm, one can check the conditions (4.50a) or measure either $\|\mathbf{x}^k - \mathbf{x}^{k+1}\|_2$ or $\|\mathbf{y}^k - \mathbf{y}^{k+1}\|_2$.

4.6.3 Bregman method

There are different versions of Bregman algorithms: original Bregman, linearized Bregman, and split Bregman. This subsection describes the original version. The linearized and split versions are given in Sections 4.6.5 and 4.7 below, respectively.

The Bregman method is based on successively minimizing the Bregman distance (also called the Bregman divergence).

Definition 8 ([184]). *The Bregman distance induced by convex function $r(\cdot)$ is defined as*

$$D_r(\mathbf{x}, \mathbf{y}; \mathbf{p}) = r(\mathbf{x}) - r(\mathbf{y}) - \langle \mathbf{p}, \mathbf{x} - \mathbf{y} \rangle, \quad \textit{where } \mathbf{p} \in \partial r(\mathbf{y}). \tag{4.54}$$

Since $\partial r(\mathbf{y})$ may include multiple subgradients, different choices of $\mathbf{p} \in \partial r(\mathbf{y})$ lead to different Bregman distances, so we include $\mathbf{p}$ in $D_r(\mathbf{x}, \mathbf{y}; \mathbf{p})$. When r is differentiable or the choice of $\mathbf{p}$ is clear from the context, we drop $\mathbf{p}$ and write $D_r(\mathbf{x}, \mathbf{y})$. Because we have $D_r(\mathbf{x}, \mathbf{y}; \mathbf{p}) \neq D_r(\mathbf{y}, \mathbf{x}; \mathbf{p})$ in general, the Bregman distance is not a distance in the mathematical sense. However, it measures the closeness between $\mathbf{x}$ and $\mathbf{y}$ in some sense. We have $D_r(\mathbf{x}, \mathbf{y}; \mathbf{p}) \geq 0$ uniformly, and fixing $\mathbf{x}$ and $\mathbf{y}$, we have $D_r(\mathbf{x}, \mathbf{y}; \mathbf{p}) \geq D_r(\mathbf{w}, \mathbf{y}; \mathbf{p})$ for any point $\mathbf{w}$ picked on the line segment $[\mathbf{x}, \mathbf{y}]$. In addition, if r is strictly convex, then $D_r(\mathbf{x}, \mathbf{y}; \mathbf{p}) = 0$ if and only if $\mathbf{x} = \mathbf{y}$. Figure 4.3 illustrates the Bregman distance.

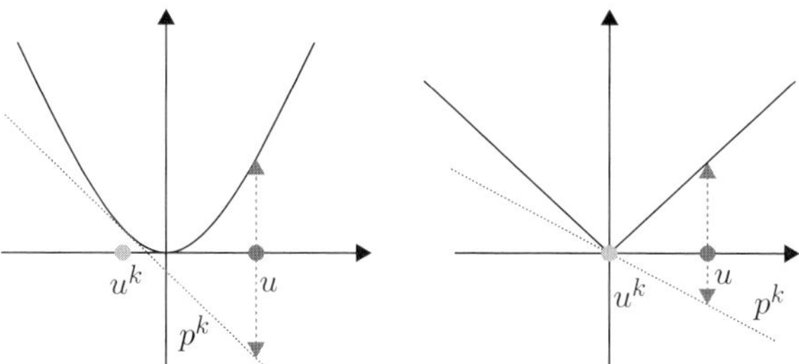

Figure 4.3 Bregman distance (the length of the dashed line) between u and u^k.

The original Bregman algorithm iteration has two steps

$$\mathbf{x}^{k+1} = \arg\min_{\mathbf{x}} \; D_r(\mathbf{x}, \mathbf{x}^k; \mathbf{p}^k) + f(\mathbf{x}), \tag{4.55a}$$

$$\mathbf{p}^{k+1} = \mathbf{p}^k - \nabla f(\mathbf{x}^{k+1}), \tag{4.55b}$$

starting at $k = 0$ and $(\mathbf{x}^0, \mathbf{p}^0) = (\mathbf{0}, \mathbf{0})$. Step (4.55b) generates $\mathbf{p}^{k+1} \in \partial r(\mathbf{x}^{k+1})$ following the first-order optimality condition of (4.55a):

$$\mathbf{0} \in \partial r(\mathbf{x}^{k+1}) - \mathbf{p}^k + \nabla f(\mathbf{x}^{k+1}), \tag{4.56}$$

so the Bregman distance $D_r(\mathbf{x}, \mathbf{x}^{k+1}; \mathbf{p}^{k+1})$ becomes well defined.

For $f(\mathbf{x}) = (\delta/2)\|\mathbf{A}\mathbf{x} - \mathbf{b}\|_2^2$, an interesting alternative form of the Bregman algorithm is the iteration of "adding back the residual":

$$\mathbf{b}^{k+1} = \mathbf{b} + (\mathbf{b}^k - \mathbf{A}\mathbf{x}^k), \tag{4.57a}$$

$$\mathbf{x}^{k+1} = \arg\min_{\mathbf{x}} \; r(\mathbf{x}) + \frac{\delta}{2}\|\mathbf{A}\mathbf{x} - \mathbf{b}^{k+1}\|_2^2, \tag{4.57b}$$

starting from $k = 0$ and $(\mathbf{x}^0, \mathbf{b}^0) = (\mathbf{0}, \mathbf{0})$.

Lemma 2. *For $f(\mathbf{x}) = \frac{\delta}{2}\|\mathbf{A}\mathbf{x} - \mathbf{b}\|_2^2$, iterations* (4.55) *and* (4.57) *yield the same sequence $\{\mathbf{x}^k\}$. In particular, $\mathbf{p}^k = -\delta\mathbf{A}^T(\mathbf{A}\mathbf{x}^k - \mathbf{b}^k)$ for all k.*

Proof. Since $\mathbf{p}^0 = 0$ and $\mathbf{b}^1 = \mathbf{b}$, both iterations yield the same $\mathbf{x}^1$ and (4.55b) gives $\mathbf{p}^1 = -\delta\mathbf{A}^T(\mathbf{A}\mathbf{x}^1 - \mathbf{b}^1)$. We prove by induction. Suppose both iterations yield the same $\mathbf{x}^k$ and $\mathbf{p}^k = -\delta\mathbf{A}^T(\mathbf{A}\mathbf{x}^k - \mathbf{b}^k)$. From (4.55a) and (4.57a), it follows

$$\begin{aligned}
\mathbf{x}^{k+1} &= \arg\min_{\mathbf{x}} r(\mathbf{x}) - \langle \mathbf{p}^k, \mathbf{x}\rangle + \frac{\delta}{2}\|\mathbf{A}\mathbf{x} - \mathbf{b}\|_2^2 \\
&= \arg\min_{\mathbf{x}} r(\mathbf{x}) + \frac{\delta}{2}\|\mathbf{A}\mathbf{x} - \big(\mathbf{b} + (\mathbf{b}^k - \mathbf{A}\mathbf{x}^k)\big)\|_2^2 \\
&= \arg\min_{\mathbf{x}} r(\mathbf{x}) + \frac{\delta}{2}\|\mathbf{A}\mathbf{x} - \mathbf{b}^{k+1}\|_2^2,
\end{aligned}$$

which is (4.57b). Hence, both iterations yield the same $\mathbf{x}^{k+1}$. Because of (4.55b) and (4.57b), we obtain

$$\begin{aligned}\mathbf{p}^{k+1} &= \mathbf{p}^k - \delta\mathbf{A}^T(\mathbf{A}\mathbf{x}^{k+1} - \mathbf{b}) \\ &= -\delta\mathbf{A}^T(\mathbf{A}\mathbf{x}^k - \mathbf{b}^k) - \delta\mathbf{A}^T(\mathbf{A}\mathbf{x}^{k+1} - \mathbf{b}) \\ &= -\delta\mathbf{A}^T\left(\mathbf{A}\mathbf{x}^{k+1} - \left(\mathbf{b} + (\mathbf{b}^k - \mathbf{A}\mathbf{x}^k)\right)\right) \\ &= -\delta\mathbf{A}^T(\mathbf{A}\mathbf{x}^{k+1} - \mathbf{b}^{k+1}).\end{aligned}$$

□

In addition, iteration (4.57) is equivalent to (4.49) by the identity $\mathbf{y}^k = \delta\mathbf{b}^{k+1}$ for all k. Therefore, we have the following theorem.

Theorem 8. *For $f(\mathbf{x}) = \frac{\delta}{2}\|\mathbf{A}\mathbf{x} - \mathbf{b}\|_2^2$, the augmented Lagrangian iteration* (4.49) *and the Bregman iteration* (4.55) *or* (4.57) *generate the same sequence* $\{\mathbf{x}^k\}$.

This result is simple yet surprising since the three equivalent iterations have very different forms and explanations. Iteration (4.49) maintains dual feasibility and works toward primal feasibility, iteration (4.55) applies successive linear relaxation to the regularization function r, and iteration (4.57) adds back the residual to $\mathbf{b}$.

If subproblems (4.49a), (4.55a), and (4.57b) are solved *inexactly*, the three iterations no longer yield the same sequence! In light of Lemma 2, the identity $\mathbf{p}^k = -\delta\mathbf{A}^T(\mathbf{A}\mathbf{x}^k - \mathbf{b}^k)$ will fail to hold. The augmented Lagrangian iteration (4.49) and the "adding-back" iteration (4.57) are more stable and, when r is a piece-wise linear function (also referred to as a polyhedral function) and the subproblems are solved with sufficient (but not necessarily high) accuracies, these two iterations enjoy the property of *error forgetting*, namely, the errors in $\mathbf{x}^k$ do not accumulate, and furthermore the errors at iterations k and $k+1$ may cancel each other, allowing $\mathbf{x}^k$ to converge to a solution to model (4.12a) up to the machine precision; see [185] for examples and analysis.

4.6.4 Bregman iterations and denoising

It is known that the augmented Lagrangian and Bregman iterations converge to a solution of the underlying equality-constrained problem (4.12a), which is a model for noiseless measurements $\mathbf{b}$. When there is noise in $\mathbf{b}$, properly stopping the iterations before convergence may yield clean solutions with better fitting than the exact solutions to problems (4.12b) and (4.12c), even with their best parameters.

We illustrate this through an example and quoting a theorem. In this example, we generated a signal $\mathbf{x}^o$ of $n = 250$ dimensions with $k = 20$ non-zero components that were i.i.d. samples of the standard Gaussian distribution. $\mathbf{x}^o$ is depicted by the block dot • in Figure 4.4. Measurement matrix $\mathbf{A}$ had $m = 100$ rows and $n = 250$ columns with its entries sampled i.i.d. from the standard Gaussian distribution. Measurements were $\mathbf{b} = \mathbf{A}\mathbf{x}^o + \mathbf{n}$, where noise $\mathbf{n}$ followed the independent Gaussian noise of standard deviation 0.1. We compared

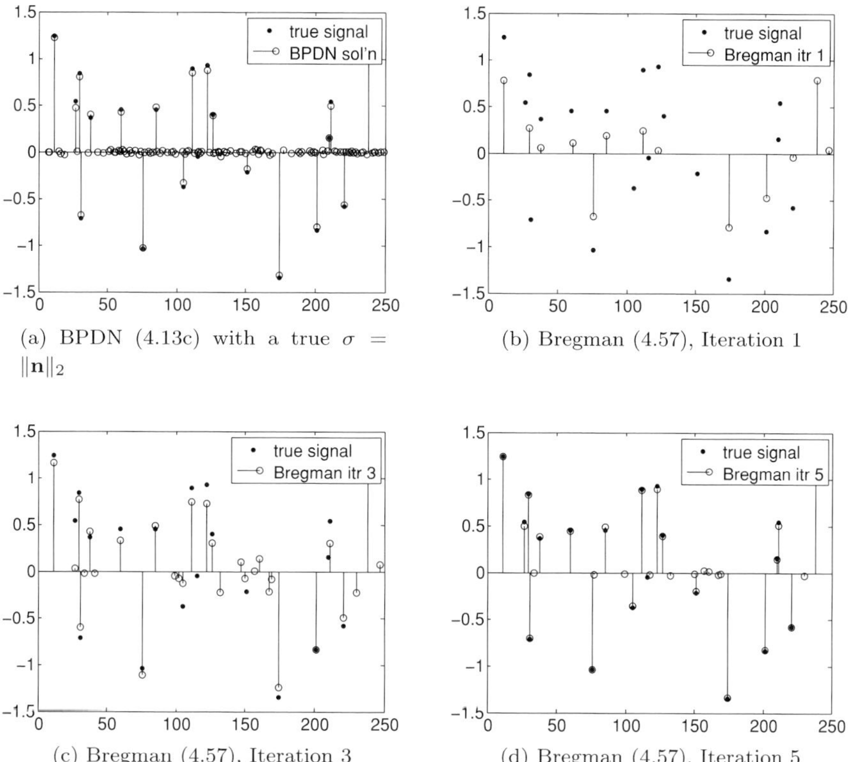

(a) BPDN (4.13c) with a true $\sigma = \|\mathbf{n}\|_2$

(b) Bregman (4.57), Iteration 1

(c) Bregman (4.57), Iteration 3

(d) Bregman (4.57), Iteration 5

Figure 4.4 BPDN versus Bregman iteration.

1. Model (4.13c) with $\sigma = \|\mathbf{n}\|_2$ in Figure 4.4(a).
2. Bregman iteration (4.57) with $r(\cdot) = \|\cdot\|_1$ and $\delta = 1/50$ at iterations 1, 3, and 5, in Figures 4.4(b), 4.4(c), and 4.4(d), respectively.

Because of the noise $\mathbf{n}$ in the measurements, neither approach gave exact reconstruction of $\mathbf{x}^0$. The BPDN solution had a large number of false non-zero components and slightly mismatched with the non-zero components of $\mathbf{x}^o$. The Bregman solutions had significantly fewer false non-zero components. Iterations 1 and 3 missed certain non-zero components of $\mathbf{x}^o$. Iteration 5 had a much better solution that better matched with $\mathbf{x}^0$, achieving a relative 2-norm error of 3.50%, than the solution of BPDN (relative 2-norm error 6.19%).

In general, as long as noise is not too large, this phenomenon arises with other dual algorithms such as the linearzied Bregman and alternating direction methods below, and it also does with the logistic loss function. When the measurements contain a moderate amount of noise, instead of solving models (4.12b) and (4.12c), solving model (4.12a) using a dual algorithm and appropriately stopping the algorithm before convergence can yield a better solution; more examples and discussions can be found in [186, 187]. We adapt a result from [186], Theorem 3.5, to our problem:

Theorem 9. *Assume that from the original signal $\mathbf{x}^o$, we obtain measurements $\mathbf{b} = \mathbf{A}\mathbf{x}^o + \mathbf{n}$, where $\mathbf{n}$ is an arbitrary noise obeying $\frac{1}{2}\|\mathbf{n}\|_2^2 \leq \sigma^2$. Let r be ℓ_1-norm. Let $\mathbf{x}^k$ and $\mathbf{p}^k$ be generated by the Bregman iteration. Then, as long as $\frac{1}{2}\|\mathbf{A}\mathbf{x}^k - \mathbf{b}\|_2^2 \leq \sigma^2$, the Bregman distance between $\mathbf{x}^o$ and $\mathbf{x}^k$ is monotonically non-increasing in k, namely,*

$$D_r(\mathbf{x}^o, \mathbf{x}^k; \mathbf{p}^k) \leq D_r(\mathbf{x}^o, \mathbf{x}^{k-1}; \mathbf{p}^{k-1}).$$

4.6.5 Linearized Bregman and augmented models

The linearized Bregman method is motivated by the challenge that step (4.55a) of Bregman iteration has no closed-form solution and needs another algorithm to solve. As the name suggests, linearized Bregman iteration applies linearization to $f(\mathbf{x})$ in (4.55a) (with an extra proximal term). The iteration is

$$\mathbf{x}^{k+1} \leftarrow \arg\min_{\mathbf{x}} D_r(\mathbf{x}, \mathbf{x}^k; \mathbf{p}^k) + \langle \nabla f(\mathbf{x}^k), \mathbf{x} \rangle + \frac{1}{2\alpha}\|\mathbf{x} - \mathbf{x}^k\|_2^2, \tag{4.58a}$$

$$\mathbf{p}^{k+1} \leftarrow \mathbf{p}^k - \nabla f(\mathbf{x}^k) - \frac{1}{\alpha}(\mathbf{x}^{k+1} - \mathbf{x}^k). \tag{4.58b}$$

Step (4.58b) generates a valid subgradient $\mathbf{p}^{k+1}$ following the first-order optimality condition of (4.58a). If we introduce the *augmented* function of r

$$\bar{r}(\mathbf{x}) := r(\mathbf{x}) + \frac{1}{2\alpha}\|\mathbf{x}\|_2^2, \tag{4.59}$$

we can write

$$D_{\bar{r}}(\mathbf{x}, \mathbf{x}^k; \bar{\mathbf{p}}^k) = D_r(\mathbf{x}, \mathbf{x}^k; \mathbf{p}^k) + \frac{1}{2\alpha}\|\mathbf{x} - \mathbf{x}^k\|_2^2$$

for $\bar{\mathbf{p}}^k = \mathbf{p}^k + (1/\alpha)\mathbf{x}^k$ using the additivity property of the Bregman distance. Therefore, iteration (4.58) can be equivalently written as

$$\mathbf{x}^{k+1} \leftarrow \arg\min_{\mathbf{x}} D_{\bar{r}}(\mathbf{x}, \mathbf{x}^k; \bar{\mathbf{p}}^k) + \langle \nabla f(\mathbf{x}^k), \mathbf{x} \rangle, \tag{4.60a}$$

$$\bar{\mathbf{p}}^{k+1} \leftarrow \bar{\mathbf{p}}^k - \nabla f(\mathbf{x}^k), \tag{4.60b}$$

where in (4.60a) the Bregman distance term and the term of $\langle \nabla f(\mathbf{x}^k), \mathbf{x} \rangle$ together remind us of the name *linearized Bregman*.

Next, we argue that if $f(\mathbf{x})$ is a quadratic function, then linearized Bregman iteration is equivalent to the Uzawa method [188], or dual gradient ascent, for the problem

$$\min_{\mathbf{x}}\{\bar{r}(\mathbf{x}) : \mathbf{A}\mathbf{x} = \mathbf{b}\}, \tag{4.61}$$

where $\bar{r}$ is defined in (4.59). Its Lagrange dual can be obtained as follows:

$$\begin{aligned}
\min_{\mathbf{x}}\{\bar{r}(\mathbf{x}) : \mathbf{A}\mathbf{x} = \mathbf{b}\} &= \min_{\mathbf{x}} \max_{\mathbf{y}} \|\mathbf{x}\|_1 + \frac{1}{2\alpha}\|\mathbf{x}\|_2^2 - \mathbf{y}^T(\mathbf{A}\mathbf{x} - \mathbf{b}) \\
&= \min_{\mathbf{x}} \max_{\mathbf{y},\mathbf{z}}\{\mathbf{x}^T\mathbf{z} + \frac{1}{2\alpha}\|\mathbf{x}\|_2^2 - \mathbf{y}^T\mathbf{A}\mathbf{x} + \mathbf{y}^T\mathbf{b} : \|\mathbf{z}\|_\infty \leq 1\} \\
&= \max_{\mathbf{y},\mathbf{z}}\{\min_{\mathbf{x}} \mathbf{x}^T\mathbf{z} + \frac{1}{2\alpha}\|\mathbf{x}\|_2^2 - \mathbf{y}^T\mathbf{A}\mathbf{x} + \mathbf{b}^T\mathbf{y} : \|\mathbf{z}\|_\infty \leq 1\} \\
&= \min_{\mathbf{y},\mathbf{z}}\{-\mathbf{b}^T\mathbf{y} + \frac{\alpha}{2}\|\mathbf{A}^T\mathbf{y} - \mathbf{z}\|_2^2 : \|\mathbf{z}\|_\infty \leq 1\},
\end{aligned}$$

where the last equality holds since the minimum is attained at $\mathbf{x}^* = \alpha(\mathbf{A}^T\mathbf{y} - \mathbf{z})$. Eliminating $\mathbf{z}$ from the last equation gives the following dual problem

$$\min_{\mathbf{y}} \ -\mathbf{b}^T\mathbf{y} + \frac{\alpha}{2}\|\mathbf{A}^T\mathbf{y} - \mathrm{Proj}_{[-1,1]^n}(\mathbf{A}^T\mathbf{y})\|_2^2. \tag{4.62}$$

Note the identify $\mathrm{shrink}(\mathbf{x}, 1) = \mathbf{x} - \mathrm{Proj}_{[-1,1]^n}(\mathbf{x})$. The Uzawa or dual gradient iteration is

$$\mathbf{y}^{k+1} \leftarrow \mathbf{y}^k - \tau\left(-\mathbf{b} + \alpha\mathbf{A}\,\mathrm{shrink}(\mathbf{A}^T\mathbf{y}^k, 1)\right), \tag{4.63}$$

where $\tau > 0$ is the step size. However, (4.63) is not the form given in [189–191]. That form can be obtained by introducing $\mathbf{x}^k := \alpha\,\mathrm{shrink}(\mathbf{A}^T\mathbf{y}^k, 1)$ and $\mathbf{v}^k = \mathbf{A}^T\mathbf{y}^k$:

$$\mathbf{x}^{k+1} \leftarrow \alpha\,\mathrm{shrink}(\mathbf{v}^{(k)}, 1), \tag{4.64}$$

$$\mathbf{v}^{k+1} \leftarrow \mathbf{v}^{(k)} + \tau\mathbf{A}^T(\mathbf{b} - \mathbf{A}\mathbf{x}^{(k+1)}). \tag{4.65}$$

The iteration looks very simple! Now, setting $\mathbf{v}^{(k)} = \mathbf{p}^{(k)} + \tau\mathbf{A}^T(\mathbf{b} - \mathbf{A}\mathbf{x}^{(k)}) + \frac{\mathbf{x}^{(k)}}{\alpha}$ and $\tau = 1$, we recover (4.58) and (4.60).

Compared to model (4.13a), model (4.61) has an extra least-squares term in the objective. A pure least-squares solution is nowhere sparse unless $\mathbf{A}\mathbf{x} = \mathbf{b}$ admits a zero solution. Hence, the solution to model (4.61) is generally not as sparse as that of (4.13a).

However, when α is sufficiently large, (4.61) has the exact regularization property [192, 193], namely, the solution of (4.61) is also a solution to (4.13a). Furthermore, [194] suggests $\alpha \geq 10\|\mathbf{x}^o\|_\infty$ and demonstrates that, for sparse optimization, this condition leads to the sparsest solution $\mathbf{x}^*$ *nearly* whenever the original model (4.13a) does.

The term $\frac{1}{2\alpha}\|\mathbf{x}\|_2^2$ in the objective leads to numerical advantages. Since it is a strongly convex function and strongly convex problems have differentiable dual problems, problem (4.62) has a differentiable objective, whose gradient is $-\mathbf{b} + \alpha\mathbf{A}\,\mathrm{shrink}(\mathbf{A}^T\mathbf{y})$. Hence, one can apply classical techniques such as Barzilai–Borwein steps, line search, limited-memory BFGS, Nesterov's acceleration, semi-smooth Newton, etc., to obtain algorithms of (4.62) that are often much faster than the aforementioned algorithms for (4.13a).

In general, we call a model with an added strongly convex term *an augmented model*. Next, we review the Lagrange dual problems of a few augmented models with $\frac{1}{2\alpha}\|\mathbf{x}\|_2^2$.

The duals of the augmented models of (4.13b) and (4.13c) are, respectively,

$$\min_{\mathbf{y}} -\mathbf{b}^T\mathbf{y} + \frac{1}{2\mu}\|\mathbf{y}\|_2^2 + \frac{\alpha}{2}\|\mathbf{A}^T\mathbf{y} - \mathrm{Proj}_{[-1,1]^n}(\mathbf{A}^T\mathbf{y})\|_2^2, \tag{4.66}$$

$$\min_{\mathbf{y}} -\mathbf{b}^T\mathbf{y} + \sigma\|\mathbf{y}\|_2 + \frac{\alpha}{2}\|\mathbf{A}^T\mathbf{y} - \mathrm{Proj}_{[-1,1]^n}(\mathbf{A}^T\mathbf{y})\|_2^2. \tag{4.67}$$

Both (4.66) and (4.67) have differentiable objectives except for the latter at $\mathbf{y} = \mathbf{0}$. However, unless the original model (4.13c) has solution $\mathbf{x}^* = \mathbf{0}$, $\mathbf{y} = \mathbf{0}$ is not a solution to (4.67) so it can be avoided in an algorithm. Furthermore, the original solutions of (4.62), (4.66), and (4.67) can be recovered, just like in (4.64), from any dual solution $\mathbf{y}^*$ via

$$\mathbf{x}^* = \alpha\,\mathrm{shrink}(\mathbf{A}^T\mathbf{y}^*, 1). \tag{4.68}$$

Finally, recovery guarantees for sparse solutions $\mathbf{x}^*$ and global linear convergence are given in [194], which exploits a certain restricted strong convexity property of the objective function of (4.62). Recovery guarantees for the augmented models for the matrix completion and robust PCA models are studied in [195].

4.6.6 Handling complex data and variables

Let $\mathrm{r}(\cdot)$, $\mathrm{i}(\cdot)$, and $|\cdot|$ denote the operators that return the real part, the imaginary part, and the magnitude of the input. Many researchers define the ℓ_1-norm and ℓ_2-norm of a complex-valued vector $\mathbf{x} \in \mathbb{C}^n$ as

$$\|\mathbf{x}\|_1 := \sum_{i=1}^{n} \left(\mathrm{r}^2(x_i) + \mathrm{i}^2(x_i)\right)^{1/2} = \|\mathbf{a}\|_1,$$

$$\|\mathbf{x}\|_2 := \left(\sum_{i=1}^{n} \mathrm{r}^2(x_i) + \mathrm{i}^2(x_i)\right)^{1/2} = \|\mathbf{a}\|_2,$$

respectively, where $\mathbf{a} = [|x_1|, \ldots, |x_n|]^T \in \mathbb{R}^n$. They reduce to the standard ℓ_1-norm and ℓ_2-norm if $\mathbf{x}$ is real-valued. In general, we define the ℓ_p-norm of $\mathbf{x}$ by $\|\mathbf{a}\|_p$, which is convex for $1 \le p \le \infty$.

The *dual norm* of $\|\mathbf{x}\|_p$ is $\|\mathbf{y}\|_d = \max\{\mathrm{r}(\mathbf{y}^*\mathbf{x}) : \|\mathbf{x}\|_p \le 1\}$, where $\mathbf{y}^*$ denotes the conjugate transpose of $\mathbf{y}$.

We are interested in developing the dual problems of

$$\min_{\mathbf{x}}\{\|\mathbf{x}\|_p : \mathbf{A}\mathbf{x} = \mathbf{b}\}, \tag{4.69a}$$

$$\min_{\mathbf{x}} \|\mathbf{x}\|_p + \frac{\mu}{2}\|\mathbf{A}\mathbf{x} - \mathbf{b}\|_2^2, \tag{4.69b}$$

$$\min_{\mathbf{x}}\{\|\mathbf{x}\|_p : \|\mathbf{A}\mathbf{x} - \mathbf{b}\|_2 \le \sigma\}. \tag{4.69c}$$

The constraints $\mathbf{A}\mathbf{x} = \mathbf{b}$ have their real and imaginary parts, namely, $\mathrm{r}(\mathbf{A}\mathbf{x}) = \mathrm{r}(\mathbf{b})$ and $\mathrm{i}(\mathbf{A}\mathbf{x}) = \mathrm{i}(\mathbf{b})$. To each of them, we shall assign a real-valued dual vector. If we assign real-valued vectors $\mathbf{p}$ and $\mathbf{q}$ to the real and imaginary parts, respectively, and introduce complex-valued vector $\mathbf{y} = \mathbf{p} + \mathbf{q}i$, then we obtain

$$\mathbf{p}^T(\mathrm{r}(\mathbf{Ax}) - \mathrm{r}(\mathbf{b})) + \mathbf{q}^T(\mathrm{i}(\mathbf{Ax}) - \mathrm{i}(\mathbf{b})) = \mathrm{r}(\mathbf{y}^*(\mathbf{Ax} - \mathbf{b})).$$

Hence, we can obtain the dual of (4.69a) via

$$\min_{\mathbf{x}}\{\|\mathbf{x}\|_p : \mathbf{Ax} = \mathbf{b}\} = \min_{\mathbf{x}}\max_{\mathbf{y}} \|\mathbf{x}\|_p - \mathrm{r}(\mathbf{y}^*(\mathbf{Ax} - \mathbf{b})) \tag{4.70a}$$

$$= \max_{\mathbf{y}}\min_{\mathbf{x}} \|\mathbf{x}\|_p - \mathrm{r}(\mathbf{y}^*(\mathbf{Ax} - \mathbf{b})) \tag{4.70b}$$

$$= \max_{\mathbf{y}}\{\mathrm{r}(\mathbf{b}^*\mathbf{y}) : \|\mathbf{A}^*\mathbf{y}\|_d \leq 1\}, \tag{4.70c}$$

where (4.70a) follows from the fact that $\max_{\mathbf{y}}\{-\mathrm{r}(\mathbf{y}^*(\mathbf{Ax} - \mathbf{b})\}$ stays bounded if and only if $\mathbf{Ax} = \mathbf{b}$, (4.70b) follows from the mini-max theorem, and (4.70c) follows from the dual relation between $\|\cdot\|_p$ and $\|\cdot\|_d$, as well as

$$\min_{\mathbf{x}} \|\mathbf{x}\|_p - \mathrm{r}(\mathbf{y}^*(\mathbf{Ax})) = \begin{cases} 0, & \|\mathbf{A}^*\mathbf{y}\|_d \leq 1, \\ -\infty, & \text{otherwise.} \end{cases} \tag{4.71}$$

Leaving the details to the reader, the dual problem of (4.69b) is

$$\max_{\mathbf{y}}\{\mathrm{r}(\mathbf{b}^*\mathbf{y}) - \frac{1}{2\mu}\|\mathbf{y}\|_2^2 : \|\mathbf{A}^*\mathbf{y}\|_d \leq 1\}.$$

One way to dualize problem (4.69c) is

$$\min_{\mathbf{x}}\{\|\mathbf{x}\|_p : \|\mathbf{Ax} - \mathbf{b}\|_2 \leq \sigma\} = \min_{\mathbf{x},\mathbf{z}}\{\|\mathbf{x}\|_p : \mathbf{Ax} + \mathbf{z} = \mathbf{b}, \|\mathbf{z}\|_2 \leq \sigma\} \tag{4.72a}$$

$$= \min_{\mathbf{x},\mathbf{z}}\max_{\mathbf{y}}\{\|\mathbf{x}\|_p - \mathrm{r}(\mathbf{y}^*(\mathbf{Ax} + \mathbf{z} - \mathbf{b})) : \|\mathbf{z}\|_2 \leq \sigma\} \tag{4.72b}$$

$$= \max_{\mathbf{y}}\min_{\mathbf{x},\mathbf{z}}\{\|\mathbf{x}\|_p - \mathrm{r}(\mathbf{y}^*(\mathbf{Ax} + \mathbf{z} - \mathbf{b})) : \|\mathbf{z}\|_2 \leq \sigma\}. \tag{4.72c}$$

The inner problem of (4.72c), given a fixed $\mathbf{y}$, is separable in the unknowns $\mathbf{x}$ and $\mathbf{z}$. The optimal $\mathbf{z}$ is given by

$$\mathbf{z} = \begin{cases} \sigma\mathbf{y}/\|\mathbf{y}\|_2, & \mathbf{y} \neq \mathbf{0}, \\ \mathbf{0}, & \text{otherwise.} \end{cases}$$

In either case, $-\mathrm{r}(\mathbf{y}^*\mathbf{z}) = -\sigma\|\mathbf{y}\|_2$. Also with (4.71), (4.72c) is equivalent to

$$\max_{\mathbf{y}}\{\mathrm{r}(\mathbf{b}^*\mathbf{y}) - \sigma\|\mathbf{y}\|_2 : \|\mathbf{A}^*\mathbf{y}\|_d \leq 1\}.$$

4.7 Alternating direction method of multipliers

The subproblems of augmented Lagrangian and original Bregman algorithms do not have closed-form solutions. For choices of $r(\mathbf{x})$ such as ℓ_1 and its variants, the algorithms in this section can have subproblems with closed-form solutions, and thus often run faster than the augmented Lagrangian and original Bregman algorithms.

4.7.1 Framework

We introduce the general framework of variable splitting and the alternating direction method of multipliers, abbreviated as ADM or ADMM. Consider the problem

$$\min_{\mathbf{x},\mathbf{y}}\{f(\mathbf{x})+h(\mathbf{y}):\mathbf{A}\mathbf{x}+\mathbf{B}\mathbf{y}=\mathbf{b}\}. \tag{4.73}$$

The augmented Lagrangian for (4.73) is

$$L(\mathbf{x},\mathbf{y};\lambda)=f(\mathbf{x})+h(\mathbf{y})-\lambda^T(\mathbf{A}\mathbf{x}+\mathbf{B}\mathbf{y}-\mathbf{b})+\frac{c}{2}\|\mathbf{A}\mathbf{x}+\mathbf{B}\mathbf{y}-\mathbf{b}\|_2^2,$$

and the augmented Lagrangian algorithm is

$$(\mathbf{x}^{k+1},\mathbf{y}^{k+1})=\arg\min_{\mathbf{x},\mathbf{y}} L(\mathbf{x},\mathbf{y},\lambda^k), \tag{4.74a}$$

$$\lambda^{k+1}=\lambda^k-\gamma c(A\mathbf{x}^{k+1}+B\mathbf{y}^{k+1}-\mathbf{b}), \tag{4.74b}$$

where $\gamma\in(0,2)$ leads to convergence. The subproblems, if inexactly solved, must be solved with increasing accuracies [196].

The ADM relaxes the joint minimization (4.74a) to individual minimization so that the subproblems are easier to solve:

$$\mathbf{x}^{k+1}=\arg\min_{\mathbf{x}} L(\mathbf{x},\mathbf{y}^k,\lambda^k), \tag{4.75a}$$

$$\mathbf{y}^{k+1}=\arg\min_{\mathbf{y}} L(\mathbf{x}^{k+1},\mathbf{y},\lambda^k), \tag{4.75b}$$

$$\lambda^{k+1}=\lambda^k-\gamma c(\mathbf{A}\mathbf{x}^{k+1}+\mathbf{B}\mathbf{y}^{k+1}-\mathbf{b}), \tag{4.75c}$$

where $\gamma\in(0,(\sqrt{5}+1)/2)$ leads to convergence; see [167, 197] for proofs and also [198] for the proof for the more general setting below.

If one of the subproblems, say, the **x**-subproblem (4.75a), has no known closed-from solution, one can apply the inexact ADM:

$$\mathbf{x}^{k+1}\approx\arg\min_{\mathbf{x}} L(\mathbf{x},\mathbf{y}^k,\lambda^k), \tag{4.76a}$$

$$\mathbf{y}^{k+1}=\arg\min_{\mathbf{y}} L(\mathbf{x}^{k+1},\mathbf{y},\lambda^k), \tag{4.76b}$$

$$\lambda^{k+1}=\lambda^k-\gamma c(\mathbf{A}\mathbf{x}^{k+1}+\mathbf{B}\mathbf{y}^{k+1}-\mathbf{b}), \tag{4.76c}$$

or

$$\mathbf{y}^{k+1} = \arg\min_{\mathbf{y}} L(\mathbf{x}^k, \mathbf{y}, \lambda^k), \tag{4.77a}$$

$$\mathbf{x}^{k+1} \approx \arg\min_{\mathbf{x}} L(\mathbf{x}, \mathbf{y}^{k+1}, \lambda^k), \tag{4.77b}$$

$$\lambda^{k+1} = \lambda^k - \gamma c(\mathbf{A}\mathbf{x}^{k+1} + \mathbf{B}\mathbf{y}^{k+1} - \mathbf{b}). \tag{4.77c}$$

The $\mathbf{y}$-subproblem can also be solved inexactly, and both subproblems can be solved inexactly at the same time. Solving subproblems inexactly improves the speed of each iteration but tends to increase the total number of iterations. The goal is to wisely balance speed and subproblem accuracies so that the entire ADM runs more efficiently.

In the ADM, the two subproblems can be solved in either order. However, when one subproblem is solved less exactly than the other, the ADM *tends* to run faster if the less exact one is solved *later* – like (4.77) as opposed to (4.76) – because at each iteration, the ADM updates the variables in the Gauss–Seidel fashion. If the less-exact one runs first, its relatively inaccurate solution will then affect the more exact step, wasting its higher accuracy. Note that this is merely intuitive and based on observations, not a rigorous statement.

One can inexactly solve the ADM subproblems in many different ways as follows:

(i) *Iteration limiter*: If one applies an iterative solver (e.g., CG or L-BFGS) for a subproblem, the number of iterations can be limited, say, 10 CG iterations.

(ii) *Linear proximal* [199]: Suppose we deal with (4.76a). A linear proximal version of (4.76a) is

$$\mathbf{x}^{k+1} = \arg\min_{\mathbf{x}} f(\mathbf{x}) + c(\mathbf{g}^k)^T(\mathbf{x} - \mathbf{x}^k) + \frac{1}{2t}\|\mathbf{x} - \mathbf{x}^k\|_2^2, \tag{4.78}$$

where $\mathbf{g}^k = \mathbf{A}^T(\mathbf{A}\mathbf{x}^k + \mathbf{B}\mathbf{y}^k - \mathbf{b} - \lambda/c)$ is the $\mathbf{x}$-gradient of the last two terms of $L(\mathbf{x}^k, \mathbf{y}^k, \lambda^k)$ and t is the step size. The middle and last terms of (4.78) are the linear and proximal terms, respectively. The benefit of (4.78) is that one no longer needs to deal with the term $\frac{c}{2}\mathbf{x}^T\mathbf{A}^T\mathbf{A}\mathbf{x}$ in $L(\mathbf{x}, \mathbf{y}^k, \lambda^k)$, and the last two terms in (4.78) are separable over x_i for all i. Therefore, (4.78) is easy to compute for $f(\mathbf{x})$ that is one of the examples in Section 4.4.

Also for $f(\mathbf{x}) = f([\mathbf{x}_1, \cdots, \mathbf{x}_L]) = f_1(\mathbf{x}_1) + \cdots + f_L(\mathbf{x}_L)$, (4.78) is separable over each block of $\mathbf{x}$.

(iii) *Gradient descent*. Suppose we deal with (4.76a). We can simply apply one or a few steps of gradient based on $L(\mathbf{x}, \mathbf{y}^k, \lambda^k)$.

(iv) *Approximating* $\mathbf{A}^T\mathbf{A}$ *or* $\mathbf{B}^T\mathbf{B}$ [198]. The penalty term $\frac{c}{2}\|A\mathbf{x} + B\mathbf{y} - \mathbf{b}\|_2$ contains $\frac{c}{2}\mathbf{x}^T\mathbf{A}^T\mathbf{A}\mathbf{x}$ and $\frac{c}{2}\mathbf{y}^T\mathbf{B}^T\mathbf{B}\mathbf{y}$. Suppose $\mathbf{A}^T\mathbf{A}$ can be approximated by $\mathbf{D}$, which is much more efficient to form and invert. Then one can apply

$$\mathbf{x}^{k+1} = \arg\min_{\mathbf{x}} L(\mathbf{x}, \mathbf{y}^k, \lambda^k) + \frac{c}{2}(\mathbf{x} - \mathbf{x}^k)^T(\mathbf{D} - \mathbf{A}^T\mathbf{A})(\mathbf{x} - \mathbf{x}^k). \tag{4.79}$$

The right-hand side objective function has $\frac{c}{2}\mathbf{x}^T\mathbf{D}\mathbf{x}$ instead of $\frac{c}{2}\mathbf{x}^T\mathbf{A}^T\mathbf{A}\mathbf{x}$.

Note that the choice of inexact approaches affects the scope of step-size parameter γ that guarantees convergence. There are only partial results available; see [187, 198, 200] for details.

4.7.2 Applications of ADM in sparse optimization

Example 9. The formulation (4.73) often arises when one "splits" the variable $\mathbf{x}$ in problem

$$\min_{\mathbf{x}} p(\mathbf{x}) + q(\mathbf{A}\mathbf{x})$$

into two variables $\mathbf{x}$ and $\mathbf{y}$, giving rise to

$$\min_{\mathbf{x},\mathbf{y}} \{p(\mathbf{x}) + q(\mathbf{y}) : \mathbf{A}\mathbf{x} - \mathbf{y} = 0\},$$

where the objective has separable terms $p(\mathbf{x})$ and $q(\mathbf{y})$.

Example 10 (Basis pursuit denoising). The penalty formulation (4.13b) can be written equivalently as

$$\min_{\mathbf{x}} \left\{ \|\mathbf{x}\|_1 + \frac{\mu}{2}\|\mathbf{y}\|_2^2 : \mathbf{A}\mathbf{x} + \mathbf{y} = \mathbf{b} \right\}. \tag{4.80}$$

With augmented Lagrangian

$$L(\mathbf{x}, \mathbf{y}; \lambda) = \|\mathbf{x}\|_1 + \frac{\mu}{2}\|\mathbf{y}\|_2^2 - \lambda^T(\mathbf{A}\mathbf{x} + \mathbf{y} - \mathbf{b}) + \frac{c}{2}\|\mathbf{A}\mathbf{x} + \mathbf{y} - \mathbf{b}\|_2^2,$$

applying the inexact ADM to (4.80) yields the iteration

$$\mathbf{y}^{k+1} = \frac{c}{\mu + c}(\lambda^k/c - (\mathbf{A}\mathbf{x}^k - \mathbf{b})), \tag{4.81a}$$

$$\mathbf{x}^{k+1} = \text{shrink}(\mathbf{x}^k - \tau_k \nabla_{\mathbf{x}} L(\mathbf{x}^k, \mathbf{y}^{k+1}; \lambda^k), \tau_k), \tag{4.81b}$$

$$\lambda^{k+1} = \lambda^k - \gamma c(\mathbf{A}\mathbf{x}^{k+1} + \mathbf{y}^{k+1} - \mathbf{b}), \tag{4.81c}$$

which is (4.77), where (4.77b) is computed by the prox-linear step (4.81b).

The basis pursuit formulation (4.13a) corresponds to (4.80) with $\mu = \infty$ and thus $\mathbf{y}^k \equiv \mathbf{0}$ in (4.81), which reduces to the inexact augmented Lagrangian iteration.

Example 11 (Basis pursuit denoising). In the reformulation (4.80), $\mathbf{x}$ is associated with both $\|\cdot\|_1$ and $\mathbf{A}$; hence, the $\mathbf{x}$-subproblem is in the form of $\min_{\mathbf{x}} \|\mathbf{x}\|_1 + \frac{c}{2}\|\mathbf{A}\mathbf{x} - (\cdots)\|_2^2$. A different splitting can separate $\|\cdot\|_1$ and $\mathbf{A}$ as follows

$$\min_{\mathbf{x}} \left\{ \|\mathbf{x}\|_1 + \frac{\mu}{2}\|\mathbf{A}\mathbf{y} - \mathbf{b}\|_2^2 : \mathbf{x} - \mathbf{y} = \mathbf{0} \right\}. \tag{4.82}$$

With augmented Lagrangian

$$L(\mathbf{x}, \mathbf{y}; \lambda) = \|\mathbf{x}\|_1 + \frac{\mu}{2}\|\mathbf{A}\mathbf{y} - \mathbf{b}\|_2^2 - \lambda^T(\mathbf{x} - \mathbf{y}) + \frac{c}{2}\|\mathbf{x} - \mathbf{y}\|_2^2,$$

applying the inexact ADM (iii) to (4.82) yields the iteration

$$\mathbf{x}^{k+1} = \text{shrink}(\mathbf{y}^k + \lambda^k/c, c^{-1}), \tag{4.83a}$$
$$\mathbf{y}^{k+1} = \mathbf{y}^k - \tau_k \nabla_{\mathbf{y}} L(\mathbf{x}^{k+1}, \mathbf{y}^k; \lambda^k), \tag{4.83b}$$
$$\lambda^{k+1} = \lambda^k - \gamma c(\mathbf{A}\mathbf{x}^{k+1} + \mathbf{y}^{k+1} - \mathbf{b}), \tag{4.83c}$$

where (4.83a) is an exact solve and (4.83b) is a gradient step.

One advantage of separate $\|\cdot\|_1$ and $\mathbf{A}$ in two subproblems is that if $\mathbf{A}\mathbf{A}^T = I$, one can replace (4.83b) by an exact solve at a relatively small cost. Specifically, minimizing $L(\mathbf{x}^{k+1}, \mathbf{y}; \lambda^k)$ over $\mathbf{y}$ amounts to solving the equation system $(\mu\mathbf{A}^T\mathbf{A} + cI)\mathbf{y} = \mu\mathbf{A}^T\mathbf{b} - \lambda^k + c\mathbf{x}^{k+1}$, and one can apply the Sherman–Morrison–Woodbury formula (or the fact that $\mathbf{A}\mathbf{A}^T$ is an idempotent matrix) to derive $(\mu\mathbf{A}^T\mathbf{A} + cI)^{-1} = c^{-1}(I - ((\mu c)^{-1} + c^{-2})\mathbf{A}^T\mathbf{A})$.

Alternatively, one can pre-compute $(\mu\mathbf{A}^T\mathbf{A} + cI)^{-1}$ or the Cholesky factorization of $(\mu\mathbf{A}^T\mathbf{A} + cI)$ before starting the ADM iterations. Then, at each iteration, instead of (4.83b), one can exactly compute

$$\mathbf{y}^{k+1} = \arg\min_{\mathbf{y}} L(\mathbf{x}^{k+1}, \mathbf{y}; \lambda^k)$$

at a low cost.

Example 12 (ℓ_2-constrained basis pursuit denoising). Next, to apply inexact ADM to (4.13c), we begin with rewriting (4.13c) as

$$\min_{\mathbf{x}} \{\|\mathbf{x}\|_1 : \mathbf{A}\mathbf{x} + \mathbf{y} = \mathbf{b}, \|\mathbf{y}\|_2 \le \sigma\}. \tag{4.84}$$

Treating the constraint $\|\mathbf{y}\|_2 \le \sigma$ as an indicator function $\chi_{\|\mathbf{y}\|_2 \le \sigma}$ in the objective, we obtain the Lagrangian

$$L(\mathbf{x}, \mathbf{y}; \lambda) = \|\mathbf{x}\|_1 + \chi_{\|\mathbf{y}\|_2 \le \sigma} - \lambda^T(\mathbf{A}\mathbf{x} + \mathbf{y} - \mathbf{b}) + \frac{c}{2}\|\mathbf{A}\mathbf{x} + \mathbf{y} - \mathbf{b}\|_2^2$$

and the inexact ADM iteration

$$\mathbf{y}^{k+1} = \text{Proj}_{\mathbf{B}_\sigma}(\lambda^k/c - (\mathbf{A}\mathbf{y}^k - \mathbf{b})), \tag{4.85a}$$
$$\mathbf{x}^{k+1} = \text{shrink}(\mathbf{x}^k - \tau_k \nabla_{\mathbf{x}} L(\mathbf{x}^k, \mathbf{y}^{k+1}; \lambda^k), \tau_k), \tag{4.85b}$$
$$\lambda^{k+1} = \lambda^k - \gamma c(\mathbf{A}\mathbf{x}^{k+1} + \mathbf{y}^{k+1} - \mathbf{b}), \tag{4.85c}$$

where $\text{Proj}_{\mathbf{B}_\sigma}$ stands for Euclidean projection to the σ-ball $\mathbf{B}_\sigma = \{\mathbf{y} : \|\mathbf{y}\|_2 \le \sigma\}$.

The exact and inexact ADM can also be applied to the dual problems of (4.13a)–(4.13c), as well as the primal and dual problems of the augmented versions of various models. It can have a large number of different forms.

Example 13 (Robust PCA). Our next example is more complicated since it has two regularization terms. *Robust PCA* [95] aims to decompose a data matrix $\mathbf{M}$ into the sum of a low-rank matrix $\mathbf{L}$ and a sparse matrix $\mathbf{S}$; i.e., $\mathbf{M} = \mathbf{L} + \mathbf{S}$, which has many applications in data mining, video processing, face recognition, and so on. The robust PCA model solves the problem

$$\min_{\mathbf{L},\mathbf{S}} \|\mathbf{L}\|_* + \alpha\|\mathbf{S}\|_1 + \frac{\mu}{2}\|\mathcal{A}(\mathbf{L}) + \mathcal{B}(\mathbf{S}) - \mathbf{b}\|_2^2, \tag{4.86}$$

where $\mathcal{A}$ and $\mathcal{B}$ are matrix linear operators, and α and μ are two weight parameters. Problem (4.86) is equivalent to

$$\min_{\mathbf{x},\mathbf{L},\mathbf{S}} \left\{\|\mathbf{L}\|_* + \alpha\|\mathbf{S}\|_1 + \frac{\mu}{2}\|\mathbf{x}\|_2^2 : \mathbf{x} + \mathcal{A}(\mathbf{L}) + \mathcal{B}(\mathbf{S}) = \mathbf{b}\right\}. \tag{4.87}$$

Introduce

$$\mathbf{y} := (\mathbf{L}, \mathbf{S})$$

and linear operator $\mathcal{C}(\mathbf{y}) := \mathcal{A}(\mathbf{L}) + \mathcal{B}(\mathbf{S})$ and $h(\mathbf{y}) = \|\mathbf{L}\|_* + \alpha\|\mathbf{S}\|_1$. Then problem (4.87) can be rewritten as

$$\min_{\mathbf{x},\mathbf{y}} \left\{\frac{\mu}{2}\|\mathbf{x}\|_2^2 + h(\mathbf{y}) : \mathbf{x} + \mathcal{C}(\mathbf{y}) = \mathbf{b}\right\}, \tag{4.88}$$

to which one can apply the inexact ADM iterations in which the $\mathbf{y}$-subproblem decouples $\mathbf{L}$ and $\mathbf{S}$ with the linear proximal step. Specifically,

$$\mathbf{x}^{k+1} = \arg\min_{\mathbf{x}} \frac{\mu}{2}\|\mathbf{x}\|_2^2 + \frac{c}{2}\|\mathbf{x} + \mathcal{C}(\mathbf{y}^k) - \mathbf{b} - \lambda^k/c\|_2^2, \tag{4.89a}$$

$$\mathbf{L}^{k+1} = \arg\min_{\mathbf{L}} \|\mathbf{L}\|_* + c\langle \mathcal{A}^*(\mathbf{x}^{k+1} + \mathcal{C}(\mathbf{y}^k) - \mathbf{b} - \lambda^k/c), \mathbf{L}\rangle + \frac{1}{2t}\|\mathbf{L} - \mathbf{L}^k\|_2, \tag{4.89b}$$

$$\mathbf{S}^{k+1} = \arg\min_{\mathbf{S}} \alpha\|\mathbf{S}\|_1 + c\langle \mathcal{B}^*(\mathbf{x}^{k+1} + \mathcal{C}(\mathbf{y}^k) - \mathbf{b} - \lambda^k/c), \mathbf{S}\rangle + \frac{1}{2t}\|\mathbf{S} - \mathbf{S}^k\|_2, \tag{4.89c}$$

$$\lambda^{k+1} = \lambda^k - \gamma c(\mathbf{x}^{k+1} + \mathcal{C}(\mathbf{y}^{k+1}) - \mathbf{b}). \tag{4.89d}$$

Example 14 (ℓ_1-ℓ_1 model). This example considers the ℓ_1-ℓ_1 model

$$\min_{\mathbf{x}} \|\mathbf{x}\|_1 + \mu\|\mathbf{A}\mathbf{x} - \mathbf{b}\|_1,$$

which can be written equivalently as

$$\min_{\mathbf{x}} \{\|\mathbf{x}\|_1 + \mu\|\mathbf{y}\|_1 : \mathbf{A}\mathbf{x} + \mathbf{y} = \mathbf{b}\}.$$

The augmented Lagrangian is

$$L(\mathbf{x}, \mathbf{y}; \lambda) = \|\mathbf{x}\|_1 + \mu\|\mathbf{y}\|_1 - \lambda^T(\mathbf{A}\mathbf{x} + \mathbf{y} - \mathbf{b}) + \frac{c}{2}\|\mathbf{A}\mathbf{x} + \mathbf{y} - \mathbf{b}\|_2^2,$$

to which one can apply the inexact ADM iteration

$$\mathbf{y}^{k+1} = \text{shrink}(\lambda^k/c - (\mathbf{A}\mathbf{x}^k - \mathbf{b}), \mu/c), \tag{4.90a}$$
$$\mathbf{x}^{k+1} = \text{shrink}(\mathbf{x}^k - \tau_k \nabla_{\mathbf{x}} L(\mathbf{x}^k, \mathbf{y}^{k+1}; \lambda^k), \tau_k), \tag{4.90b}$$
$$\lambda^{k+1} = \lambda^k - \gamma c(\mathbf{A}\mathbf{x}^{k+1} + \mathbf{y}^{k+1} - \mathbf{b}), \tag{4.90c}$$

which is (4.77), where (4.77b) is computed by the prox-linear step (4.81b).

Example 15 (Group basis pursuit). Group basis pursuit is problem (4.16), for which primal and dual ADM algorithms are presented in [179]. Problem (4.16) is equivalent to

$$\min_{\mathbf{x},\mathbf{z}} \left\{ \|\mathbf{z}\|_{\mathcal{G},2,1} : \mathbf{A}\mathbf{x} = \mathbf{b}, \mathbf{x} = \mathbf{z} \right\}, \tag{4.91}$$

whose augmented Lagrangian is

$$L(\mathbf{x}, \mathbf{z}; \lambda_1, \lambda_2) = \|\mathbf{z}\|_{\mathcal{G},2,1} - \lambda_1^T(\mathbf{z} - \mathbf{x}) + \frac{c_1}{2}\|\mathbf{z} - \mathbf{x}\|_2^2 - \lambda_2^T(\mathbf{A}\mathbf{x} - \mathbf{b}) + \frac{c_2}{2}\|\mathbf{A}\mathbf{x} - \mathbf{b}\|_2^2.$$

The exact ADM subproblems are

$$\mathbf{x}^{k+1} = (c_1\mathbf{I} + c_2\mathbf{A}^T\mathbf{A})^{-1}(c_1\mathbf{z}^k - \lambda_1 + c_2\mathbf{A}^T\mathbf{b} + \mathbf{A}^T\lambda_2), \tag{4.92a}$$
$$\mathbf{z}_{\mathcal{G}_s}^{k+1} = \text{(4.26) with } \mathbf{x} \leftarrow (\mathbf{x}_{\mathcal{G}_s}^{k+1} + c_1^{-1}(\lambda_1)_{\mathcal{G}_s}) \text{ and } \tau \leftarrow (w_s/c_1),\ s = 1, \ldots, S \tag{4.92b}$$
$$\lambda^{k+1} = \lambda^k - \gamma c(\mathbf{A}\mathbf{x}^{k+1} + \mathbf{y}^{k+1} - \mathbf{b}). \tag{4.92c}$$

If it is easy to invert $(\mathbf{I} + \mathbf{A}\mathbf{A}^T)$, then one can consider the Sherman–Morrison–Woodbury formula to reduce the cost of (4.92a); otherwise, one can consider an inexact ADM step such as the gradient-descent type, or cache the inverse or the Cholesky factorization of $(\mathbf{I} + \mathbf{A}\mathbf{A}^T)$.

The group LASSO problem also has an ADM algorithm, which is not very different from (4.92); see [179].

Example 16 (Overlapping group basis pursuit). In the last example or problem (4.16), the objective is defined with non-overlapping groups, namely, $\mathcal{G}_i \cap \mathcal{G}_j = \emptyset$, $\forall i \neq j$. If the groups *overlap*, then only an easy modification is needed. For $s = 1, \ldots, S$, define $\mathbf{z}_s \in \mathbb{R}^{|\mathcal{G}_s|}$ and introduce constraints $\mathbf{x}_{\mathcal{G}_s} = \mathbf{z}_s$. This set of constraints over all s can be written compactly as $\mathbf{G}\mathbf{x} = \mathbf{z}$, where $\mathbf{z} = [\mathbf{z}_1, \ldots, \mathbf{z}_S]$ and $\mathbf{G}$ is a certain matrix, and it replaces constraints $\mathbf{x} = \mathbf{z}$ in problem (4.91). Then, one can just follow the previous example and derive the modified ADM iterations. Note that since $\mathbf{z}_s$, $s = 1, \ldots, S$, do not share variables, the $\mathbf{z}$-subproblem still decomposes to S $\mathbf{z}_s$-subproblems in parallel.

4.7.3 Applications in distributed optimization

Recent work [201] describes how ADM solves certain problems in a distributed environment. Consider in a cluster of N computing nodes, the problem of minimizing the sum of N convex functions, one for each node, over a common variable $\mathbf{x} \in \mathbb{R}^n$. We formulate this problem as

$$\min_{\mathbf{x}} \sum_{i=1}^{N} f_i(\mathbf{x}). \tag{4.93}$$

The objective function of this form is called partially separable. Let each node i keep a local copy of $\mathbf{x}$ as $\mathbf{x}^i \in \mathbb{R}^n$. Note that $\mathbf{x}^i$ is itself an n-dimensional vector, not a subvector or $\mathbf{x}$. For all the nodes to reach a consensus among $\mathbf{x}^i$, $i = 1, \ldots, N$, a common approach is to introduce a global common variable $\mathbf{y}$ and reformulate (4.93) as

$$\min_{\{\mathbf{x}^i\},\mathbf{y}} \left\{ \sum_{i=1}^{N} f_i(\mathbf{x}^i) : \mathbf{x}^i - \mathbf{y} = \mathbf{0}, i = 1, \ldots, N \right\}, \tag{4.94}$$

which is referred to as the *global consensus problem*. The same problem with regularization $r(\mathbf{x})$ over the common variable $\mathbf{x}$ is

$$\min_{\{\mathbf{x}^i\},\mathbf{y}} \left\{ \sum_{i=1}^{N} f_i(\mathbf{x}^i) + r(\mathbf{y}) : \mathbf{x}^i - \mathbf{y} = \mathbf{0}, i = 1, \ldots, N \right\}. \tag{4.95}$$

As an example, problem (4.13b) can be turned into a special case of (4.95) if one decomposes $\mathbf{A}$ and $\mathbf{b}$ into blocks of rows as

$$\mathbf{A} = \begin{bmatrix} \vdots \\ \mathbf{A}^i \\ \vdots \end{bmatrix}, \quad \mathbf{b} = \begin{bmatrix} \vdots \\ \mathbf{b}^i \\ \vdots \end{bmatrix},$$

and posts (4.13b) as

$$\min_{\{\mathbf{x}^i\},\mathbf{y}} \left\{ \|\mathbf{y}\|_1 + \sum_{i=1}^{N} \frac{\mu}{2} \|\mathbf{A}^i\mathbf{x}^i - \mathbf{b}^i\|_2^2 : \mathbf{x}^i - \mathbf{y} = \mathbf{0}, i = 1, \ldots, N \right\}. \tag{4.96}$$

When the ADM is applied to problem (4.95), all the $\mathbf{x}^i$-subproblems are independent and can be solved by individual computing nodes *in parallel*, and the $\mathbf{y}$-subproblem and λ-update are also easy as long as $r(\mathbf{y})$ is simple. For problem (4.96), it boils down to taking a vector sum over all the computing nodes and performing a shrinkage operation. Operations such as computing the sum, average, maximum, minimum, etc. are predefined reduce operations in MPI (message passing interface). They have very efficient implementations and are simple to code. Note that $\mathbf{A}^i$ only appears in the $\mathbf{x}^i$-subproblem, so it can be stored locally on node i.

In the application of (4.13b) with a "fat" matrix $\mathbf{A}$, which has more columns than rows, one may prefer to decompose it by column like

$$\mathbf{A} = \begin{bmatrix} \cdots & \mathbf{A}_j & \cdots \end{bmatrix}. \tag{4.97}$$

To distribute $\mathbf{A}_j$s to different computing nodes. We take advantage of duality. The dual of (4.13b) is

$$\min_{\mathbf{y}} \left\{ -\mathbf{b}^T\mathbf{y} + \frac{\mu}{2}\|\mathbf{y}\|_2^2 : \|\mathbf{A}^T\mathbf{y}\|_\infty \leq 1 \right\},$$

to which one can apply variable splitting and obtain the equivalent problem

$$\min_{\mathbf{y}} \left\{ -\mathbf{b}^T\mathbf{y} + \frac{\mu}{2}\|\mathbf{y}\|_2^2 + \chi_{\|\mathbf{z}\|_\infty \leq 1} : \mathbf{A}^T\mathbf{y} + \mathbf{z} = \mathbf{0} \right\}. \tag{4.98}$$

Let $\mathbf{z} = [\mathbf{z}_1; \mathbf{z}_2; \cdots; \mathbf{z}_n]$. Since $\chi_{\|\mathbf{z}\|_\infty \leq 1} = \sum_{i=1}^n \chi_{\|\mathbf{z}_i\|_\infty \leq 1}$, (4.98) is in the form of

$$\min_{\mathbf{y},\mathbf{z}} \left\{ f(\mathbf{y}) + \sum_{i=1}^n g_i(\mathbf{z}_i) : \mathbf{A}^T\mathbf{y} + \mathbf{z} = \mathbf{c} \right\}. \tag{4.99}$$

Decompose $\mathbf{A}^T\mathbf{y} + \mathbf{z} = \mathbf{c}$ into blocks of rows as

$$\mathbf{A}^T\mathbf{y} = \begin{bmatrix} \mathbf{A}_1^T \\ \mathbf{A}_2^T \\ \vdots \\ \mathbf{A}_n^T \end{bmatrix} \mathbf{y}, \quad \mathbf{z} = \begin{bmatrix} \mathbf{z}_1 \\ \mathbf{z}_2 \\ \vdots \\ \mathbf{z}_n \end{bmatrix}, \quad \mathbf{c} = \begin{bmatrix} \mathbf{c}_1 \\ \mathbf{c}_2 \\ \vdots \\ \mathbf{c}_n \end{bmatrix},$$

where $\mathbf{A}$ is decomposed according to (4.97). Introduce n *identical local copies* of $\mathbf{y}$: $\mathbf{y}_1, \ldots, \mathbf{y}_n$ and obtain

$$\min_{\mathbf{y},\{\mathbf{y}_i\},\mathbf{z}} \left\{ \sum_{i=1}^n \left(\bar{f}(\mathbf{y}_i) + g_i(\mathbf{z}_i) \right) : \mathbf{A}_i^T\mathbf{y}_i + \mathbf{z}_i = \mathbf{c}_i,\ \mathbf{y}_i - \mathbf{y} = \mathbf{0},\ \forall i = 1, \ldots, n \right\}, \tag{4.100}$$

where $\bar{f}(\cdot) = \frac{1}{n} f(\cdot)$, which is equivalent to (4.99). To see the constraint structure better, we can rewrite (4.100) as

$$\min_{\mathbf{y},\mathbf{z}} \sum_{i=1}^n \left(\bar{f}(\mathbf{y}_i) + g_i(\mathbf{z}_i) \right),$$

$$\text{s.t.} \begin{bmatrix} \mathbf{A}_1^T & & \\ & \ddots & \\ & & \mathbf{A}_n^T \\ I & & \\ & \ddots & \\ & & I \end{bmatrix} \begin{bmatrix} \mathbf{y}_1 \\ \vdots \\ \mathbf{y}_n \end{bmatrix} + \begin{bmatrix} I & & & \mathbf{0} \\ & \ddots & & \vdots \\ & & I & \mathbf{0} \\ \mathbf{0} & \cdots & \mathbf{0} & -I \\ & \ddots & & \vdots \\ \mathbf{0} & \cdots & \mathbf{0} & -I \end{bmatrix} \begin{bmatrix} \mathbf{z}_1 \\ \vdots \\ \mathbf{z}_n \\ \mathbf{y} \end{bmatrix} = \begin{bmatrix} \mathbf{c}_1 \\ \vdots \\ \mathbf{c}_n \\ \mathbf{0} \\ \vdots \\ \mathbf{0} \end{bmatrix},$$

which has an ADM-ready structure. All the $\mathbf{y}_i$-subproblems and $\mathbf{z}_i$-subproblems are independent and can be solved by individual computing nodes *in parallel*, and the $\mathbf{y}$-subproblem and λ-update are also easy based on the reduce operations. Submatrix $\mathbf{A}_i$ only shows up in the $\mathbf{y}_i$-subproblem, so it can be stored locally.

In fact, one can also decompose $\mathbf{A}$ by both row and column like

$$\mathbf{A} = \begin{bmatrix} \mathbf{A}_{11} & \mathbf{A}_{12} & \cdots & \mathbf{A}_{1n} \\ \mathbf{A}_{21} & \mathbf{A}_{22} & \cdots & \mathbf{A}_{2n} \\ & & \cdots & \\ \mathbf{A}_{m1} & \mathbf{A}_{m2} & \cdots & \mathbf{A}_{mn} \end{bmatrix},$$

and derive an ADM-ready problem equivalent to (4.13b) so that each $\mathbf{A}_{ij}$ is only used in one subproblem, and all the subproblems involving with $\mathbf{A}_{ij}$s can be solved in parallel. We leave the detail to the reader.

4.7.4 Applications in decentralized optimization

One can even go one step further and eliminate taking the vector sum over all the nodes (the reduce operation of sum) as in the distributed optimization case; instead, one can rely solely on exchanging information among neighboring nodes. This is extremely useful for collaborative computation in a multiagent network environment without a data fusion center. To do so, let us define a graph $(\mathcal{N}, \mathcal{E})$, where $\mathcal{N}$ is the set of nodes and $\mathcal{E}$ is the set of undirected arcs connecting two nodes. Assume that if $(i, j) \in \mathcal{E}$, then nodes i and j can directly exchange data without the need of any third node. We introduce a *bridge variable* $\mathbf{y}^{ij}$ for each $(i, j) \in \mathcal{E}$. We can reformulate problem (4.95) as

$$\min_{\{\mathbf{x}^i\},\{\mathbf{y}^{ij}\}} \left\{ \sum_{i=1}^{N} f_i(\mathbf{x}^i) + \frac{1}{|\mathcal{E}|} r(\mathbf{y}^{ij}) : \mathbf{x}^i - \mathbf{y}^{ij} = \mathbf{0}, \mathbf{x}^j - \mathbf{y}^{ij} = \mathbf{0}, \forall (i, j) \in \mathcal{E} \right\}. \quad (4.101)$$

As long as the graph $(\mathcal{N}, \mathcal{E})$ is connected, the constraints of (4.101) enforce $\mathbf{x}^i = \mathbf{y}^{ij} = \mathbf{x}$ for all $i \in \mathcal{N}$ and $(i, j) \in \mathcal{E}$. The ADM applied to (4.101) is an iterative process consisting of $\mathbf{x}^i$-subproblems, $\mathbf{y}^{ij}$-subproblems, as well as the updates of $\lambda^{i,ij}$ and $\lambda^{j,ij}$. The $\mathbf{x}^i$-subproblems are independent across the nodes, and the other two are independent across all pairs of neighbors. Therefore, there is no global data exchange or centralized computation of any sort; all data and operations are decentralized. Such algorithms are called *decentralized algorithms*. Textbook [202] is a good resource of such algorithms.

4.7.5 Convergence rates

Although there is rich literature on the ADM and its applications, there have been very few results on its rate of convergence until the very recent past. The work [203] shows that for a Jacobi version of the ADM applied to smooth functions with Lipschitz continuous gradients, the objective value descends at the rate of $O(1/k)$ and that of an accelerated version descends at $O(1/k^2)$. Then, [203] establishes the same rates on a Gauss–Seidel version and requires only one of the two objective functions to be smooth with Lipschitz continuous gradient. Lately, [204] shows that $\|\mathbf{u}^k - \mathbf{u}^{k+1}\|$, where $\mathbf{u}^k := (\mathbf{x}^k, \mathbf{y}^k, \lambda^k)$, of the ADM converges at $O(1/k)$ assuming at least one of the subproblems is exactly solved. Reference [205] proves that the dual objective value of a modification to the ADM descends at $O(1/k^2)$ under the assumption that the objective functions are strongly convex, one of them is quadratic, and both subproblems are solved exactly. Reference [198] shows the linear rate of convergence $O(1/c^k)$ for some $c > 1$ under a variety of scenarios in which at least one of the two objective functions is strongly convex and has Lipschitz continuous gradient. The linear rate $O(1/c^k)$ is stronger than the sublinear rates such as $O(1/k)$ and $O(1/k^2)$ and is given in terms of the solution error, which is stronger than those given in terms of the objective error in [203, 205] and the solution relative change in [204]. On the other hand, sublinear-rate results [203, 204] do not require any strongly convex functions. In addition, the ADM applied to linear programming is known to converge at a globally linear rate [206].

4.8 (Block) coordinate minimization and gradient descent

The (block) coordinate descent (BCD) method decomposes the decision variables $\mathbf{x}$ into S non-overlapping blocks $\mathbf{x}_1, \mathbf{x}_2, \ldots, \mathbf{x}_S$. To solve a problem in the form of

$$\min_{\mathbf{x}} F(\mathbf{x}) \equiv F(\mathbf{x}_1, \mathbf{x}_2, \ldots, \mathbf{x}_S), \tag{4.102}$$

in every step BCD updates only one block of variable while fixing all the other blocks. There are different rules to select which block to update, and when a block is being updated, the values of remaining blocks may or may not be up-to-date. There are a few popular rules:

- *Jacobian*: at every iteration k, all S blocks are updated in parallel; $\mathbf{x}_s$ is updated using the values of the remaining $(s-1)$ blocks at iteration $k-1$. For example,

$$\mathbf{x}_s^k \leftarrow \arg\min_{\mathbf{x}_s} f(\mathbf{x}_s) \equiv F(\mathbf{x}_1^{k-1}, \ldots, \mathbf{x}_{s-1}^{k-1}, \mathbf{x}_s, \mathbf{x}_{s+1}^{k-1}, \ldots, \mathbf{x}_S^{k-1}), \tag{4.103}$$

  for $s = 1, \ldots, S$ in parallel.
- *Gauss–Seidel*: at every iteration k, the blocks are cyclically and sequentially selected to update; $\mathbf{x}_s$ is updated using the most recent values of the remaining $(S-1)$ blocks. For example,

$$\mathbf{x}_s^k \leftarrow \arg\min_{\mathbf{x}_s} f(\mathbf{x}_s) \equiv F(\mathbf{x}_1^{k}, \ldots, \mathbf{x}_{s-1}^{k}, \mathbf{x}_s, \mathbf{x}_{s+1}^{k-1}, \ldots, \mathbf{x}_S^{k-1}), \tag{4.104}$$

  for $s = 1, \ldots, S$. (4.104) uses $\mathbf{x}_{s+1}^{k-1}, \ldots, \mathbf{x}_S^{k-1}$ since they have not been updated to $\mathbf{x}_{s+1}^{k}, \ldots, \mathbf{x}_S^{k}$ yet.
- *Gauss–Southwell*: define merit values $m_1, m_2, \ldots, m_S \geq 0$ in the sense that a larger merit value means that updating the corresponding block tends to be more rewarding; fix a constant $0 < \nu \leq 1$, the rule selects any $\mathbf{x}_s$ satisfying

$$m_s \geq \nu \cdot \max\{m_1, m_2, \ldots, m_S\}; \tag{4.105}$$

  if $\nu = 1$, the most rewarding one is selected; the selected $\mathbf{x}_s$ is updated using the most recent values of the remaining $(S-1)$ blocks. The *Gauss–Southwell-r* rule uses a certain block-wise step length as the merit value. The *Gauss–Southwell-q* rule uses a certain block-wise objective-value improvement as the merit value.

Obviously, updating one block at a time gives rise to relatively simple subproblems such as (4.103) and (4.104). For this reason, BCD has been a popular method used widely for solving both convex and non-convex problems. The interesting question is when BCD converges and why BCD can outperform other algorithms on certain sparse optimization problems.

The work of BCD dates back to methods in [207] for solving equation systems and to references [208–211], which analyze the method assuming F to be convex (or quasiconvex or hemivariate), differentiable, and have bounded level sets except for certain classes of convex functions. If F is non-differentiable, like in various sparse optimization problems, BCD can generally get stuck at a non-stationary point; see [210], p. 94, for an example. However, subsequent convergence can be obtained if the non-differentiable part is separable; see [212–215] for results on different forms of F. The recent work [216] describes a very efficient method for non-smooth separable minimization, especially, problems (4.13b) and

$$\min \sum_{s=1}^{S} \|\mathbf{x}_s\|_2 + f(\mathbf{x}), \tag{4.106}$$

where f is a differentiable function. Closely related works are [217–219]. Since (4.106) is more general than (4.13b), where the latter is a special case of each scalar x_i being a block, we describe the algorithm of [216] for (4.106), which is called the block coordinate gradient descent (BCGD) method.

For each block s, define

$$\mathbf{d}(\bar{\mathbf{x}}; s) = \arg\min_{\mathbf{d}} \|\bar{\mathbf{x}}_s + \mathbf{d}\|_2 + \nabla_{\mathbf{x}_s} f(\bar{\mathbf{x}})^T \mathbf{d} + \frac{1}{\delta}\|\mathbf{d}\|_2^2, \tag{4.107}$$

which can be viewed as the prox-linear step of block s. Given a current point $\mathbf{x}^k$, the standard prox-linear step is

$$\mathbf{x}^{k+1} = \arg\min_{\mathbf{x}} \sum_{s} \left(\|\mathbf{x}_s\|_2 + \nabla_{\mathbf{x}_s} f(\mathbf{x}^k)^T (\mathbf{x}_s - \mathbf{x}_s^k) + \frac{1}{\delta}\|\mathbf{x}_s - \mathbf{x}_s^k\|_2^2 \right). \tag{4.108}$$

If we focus just on block s, it is easy to see $\mathbf{x}_s^{k+1} - \mathbf{x}_s^k = \mathbf{d}(\mathbf{x}^k, s)$.

The algorithm BCGD is based on iterating the following steps:

1. choose a block s;
2. compute $\mathbf{d}(\mathbf{x}^k; s)$;
3. set a step-size $\alpha_k > 0$;
4. set $\mathbf{x}_s^{k+1} \leftarrow \mathbf{x}_s^k + \alpha_k \mathbf{d}(\mathbf{x}^k; s)$ and $\mathbf{x}_t^{k+1} \leftarrow \mathbf{x}_t^k$ for $t \neq s$.

Among the three update rules, the Gauss–Southwell rule is most effective for sparse optimization. The merit values can be set as $m_s := \|\mathbf{d}(\mathbf{x}^k; s)\|_\infty$, the Gauss–Southwell-r rule, or $m_s := \|\mathbf{x}_s^k\|_2 - (\|\mathbf{x}_s^k + \mathbf{d}\|_2 + \nabla_{\mathbf{x}_s} f(\mathbf{x}^k)^T \mathbf{d} + \frac{1}{\delta}\|\mathbf{d}\|_2^2)$, the Gauss–Southwell-q rule.

Roughly speaking, the relatively superior performance of the Gauss–Southwell rule on sparse optimization over the other two rules is because of its greed – it selects the most or nearly most potential blocks to update, which tend to be the true non-zero blocks in the solution. Take $f(\mathbf{x}) = \frac{\mu}{2}\|\mathbf{A}\mathbf{x} - \mathbf{b}\|_2^2$ as an example. Let $\mathcal{S}^o$ denote the set of non-zero blocks in $\mathbf{x}^o$, the original group-sparse vector. Suppose $\mathbf{b} = \mathbf{A}\mathbf{x}^o + \mathbf{w}$, where $\mathbf{w}$ is noise, and $\{s : \mathbf{x}_s^k \neq \mathbf{0}\} \subset \mathcal{S}$, namely, the current active blocks are among the non-zero blocks and thus $\mathbf{A}\mathbf{x}^k = \mathbf{A}_{\mathcal{S}^o}\mathbf{x}^k_{\mathcal{S}^o}$. Then, $\nabla_{\mathbf{x}_s} f(\mathbf{x}^k) = \mu\mathbf{A}_s^T(\mathbf{A}\mathbf{x}^k - \mathbf{b}) = \mu\mathbf{A}_s^T(\mathbf{A}_{\mathcal{S}^o}\mathbf{x}^k_{\mathcal{S}^o} - \mathbf{A}_{\mathcal{S}^o}\mathbf{x}^o_{\mathcal{S}^o} - \mathbf{w}) = \mu\mathbf{A}_s^T\mathbf{A}_{\mathcal{S}^o}(\mathbf{x}^k_{\mathcal{S}^o} - \mathbf{x}^o_{\mathcal{S}^o}) - \mu\mathbf{A}_s^T\mathbf{w}$. Recall that matrices satisfying RIP (3.13) tend to have orthogonal columns when restricted to sparse vectors. If $s \in \mathcal{S}^o$, then $\mathbf{A}_s^T\mathbf{A}_{\mathcal{S}^o}$ tends to have large entries; otherwise, $\mathbf{A}_s^T\mathbf{A}_{\mathcal{S}^o}$ tends to have smaller entries. As a result, $\nabla_{\mathbf{x}_s} f(\mathbf{x}^k)$ tends to be larger for $s \in \mathcal{S}^o$, and the same blocks tend to give rise to larger m_s. In short, the stronger correlation between the residual $\mathbf{A}\mathbf{x}^k - \mathbf{b}$ and the submatrices $\mathbf{A}_s$ over $s \in \mathcal{S}^o$ is taken advantage of by greedy selections of blocks, and this leads to a high chance of picking blocks $s \in \mathcal{S}^o$ for update. Hence, blocks in $(\mathcal{S}^o)^c$ are seldom chosen for update; under some conditions, one can even show that they are never chosen. Of course, we have assumed $f(\mathbf{x}) = \frac{\mu}{2}\|\mathbf{A}\mathbf{x} - \mathbf{b}\|_2^2$ and $\mathbf{A}$ to have good RIP, which is not the most general case. However, in many sparse optimization problems, greed is good to varying extents. Some interesting algorithms and analysis can be found in [218, 219].

4.9 Homotopy algorithms and parametric quadratic programming

For model (4.13b), there is a method to compute its solutions corresponding to all values of $\mu > 0$ since, assuming the uniqueness of solution $\mathbf{x}^*$ for each μ, the solution path $\mathbf{x}^*(\mu)$ is continuous and piece-wise linear in μ. To see this, let us fix μ and study the optimality condition of (4.13b) satisfied by $\mathbf{x}^*$:

$$\mathbf{0} \in \partial\|\mathbf{x}^*\|_1 + \mu\mathbf{A}^T(\mathbf{A}\mathbf{x}^* - \mathbf{b}), \tag{4.109}$$

where the subdifferential of ℓ_1 is given by

$$\partial\|\mathbf{x}\|_1 = \partial|x_1| \times \ldots \times \partial|x_n| \subset \mathbb{R}^n, \quad \partial|x_i| = \begin{cases} \{1\}, & x_i > 0, \\ [-1, 1], & x_i = 0, \\ \{-1\}, & x_i < 0. \end{cases}$$

Since $\partial\|\mathbf{x}\|_1$ is component-wise separable, the condition reduces to

$$0 \in \partial|x_i^*| + \mu\mathbf{a}_i^T(\mathbf{A}\mathbf{x}^* - \mathbf{b}), \quad i = 1, \ldots, n,$$

or equivalently,

$$\mu\mathbf{a}_i^T(\mathbf{A}\mathbf{x}^* - \mathbf{b}) \in -\partial|x_i^*|, \quad i = 1, \ldots, n.$$

By definition, we know $\mu\mathbf{a}_i^T(\mathbf{A}\mathbf{x}^* - \mathbf{b}) \in [-1, 1]$ and

$$x_i^* \begin{cases} \geq 0, & \text{if } \mu\mathbf{a}_i^T(\mathbf{A}\mathbf{x}^* - \mathbf{b}) = -1, \\ = 0, & \text{if } \mu\mathbf{a}_i^T(\mathbf{A}\mathbf{x}^* - \mathbf{b}) \in (-1, 1), \\ \leq 0, & \text{if } \mu\mathbf{a}_i^T(\mathbf{A}\mathbf{x}^* - \mathbf{b}) = 1. \end{cases}$$

This offers a way to rewrite the optimality conditions. If we let

$$\begin{aligned}
\mathcal{S}_+ &:= \{i : \mu \mathbf{a}_i^T(\mathbf{A}\mathbf{x}^* - \mathbf{b}) = -1\},\\
\mathcal{S}_0 &:= \{i : \mu \mathbf{a}_i^T(\mathbf{A}\mathbf{x}^* - \mathbf{b}) \in (-1, 1)\},\\
\mathcal{S}_- &:= \{i : \mu \mathbf{a}_i^T(\mathbf{A}\mathbf{x}^* - \mathbf{b}) = +1\},\\
\mathcal{S} &:= \mathcal{S}_+ \cup \mathcal{S}_-,
\end{aligned}$$

then we have $\text{supp}(\mathbf{x}^*) \subseteq \mathcal{S}$ and can rewrite (4.109) equivalently as

$$\mathbf{1} + \mu \mathbf{A}_{\mathcal{S}_+}^T(\mathbf{A}_{\mathcal{S}}\mathbf{x}_{\mathcal{S}}^* - \mathbf{b}) = \mathbf{0}, \tag{4.110a}$$

$$-\mathbf{1} + \mu \mathbf{A}_{\mathcal{S}_-}^T(\mathbf{A}_{\mathcal{S}}\mathbf{x}_{\mathcal{S}}^* - \mathbf{b}) = \mathbf{0}, \tag{4.110b}$$

$$\mu \mathbf{a}_i^T(\mathbf{A}_{\mathcal{S}}\mathbf{x}_{\mathcal{S}}^* - \mathbf{b}) \in (-1, 1), \quad \forall i \notin \mathcal{S}, \tag{4.110c}$$

where $\mathbf{A}_{\mathcal{T}}$ denotes the submatrix of $\mathbf{A}$ formed by the columns of $\mathbf{A}$ in $\mathcal{T}$ and $\mathbf{a}_i$ is the ith column of $\mathbf{A}$. Note that the composition is *not* defined by the sign of x_i^* but by the values of $\mu \mathbf{a}_i^T(\mathbf{A}\mathbf{x}^* - \mathbf{b})$. It is *possible* to have $x_i^* = 0$ for some $i \in (\mathcal{S}_+ \cup \mathcal{S}_-)$. However, if $i \in \mathcal{S}_0$, then $x_i^* = 0$ must hold. Combining (4.110a) and (4.110b), we obtain

$$(\mathbf{A}_{\mathcal{S}}^T\mathbf{A}_{\mathcal{S}})\mathbf{x}_{\mathcal{S}}^* = \mathbf{A}_{\mathcal{S}}^T\mathbf{b} + \mu^{-1}\begin{bmatrix}-\mathbf{1}\\ \mathbf{1}\end{bmatrix}, \tag{4.111}$$

which uniquely determines $\mathbf{x}_{\mathcal{S}}^*$ if $(\mathbf{A}_{\mathcal{S}}^T\mathbf{A}_{\mathcal{S}})$ is non-singular. The solution $\mathbf{x}_{\mathcal{S}}^*$ is piece-wise linear in μ^{-1}. If $(\mathbf{A}_{\mathcal{S}}^T\mathbf{A}_{\mathcal{S}})$ is singular, $\mathbf{A}_{\mathcal{S}}\mathbf{x}_{\mathcal{S}}^*$ is unique and piece-wise linear in μ^{-1}; we leave the proof to the interested reader. Condition (4.110c) joins (4.110a) and (4.110b) in determine $\mathcal{S}$ for each given $\mu > 0$ so that there exists $\mathbf{x}_{\mathcal{S}}^*$ satisfying (4.110) and consistent with $\mathcal{S}$; (4.110c) must be satisfied but it does not directly determine the values of $\mathbf{x}_{\mathcal{S}}^*$.

For each $\mu > 0$, both $\mathbf{A}\mathbf{x}^* - \mathbf{b}$ and $\mathcal{S}$ are unique. This is because any point on the line segment between two solutions of a convex program is also a solution. If there are two distinct solutions, they must give the same value of $\|\mathbf{x}\|_1 + \frac{\mu}{2}\|\mathbf{A}\mathbf{x} - \mathbf{b}\|_2^2$. Over the line segment between the two points, $\|\mathbf{x}\|_1$ is piece-wise linear, but $\frac{\mu}{2}\|\mathbf{A}\mathbf{x} - \mathbf{b}\|_2^2$ cannot be piece-wise linear unless $\mathbf{A}\mathbf{x} - \mathbf{b}$ is constant over the line segment. So, not only is $\mathbf{A}\mathbf{x}^* - \mathbf{b}$ unique for each μ, so is $\|\mathbf{x}^*\|_1$. $\mathcal{S}$ is defined by $\mathbf{A}^T(\mathbf{A}\mathbf{x}^* - \mathbf{b})$, so it is also unique. So far, we have shown that μ uniquely determines $\mathcal{S}$ and $\mathbf{x}_{\mathcal{S}}^*$ (or $\mathbf{A}_{\mathcal{S}}\mathbf{x}_{\mathcal{S}}^*$ if $(\mathbf{A}_{\mathcal{S}}^T\mathbf{A}_{\mathcal{S}})$ is singular), the latter of which is linear in μ^{-1}.

On the other hand, each $\mathcal{S}$ is associated with a (possibly empty) *interval* of μ, which can be verified by plugging the solution of (4.111) into (4.110c). Therefore, $\mathbf{x}^*$ (or $\mathbf{A}\mathbf{x}^*$ if $(\mathbf{A}_{\mathcal{S}}^T\mathbf{A}_{\mathcal{S}})$ is singular) is piece-wise linear in μ^{-1}. Not only is it piece-wise linear, the entire solution path is continuous in μ. If at μ the solution path is discontinuous, consider the left line segment $\{\mathbf{A}\mathbf{x}^*(\mu - \epsilon) : 0 < \epsilon < \epsilon^0\}$ and the right one $\{\mathbf{A}\mathbf{x}^*(\mu + \epsilon) : 0 < \epsilon < \epsilon^0\}$, where $\epsilon^0 > 0$ is sufficiently small. Since the objective is continuous, taking $\epsilon \downarrow 0$, we get $\mathbf{A}\mathbf{x}^*(\mu_-)$ and $\mathbf{A}\mathbf{x}^*(\mu_+)$ that are different yet both optimal at μ, which contradicts the uniqueness of $\mathbf{A}\mathbf{x}^*$ at μ! To summarize, we have argued that $\mathbf{x}^*$, if it is unique for all μ, and $\mathbf{A}\mathbf{x}^*$ are continuous and piece-wise linear in μ^{-1}.

All the discussions above hint at a μ-homotopy method, starting from $\mu = 0$ with the corresponding solution $\mathbf{x}^* = \mathbf{0}$ and increasing μ to compute the piece-wise linear solution path [220]. This is also known as parametric quadratic programming [221]. From μ equals 0 through the minimum value such that $\mu \mathbf{a}_i^T \mathbf{b} \in (-1, 1)$ for all i, $\mathcal{S} = \emptyset$ and $\mathbf{x}^* = \mathbf{0}$ satisfy the optimality condition (4.110). If one keeps $\mathcal{S} = \emptyset$ but further increases μ, then either (4.110c) is always satisfied (possible but not likely in practice) or (4.110c) is eventually violated for some i. Right at the violating point of μ, add that i into $\mathcal{S}$ and compute $\mathbf{x}_{\mathcal{S}}^*$ from (4.111), which is in a μ-parametric form as

$$\mathbf{x}_{\mathcal{S}}^* = (\mathbf{A}_{\mathcal{S}}^T \mathbf{A}_{\mathcal{S}})^{-1} \mathbf{A}_{\mathcal{S}}^T \mathbf{b} + \mu^{-1} (\mathbf{A}_{\mathcal{S}}^T \mathbf{A}_{\mathcal{S}})^{-1} \begin{bmatrix} -\mathbf{1} \\ \mathbf{1} \end{bmatrix}, \tag{4.112}$$

or

$$\mathbf{A}_{\mathcal{S}} \mathbf{x}_{\mathcal{S}}^* = \mathbf{A}_{\mathcal{S}} (\mathbf{A}_{\mathcal{S}}^T \mathbf{A}_{\mathcal{S}})^{\dagger} \mathbf{A}_{\mathcal{S}}^T \mathbf{b} + \mu^{-1} \mathbf{A}_{\mathcal{S}} (\mathbf{A}_{\mathcal{S}}^T \mathbf{A}_{\mathcal{S}})^{\dagger} \begin{bmatrix} -\mathbf{1} \\ \mathbf{1} \end{bmatrix}, \tag{4.113}$$

if $\mathbf{A}_{\mathcal{S}}^T \mathbf{A}_{\mathcal{S}}$ is singular. Note at the moment, $\mathcal{S}$ has a single entry, so one of $\mathcal{S}_+$ and $\mathcal{S}_-$ is empty and the other has a single entry. Therefore, all quantities in the above equation are scalars, and in particular, $\begin{bmatrix} -\mathbf{1} \\ \mathbf{1} \end{bmatrix}$ reduces to either 1 or -1. Now one further increases μ, until either (4.110c) is violated for another i at a point of μ (if it exists, this violating point can be easily computed by plugging (4.112) and (4.113) into (4.110c)) or $\text{sign}(\mathbf{x}_i^*)$ violates $\mathcal{S}$. In the former case, one again adds that i to $\mathcal{S}$, and in the latter case, one drops the violating i from $\mathcal{S}$. This process of augmenting and shrinking $\mathcal{S}$ is continued until μ reaches a target value or μ can increase to ∞ without any violations. This is a finite process since there are only finitely many possibilities of $\mathcal{S}$. Although we have omitted the computation detail, this is roughly the algorithm in [220, 221].

The homotopy algorithm is extremely efficient when $\mathcal{S}$ only needs to be updated a small number of times before reaching the target μ since the computing cost for each linear piece is low when $\mathcal{S}$ is small. Therefore, the homotopy algorithm suits a rather small value of μ and problems with highly sparse solutions. For larger μ though one cannot make much saving on computing (4.112), checking the condition (4.110c) is time consuming. Fortunately, the check can be skipped on certain sets of i. This is referred to as the elimination of coordinates. A safe elimination rule is given in [222], and a more efficient heuristic is presented in [223].

4.10 Continuation, varying step sizes, and line search

The convergence of the prox-linear iterations for (4.12b) depends highly on the parameter μ and the step-size δ_k. The discussion begins with μ. A smaller μ gives more emphasis to the regularization function $r(\mathbf{x})$ and often leads to more structured, or sparser, solutions. Such solutions are simpler, and for technical reasons that we skip here, are quicker to obtain. In fact, a large μ can lead to prox-linear iterations that are extremely slow.

Continuation on μ is a simple technique used in several codes such as FPC [173] and SPARSA [224] to accelerate the convergence of prox-linear iterations. To apply this technique, one does not use the μ given in (4.12b) (which we call the target μ) to start the iterations; instead, use a small $\bar{\mu}$ and gradually increase $\bar{\mu}$ over the iterations until it reaches the target μ. This lets the iteration produce highly structured (low $r(\mathbf{x})$) yet less-fitted (large $f(\mathbf{Ax}-\mathbf{b})$) solutions first, corresponding to smaller $\bar{\mu}$ values, and use them to "warm start" the iterations for producing solutions with more balanced structure and fitting, corresponding to larger $\bar{\mu}$ values. Continuation does not have rigorous theories to give the optimal sequence of $\bar{\mu}$ and how many iterations are needed for each $\bar{\mu}$. It is more regarded as a very effective heuristic. One can geometrically increase $\bar{\mu}$ and, since warm starts do not require high accuracies, one just needs a moderate number of iterations for each $\bar{\mu}$ except after it reaches the target μ given in (4.12b).

It turns out even a simple continuation strategy leads to much faster convergence. In [173], an initial $\bar{\mu}$ is chosen simply to have a value large enough to avoid a zero solution for (4.12b) with $\mu = \bar{\mu}$. For $r(\cdot) = \|\cdot\|_2$ and $f(\cdot) = \frac{1}{2}\|\cdot\|_2^2$, the choice is $\|\mathbf{A}^T\mathbf{b}\|_\infty^{-1}$. During the iterations, once $\mathbf{x}^k$ and $\mathbf{x}^{k+1}$ become very close or the $\nabla f(\mathbf{x}^k)$ has a small size, $\bar{\mu}$ is increased by multiplying with a constant factor.

Line search is widely used in non-linear optimization to speed up objective descent and, in some cases, to guarantee convergence. It allows one to vary the step size δ_k in a systematic way. Let

$$\psi(\mathbf{x}) = r(\mathbf{x}) + \mu f(\mathbf{x}).$$

One approach [225] replaces (4.34) by the following modified Armijo–Wolfe line search iteration

$$\text{trial point:} \quad \bar{\mathbf{x}}^{k+1} = \arg\min_{\mathbf{x}} \; r(\mathbf{x}) + \mu f(\mathbf{x}^k) + \langle \mu \nabla f(\mathbf{x}), \mathbf{x} - \mathbf{x}^k \rangle + \frac{1}{2\delta}\|\mathbf{x} - \mathbf{x}^k\|_2^2, \tag{4.114a}$$

$$\text{search direction:} \quad \mathbf{d} = \bar{\mathbf{x}}^{k+1} - \mathbf{x}^k, \tag{4.114b}$$

$$\text{decay rate:} \quad \Delta = \mu \nabla f(\mathbf{k}^k)^T \mathbf{d} + r(\bar{\mathbf{x}}^{k+1}) - r(\mathbf{x}^k), \tag{4.114c}$$

$$\text{backtracking:} \quad h^* = \arg\min\{h \in \mathbb{Z}_+ : \psi(\mathbf{x}^k + \alpha\rho^h \mathbf{d}) \le \psi(\mathbf{x}^k) + \eta\alpha\rho^h \Delta\}, \tag{4.114d}$$

$$\text{new point:} \quad \mathbf{x}^{k+1} = \mathbf{x}^k + \alpha\rho^{h^*}\mathbf{d}. \tag{4.114e}$$

There is no universally best values for α, η, and ρ, but one can start with $\alpha = 1, \rho = 0.85$, and $\eta = 0.001$. The idea is to exploit the trial point $\bar{\mathbf{x}}^{k+1}$ by making a large step along the direction $\mathbf{d}$ while guaranteeing that the new point $\mathbf{x}^{k+1}$ gives a sufficient descent in the sense of (4.114d). See [226] for a nice explanation of this "sufficient descent" condition. The main computing cost of line search is "backtracking" when the condition (4.114d) is not satisfied by $h = 0$. For one to benefit from line search, the cost of backtracking must be significantly smaller than that of the trial point computation. See [225] on how to efficiently compute h^*.

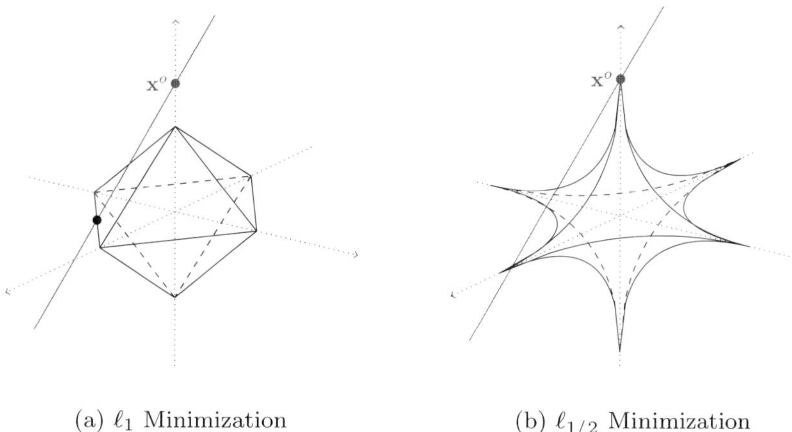

Figure 4.5 ℓ_1 vs. $\ell_{1/2}$ minimization.

4.11 Non-convex approaches for sparse optimization

Non-convex optimization, including ones based on minimizing the non-convex ℓ_q quasi-norm

$$\|\mathbf{x}\|_q = (\sum_i |x_i|^q)^{1/q}, \quad 0 < q < 1,$$

and its variants, has been used to develop algorithms for recovering sparse vectors in [227–229] and low-rank matrices in [230–232]. Compared to the norm $\|\mathbf{x}\|_1$, $\|\mathbf{x}\|_q^q$ for $0 < q < 1$ makes a closer approximate to the "counting norm" $\|\mathbf{x}\|_0$. It is shown in [233] that assuming certain restricted isometry properties (RIPs) of the sensing matrix $\mathbf{A}$, a sparse vector $\mathbf{x}^0 \in \mathbb{R}^N$ is the (global) ℓ_q minimizer of $\mathbf{A}\mathbf{x} = \mathbf{b}$, where $\mathbf{b}$ can have fewer observations than needed by convex ℓ_1 minimization. References [110, 234, 235] derive sufficient RIPs for ℓ_q minimization to recover sparse vectors that are weaker than those known for ℓ_1 minimization.

Figure 4.5 illustrates an example with unsuccessful recovery of $\mathbf{x}^0$ with ℓ_1 minimization but successful recovery with $\ell_{1/2}$ minimization. In Figure 4.5(a), the line represents the set of points satisfying $\mathbf{A}\mathbf{x} = \mathbf{b}$, and the smallest ℓ_1-ball touching the line is the lower point, which has two non-zero components. In Figure 4.5(b), the smallest $\ell_{1/2}$-ball touches the same line at $\mathbf{x}^0$, which has just one non-zero component.

However, the ℓ_q quasi-norm is non-convex, and such ℓ_q minimization is generally NP-hard [236]. Directly minimizing the ℓ_q quasi-norm most likely ends up with one of its many local minimizers, which are the intersections of the dark lines in Figure 4.6(a). Algorithms [227–229] solve a sequence of approximate subproblems. Specifically, [227] solves a sequence of reweighted ℓ_1 subproblems: given an existing iterate $\mathbf{x}^k$, the algorithm generates a new point $\mathbf{x}^{k+1}$ by minimizing $\sum_i w_i|x_i|$ with weights $w_i := (\epsilon + |x_i^k|)^{q-1}$ together with other objective terms or constraints, starting from the previous point. To see how it relates to ℓ_q quasi-norm, one can let $\mathbf{x}^k = \mathbf{x}, \epsilon = 0$, and $0/0$ be 0 and thus obtain $\sum_i w_i|x_i| = \sum_i |x_i|^q = \|\mathbf{x}\|_q^q$. Alternatively, [228, 229] solve reweighted least-squares (also called reweighted ℓ_2) subproblems: at each iteration, they approximate $\|\mathbf{x}\|_q^q$ by $\sum_i w_i|x_i|^2$ with weights $w_i := (\epsilon^2 + |x_i^k|^2)^{q/2-1}$.

In the reweighted ℓ_1/ℓ_2 iterations, a fixed $\epsilon > 0$ not only avoids division by zero but also often leads to a limit $\mathbf{x}^\epsilon = \lim_{k\to\infty} \mathbf{x}^k$ with very few entries larger than $O(|\epsilon|)$ in magnitude. In this sense, $\mathbf{x}^\epsilon$ is often a good approximate of $\mathbf{x}^0$ up to $O(|\epsilon|)$. To recover a sparse vector $\mathbf{x}$ from $\mathbf{b} = \mathbf{A}\mathbf{x}$, these algorithms must vary ϵ, starting from a large value and gradually reducing it. In particular, [229] sets ϵ in terms of the $(r+1)$th largest entry of the latest iterate, where r is the sparsity guesstimate. Empirical results show that to recover vectors with entries in decaying magnitudes, the reweighted ℓ_1/ℓ_2 algorithms require significantly fewer measurements than convex ℓ_1 minimization, and this measurement reduction translates to savings in the sensing time and cost. Figure 4.6 illustrates that the reweighted ℓ_2 algorithm [228] easily locates the global minimizer corresponding to $\epsilon = 1$, then successively decreases ϵ as 10^{-1}, 10^{-2}, 10^{-3} and obtains a sequences of global minimizers, whose limit is $\mathbf{x}^0$. Since ϵ decreases geometrically, one can start from a quite large ϵ, which will turn the problem to nearly a least-squares problem near the origin.

4.12 Greedy algorithms

4.12.1 Greedy pursuit algorithms

Strictly speaking, the greedy algorithms for sparse optimization are not optimization algorithms since they are not driven by an overall objective. Instead, they build a solution – or the support of the solution – part by part, yet in many cases, the solution is indeed the sparsest solution.

A basic greedy algorithm is called the Orthogonal Matching Pursuit (OMP) [124]. With an initial point $\mathbf{x}^0 = 0$ and empty initial support $\mathcal{S}^0$, starting at $k = 1$, it iterates

$$\mathbf{r}^k = \mathbf{b} - \mathbf{A}\mathbf{x}^{k-1}, \tag{4.115a}$$

$$\mathcal{S}^k = \mathcal{S}^{k-1} \cup \arg\min_i\{\|\phi_i\alpha - \mathbf{r}^k\|_2 : i \notin \mathcal{S}^{k-1}, \alpha \in \mathbb{R}\}, \tag{4.115b}$$

$$\mathbf{x}^k = \min_{\mathbf{x}}\{\|\mathbf{A}\mathbf{x} - \mathbf{b}\|_2 : \text{supp}(\mathbf{x}) \subseteq \mathcal{S}^k\}, \tag{4.115c}$$

until $\|\mathbf{r}^k\|_2 \leq \epsilon$ is satisfied. Step (4.115a) computes the measurement residual, step (4.115b) adds the columns of $\mathbf{A}$ that best explains the residual to the running support, and step (4.115c) updates the solution to one that best explains the measurements while confined to the running support.

In (4.115b), for each i, the minimizer is $\alpha = \phi_i^T\mathbf{r}^k/\|\phi_i\|_2$. Hence, the selected i has the smallest $\|(\phi_i^T\mathbf{r}^k/\|\phi_i\|_2)\phi_i - \mathbf{r}^k\|_2$ over all $i \in \mathcal{S}^{k-1}$. Step (4.115c) solves a least-squares

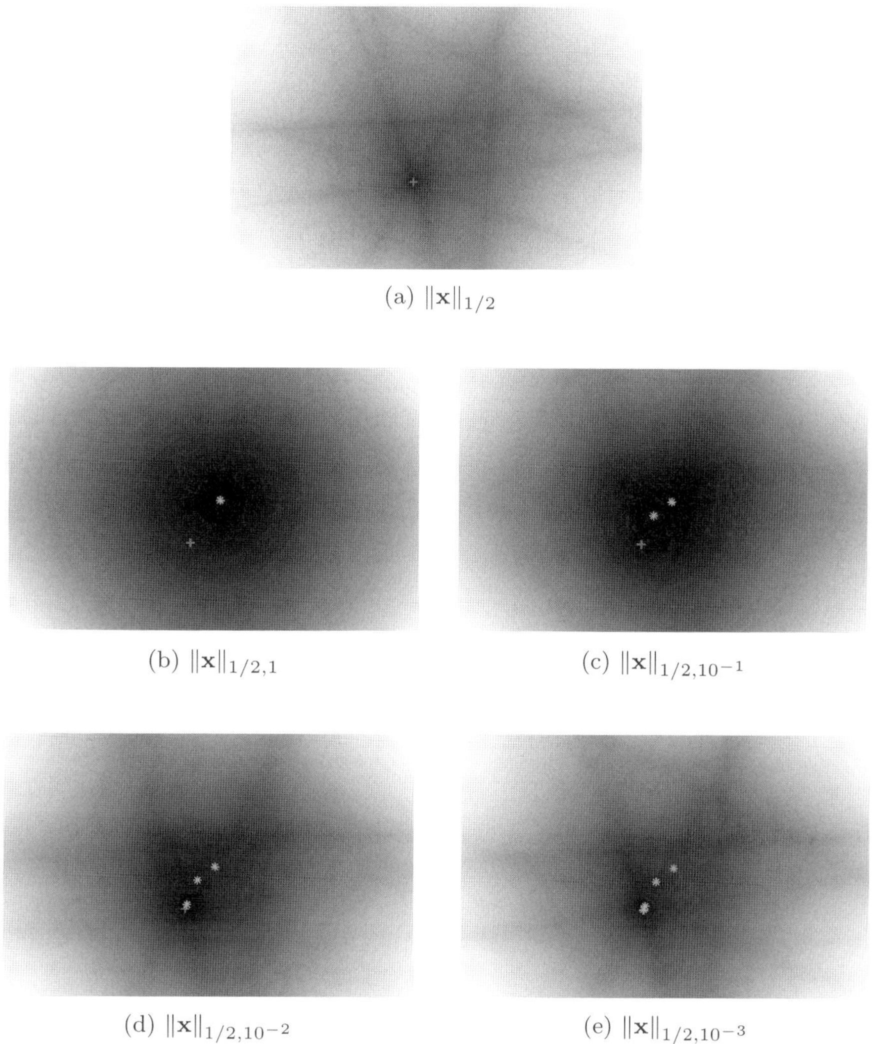

Figure 4.6 $\|\mathbf{x}\|_{1/2,\epsilon}$ over $\{\mathbf{x} : \mathbf{A}\mathbf{x} = \mathbf{b}\}$. Cross: $\mathbf{x}^0$; solid points: minimizer sequence. Courtesy of Rick Chartrand.

problem, restricting $\mathbf{x}$ to the support $\mathcal{S}^k$. There are numerical skills to speed up these steps, but we do not review them here.

Many further developments to OMP lead to algorithms with improved solution quality and/or speed such as stage-wise OMP (StOMP) [125], regularized OMP [126], subspace pursuit [127], CoSaMP [128], as well as HTP [237].

Subspace pursuit and CoSaMP have better performance than many other greedy algorithms. Because they are similar, we describe only CoSaMP. There are more steps in each iteration to predict and refine the intermediate solution and support. With an initial point $\mathbf{x}^0$ and an estimate sparsity level s, starting at $k = 1$, CoSaMP iterates

$$\mathbf{r}^k \leftarrow \mathbf{b} - \mathbf{A}\mathbf{x}^{k-1} \qquad \text{(residual)} \qquad (4.116a)$$

$$\mathbf{a} \leftarrow \mathbf{A}^*\mathbf{r}^k \qquad \text{(correlation)} \qquad (4.116b)$$

$$\mathcal{T} \leftarrow \operatorname{supp}(\mathbf{x}^{k-1}) \cup \operatorname{supp}(\mathbf{a}_{2s}) \qquad \text{(merge supports)} \qquad (4.116c)$$

$$\mathbf{c} \leftarrow \min_{\mathbf{x}}\{\|\mathbf{A}\mathbf{x} - \mathbf{b}\|_2 : \operatorname{supp}(\mathbf{x}) \subseteq \mathcal{T}\} \qquad \text{(least-squares)} \qquad (4.116d)$$

$$\mathbf{x}^k \leftarrow \mathbf{c}_s, \qquad \text{(pruning)} \qquad (4.116e)$$

until $\|\mathbf{r}^k\|_2 \le \epsilon$ is satisfied, where $\mathbf{a}_{2s} = \arg\min\{\|\mathbf{x} - \mathbf{a}\|_2 : \|\mathbf{x}\|_0 \le 2s\}$ is the best $2s$-approximate of $\mathbf{a}$ and similarly, $\mathbf{c}_s$ is the best s-approximate of $\mathbf{c}$. Over the iterations, $\operatorname{supp}(\mathbf{x}^k)$ are updated but kept to contain no more than s components. Hence, $\mathcal{T}$ has no more than $3s$ components. Unlike OMP, CoSaMP's support is updated batch by batch.

4.12.2 Iterative support detection

In the previous subsection, greedy algorithms find a solution by iteratively growing (in OMP) or updating (in subspace pursuit and CoSaMP) the solution support. To add or update the support, greedy algorithms measure the column-residual correlation $\mathbf{A}^*\mathbf{r}^k$ or normalized correlation $\min\{\|\phi_i\alpha - \mathbf{r}^k\|_2 : \alpha \in \mathbb{R}\}$ for each $i \notin \mathcal{S}^{k-1}$. Given the support, a least-squares restricted to the support estimates the solution value. In this section, we overview an algorithm that is based on a similar support-value alternating iteration, but it uses different ways.

The iterative support detection (ISD) framework is the iteration with initial $\mathcal{T}$ set to $\{1, 2, \ldots, n\}$:

$$\mathbf{x}^k \leftarrow \arg\min\{\|\mathbf{x}_{\mathcal{T}}\|_1 : \mathbf{A}\mathbf{x} = \mathbf{b}\}, \qquad (4.117a)$$

$$\mathcal{T} \leftarrow \{|x_i^k| < \rho^k\}, \qquad (4.117b)$$

while $\|\mathbf{x}^k\|_0 > m$ and $|\mathcal{T}| \ge n - m$, where $\{\rho^k\}$ is a monotonically decreasing sequence of scalars and $\|\mathbf{x}_{\mathcal{T}}\|_1 = \sum_{i\in\mathcal{T}} |x_i|$. The iterative update of the support $\mathcal{T}$ is not like any greedy algorithms in the sense that it is typically reducing yet is not always monotonic in k.

The iteration (4.117) is motivated by the fact that if the entire correct support – $\operatorname{supp}(\mathbf{x}^o)$ – is out of $\mathcal{T}$, then (4.117a) will return the correct recovery, and a larger $\mathcal{T}^C \cap \operatorname{supp}(\mathbf{x}^o)$ tends to let (4.117a) give a better estimate of $\mathbf{x}^o$.

Since at $k = 1$ step (4.117a) is equivalent to (4.13a). If it does not return a faithful reconstruction, ISD with $k \ge 2$ addresses failed reconstructions of (4.13a) by hopefully discovering a partial correct support from $\mathbf{x}^k$. Analysis in [238] gives conditions on $\mathcal{T}$ that will guarantee the recovery of $\mathbf{x}^o$.

There are two existing rules for choosing ρ^k. The simple one uses

$$\rho^k = \beta^{-k}\|\mathbf{x}^k\|_\infty, \qquad (4.118)$$

where $\beta > 1$ can be chosen as, for example, 3. Since $\|\mathbf{x}^k\|_\infty$ is quite stable in k, ρ^k will become sufficiently small to make $|\mathcal{T}| < n - m$ in a small number of iterations. Hence, the number of iterations of (4.117) is typically small.

A less-simple rule for ρ^k is based on the "first significant jump," which is effective for $\mathbf{x}^o$ with non-zero entries following a fast-decaying distribution. The idea is that with such $\mathbf{x}^o$, $\mathbf{x}^k$ tends to be fast decaying as well, yet the false non-zero entries in $\mathbf{x}^k$ are often clustered and form the tail of $\mathbf{x}^k$. See Figure 2 of [238] for an illustration. When the components of $\mathbf{x}^k$ are sorted in the increasing order like $|x^k_{[1]}| \leq |x^k_{[2]}| \leq \cdots \leq |x^k_{[n]}|$, the tail and head are often separated by the first significant jump, which is the smallest i such that

$$|x^k_{[i+1]}| - |x^k_{[i]}| > \beta^{-k}. \tag{4.119}$$

Both rules were tested empirically, and the results show that ISD and non-convex approaches have roughly equal recovery quality. Yet, it is more versatile to incorporate model sparsity into the selection of $\mathcal{T}$.

4.12.3 Hard thresholding

Hard thresholding refers to the operator that keeps the K largest entries of a vector in magnitude for a given K. We write it as hardthr($\mathbf{x}$). For vector $\mathbf{x}$ and the index set $\mathcal{K} = \{K \text{ largest entries of } \mathbf{x} \text{ in magnitude}\}$, hardthr($\mathbf{x}$) obeys

$$(\text{hardthr}(\mathbf{x}))_i = \begin{cases} x_i, & i \in \mathcal{K}, \\ 0, & \text{otherwise}. \end{cases} \tag{4.120}$$

Unlike soft thresholding, the values of the K largest entries are not changed. The basic iterative hard-thresholding iteration [129] is

$$\mathbf{x}^{k+1} \leftarrow \text{hardthr}\left(\mathbf{x}^k + \mathbf{A}^T(\mathbf{b} - \mathbf{A}\mathbf{x}^k)\right), \tag{4.121}$$

which differs from (4.41) only by the use of the hard, as opposed to the soft, thresholding, ignoring the difference in the step lengths. The iteration (4.121) can be used to solve the problem

$$\min_{\mathbf{x}} \left\{ \|\mathbf{A}\mathbf{x} - \mathbf{b}\|_2^2 : \|\mathbf{x}\|_0 \leq K \right\}, \tag{4.122}$$

with the guarantee to converge to one of its local minima provided that $\|\mathbf{A}\|_2 \leq 1$. Furthermore, under the RIP assumption of $\mathbf{A}$, $\delta_{3K}(\mathbf{A}) \leq 32^{-1/2}$, this local minimum $\mathbf{x}^*$ has bounded distance to the original unknown signal $\mathbf{x}_0$. Specifically, $\|\mathbf{x}^* - \mathbf{x}^0\|_2 \leq \|\mathbf{x}^0 - \text{hardthr}(\mathbf{x}^0)\|_2 + \|\mathbf{x}^0 - \text{hardthr}(\mathbf{x}^0)\|_1/\sqrt{k} + \|\mathbf{n}\|_2$, where $\mathbf{n} = \mathbf{b} - \mathbf{A}\mathbf{x}^0$.

The iteration (4.121) is not very competitive compared to ℓ_1 minimization, greedy algorithms, especially to non-convex ℓ_q and iterative support detection algorithms. However, the normalized iterative hard-thresholding iteration

$$\mathbf{x}^{k+1} \leftarrow \text{hardthr}\left(\mathbf{x}^k + \mu^k\mathbf{A}^T(\mathbf{b} - \mathbf{A}\mathbf{x}^k)\right), \tag{4.123}$$

where μ^k is a changing step size, has much better performance. The rule of choosing μ^k is detailed in [130], and it depends whether the resulting supp($\mathbf{x}^{k+1}$) equals supp($\mathbf{x}^k$) or not. Under this rule, (4.123) also converges to one of the local minima of (4.122) and obeys the bounded distance to $\mathbf{x}^0$ under a certain RIP assumption on $\mathbf{A}$. On the other hand, the recovery performance of (4.123) is significantly stronger than that of (4.121).

The work [237] proposes hard-thresholding pursuit (HTP) as a combination of the (original and normalized) IHT with the idea of matching pursuit. In (4.121), instead of applying the hard-thresholding operator to return $\mathbf{x}^{k+1}$, HTP performs

$$\mathcal{T} \leftarrow \{s \text{ largest entries of } \left(\mathbf{x}^k + \mathbf{A}^T(\mathbf{b} - \mathbf{A}\mathbf{x}^k)\right)\}, \tag{4.124a}$$

$$\mathbf{x}^{k+1} \leftarrow \underset{\{\text{supp}(\mathbf{x})\in\mathcal{T}\}}{\arg\min} \|\mathbf{b} - \mathbf{A}\mathbf{x}\|_2. \tag{4.124b}$$

Reference [237] also applies the same change to (4.123) and introduces a fast version of (4.123). Exact and stable recoveries are guaranteed if $\mathbf{A}$ satisfies the RIP with $\delta_{3K}(\mathbf{A}) \leq 3^{-1/2}$.

4.13 Algorithms for low-rank matrices

When the recovery of a low-rank matrix $\mathbf{X}^o \in \mathbb{R}^{n_1\times n_2}$ is based on minimizing the nuclear-norm $\|\cdot\|_*$ and solving a convex optimization problem, most convex optimization algorithms above for ℓ_1 minimization can be extended for nuclear-norm minimization. They include, but are not limited to, the singular value thresholding (SVT) algorithm [149] based on the linearized Bregman algorithm [189, 190], fixed-point continuation code FPCA [150] extending FPC [173], the code APGL [239] extending [177], and the code [240] based on the alternating direction method [165]. Refer to Section 4.4.1 for the shrinkage operation for the nuclear-norm. References [230–232] are related to non-convex ℓ_q minimization in Section 4.11. There are also algorithms with no vector predecessors.

Suppose that the rank of the underlying matrix is given as $s \ll \min\{n_1, n_2\}$ either exactly or as the current overestimate of the true rank, the underlying matrix can be factorized as $\mathbf{X} = \mathbf{USV}^*$ or $\mathbf{X} = \mathbf{PQ}$, where $\mathbf{U}$ and $\mathbf{P}$ have s columns, $\mathbf{V}^*$ and $\mathbf{Q}$ have s rows, and $\mathbf{S}$ is an $s \times s$ diagonal matrix. Instead of $\mathbf{X}$, if one solves for $(\mathbf{U}, \mathbf{S}, \mathbf{V})$ or $(\mathbf{P}, \mathbf{Q})$, then $\mathbf{X}$ can be recovered from these variables and has its rank no more than s. Hence, these factorizations and change of variable replace the role of minimizing the nuclear-norm $\|\mathbf{X}\|_*$ and give rise to a low-rank matrix. Solvers based on these factorizations include OptSpace [241], LMaFit [242], and an SDPLR-based algorithm [89]. OptSpace [241] and LMaFit [242] are based on minimizing $\|\mathcal{P}_\Omega(\mathbf{USV}^*) - \mathbf{b}\|_2$ and $\|\mathcal{P}_\Omega(\mathbf{PQ}) - \mathbf{b}\|_2$ for recovering a low-rank matrix from a subset of its entries $\mathbf{b} = \mathbf{X}^o_\Omega$, where Ω is the set of entries observed, and the algorithm in [89] solves $\min\{\frac{1}{2}(\|\mathbf{P}\|_F^2 + \|\mathbf{Q}\|_F^2) : \mathcal{A}(\mathbf{PQ}) = \mathbf{b}\}$, which is a non-convex problem whose global solution $(\mathbf{P}^*, \mathbf{Q}^*)$ generates $\mathbf{X}^* = \mathbf{P}^*\mathbf{Q}^*$ that is the solution to $\min\{\|\mathbf{X}\|_* : \mathcal{A}(\mathbf{X}) = \mathbf{b}\}$. The common advantage of these algorithms is the many fewer unknown variables, a point easy to see by comparing $n_1 \times n_2$ to $n_1 \times s + s \times n_2$. On the other hand, these methods require judiciously selecting s, which can be known, approximately known, or completely unknown based on applications. (Arguably, the nuclear-norm convex models in the form of (4.12b) and (4.12c) also have parameters that determine the rank of their solutions. However, their parameters are affected by the noise in $\mathbf{b}$ and how close $\mathbf{X}^o$ is from its best low-rank approximation, while s in the above algorithms is determined by the unknown solution and is critical to the success of those algorithms.) To determine s, a dynamic update rule for s is included in LMaFit [242], which is reportedly very effective.

4.14 How to choose an algorithm

This chapter has gone through several algorithms and various techniques for sparse optimization. Because of the space limit, as well as our limited knowledge and the fast-growing nature of this active field of research, many efficient algorithms are not covered. But already, to recover sparse vectors and low-rank matrices, one has multiple choices. In the previous sections, when each approach is introduced or discussed, we have mentioned their strengths (also weaknesses for some). Here, we attempt to give a more direct comparison and provide some guidance on selecting algorithms. Before we start, we would like to stress that the judgement on different algorithms may have become out of date due to new developments after this chapter is written. After all, sparse optimization is an amazingly common field where a very wide spectrum of optimization, as well as non-optimization, techniques can play active roles by providing new approaches or improving the existing algorithms. Another reason is that, as this part is being written, novel sparse optimization applications are being discovered each day! An algorithm that is less competitive today might have the features that make it one of the best choices tomorrow under a new setting. We have witnessed such shifts happening on, for example, the ADM algorithms – it was introduced a long time ago, then it lost its favor in non-linear optimization for a while, but it has recently re-emerged and led to a large number of successful applications. Furthermore, although numerical experiments are a good means to assess the efficiencies of algorithms, it is very difficult to be comprehensive and nearly impossible to reveal the potentials of different algorithmic approaches. The IHT approach in its original form (with a fixed thresholding parameter), as an example, has very poor performance, yet in just three years, works [130, 237] have released its power and made it competitive. The same has happened on many other approaches covered in this chapter to different extents. These make it very tricky to compare different approaches and make judgements. The reader should be aware of our limitations and is advised to check the literature for the latest developments.

Greedy versus optimization algorithms. First, both approaches enjoy exact and stable recovery guarantees, or in other words, they can be trusted for recovering sparse solutions given sufficient measurements. Secondly, because of the nature of greedy algorithms that progressively build (and correct) the solution support, their performance on signals with entries following a *faster* decaying distribution is better – namely, fewer measurements are required when the decay is *faster*. On the other hand, ℓ_1 minimization tends to have more consistent recovery performance – the solution quality tends to be less affected by the decay speed. That said, the speed and accuracies specific ℓ_1 algorithms are still affected by the decay speed. Some algorithms prefer faster decays, and the other prefer slower decays. Notably, the algorithm in Section 4.12.2 integrates ℓ_1 minimization with greedy support selection, so it has the advantages of the two. Thirdly, the two approaches are extended in different ways, which to a large extent determine which one is more suitable for a specific application. The greedy approach can be extended to model-base CS [96], in which the underlying signals are not only sparse but their supports are restricted to certain models, e.g., the tree structure. Some of such models are difficult to express as optimization models, not to mention as convex optimization models. On the other hand, it is difficult to extend greedy approaches to handle general objective or energy functions, when they are present along with the sparse objective in a model. Sparse optimization naturally accepts objective functions and constraints of many kinds, especially if they are convex. Algorithms based on prox-linear, variable splitting, and block coordinate descent are very flexible, so they are among the best choices for engineering models with "side" objectives and constraints. Overall, the two approaches are very complementary to each other. When a problem can be solved by either one, one should examine the decay speed. But after all, they are both quite easy to implement and try out.

Convex versus non-convex. There is no doubt that on some problems that convex optimization (for example, ℓ_1 minimization) fails to return faithful reconstructions, non-convex optimization succeeds. Currently, such problems are those with fast-decaying signals and very few measurements. On sparse signals with 1 or ± 1 non-zero entries, no practical non-convex algorithms are known to perform significantly better than ℓ_1 minimization. Nevertheless, in a large number of (perhaps the majority of) applications of spare optimization, the underlying signals do indeed have a reasonably fast-decaying distribution of non-zero entries, so non-convex algorithms have the performance advantages. On the other hand, the theory and generalizations of non-convex algorithms are well behind. Their recovery guarantees typically assume global optimality. In convex optimization, no matter what algorithms are used, fast or slow, the solution quality is quite predictable. However, when existing algorithms for minimizing ℓ_q, $0 \leq q < 1$, or other non-convex objectives are extended to problems with *additional* objectives and constraints or applied to subproblems in the splitting or block coordinate framework, there is no guarantee on their performance. It is fair to say that on a given application of sparse optimization, one should perform extensive tests to confirm the performance and reliability of non-convex optimization algorithms.

Very large scale sparse optimization. To solve very large scale optimization problems, we are facing memory and computing bottlenecks. Many existing algorithms will fail to run if the problem data, such as the matrix $\mathbf{A}$, or the solution, or both, does not fit in the available memory. There are a few approaches to address this issue. First, if options are available, try to avoid general data and use data that have computation-friendly structures. In CS, examples of structured sensing matrices $\mathbf{A}$ include the discrete Fourier ensemble, a partial Toeplitz or circulant matrix, and so on, which allow fast computation of $\mathbf{Ax}$ and $\mathbf{A}^T\mathbf{y}$ (or complex-valued $\mathbf{A}^*\mathbf{y}$) (or even $(\mathbf{AA}^T)^{-1}$) without storing the entire matrix in memory. When an algorithm is based on such matrix-vector operations, it can take advantages of faster computation and be *matrix-free*. After all, if an algorithm must perform large-scale dense matrix-vector multiplications or solve dense linear systems, these bottlenecks will seriously hamper its ability to solve huge problems.

When the data are given and fast operations are not available, a viable approach is distributed computation (such as the one based on ADM in Section 4.7.3). This approach divides the original problem into a series of smaller subproblems, each involving just a subset of data, or solution, or both. The smaller subproblems are solved, typically in parallel, on distributed computing nodes. There is a trade-off between the acceleration due to parallel computing and the overheads of synchronization and splitting.

Another very effective approach, which is popular among the community of large-scale machine learning, is based on stochastic approximation. It takes advantage of the fact that in many problems, from a small (random) set of data (even a single point), one can generate a gradient (or Hessian) that approximates the actual one, whose computation would require the entire batch of data. Between using just one point and the entire batch of data, there lies an optimal trade-off among memory, speed and accuracy. Stochastic approximation methods use a very small amount of (random chosen) data to guide its iterations and are usually very quick at decreasing the objective until getting close to the solution, where further improvement would require more accurate function evaluations, gradients, (and Hessians). If higher solution accuracies are needed, one can try the stochastic approximation methods with dynamic sample sizes.

Solution structure also provides us with opportunities to address the memory and computing bottlenecks. When the solution is a highly sparse vector, algorithms such as BCD with the Gauss–Southwell update rule in Section 4.8, homotopy algorithms in Section 4.9, and some dual algorithms can keep all the intermediate solutions sparse, which effectively reduces the data/solution storage and access. When the solution is a low-rank matrix (or tensor) that has a very large size, algorithms such as those in Section 4.13 that store and update the factors, which have much smaller sizes, instead of the matrix (or tensor) have significant advantages.

5 CS analog-to-digital converter

An analog-to-digital converter (ADC) is a device that uses sampling to convert a continuous quantity to a discrete-time representation in digital form. The reverse operation is performed by a digital-to-analog converter (DAC). ADC and DAC are the gateways between the analog world and digital domain. Most signal processing tasks are implemented in the digital domain. Therefore, ADC and DAC are the key enablers for the digital signal processing and have significant impacts on the system performances.

In this chapter, we study the literature of CS-based ADCs. We introduce the traditional ADC and its concepts first. Then, we study two major types of CS-ADC architectures, Random Demodulator and Modulated Wideband Converter. Next, we investigate the unified framework, Xampling, and briefly explain the other types of implementation. Finally, we present a summary.

5.1 Traditional ADC basics

In this section, we study the basics of ADC by the sampling theorem, quantization rule and practical ADC implementation in the following subsections.

5.1.1 Sampling theorem

In digital processing, it is useful to represent a signal in terms of sample values taken at appropriately spaced intervals as illustrated in Figure 5.1. The signal can be reconstructed from the sampled waveform by passing it through an ideal low-pass filter. In order to ensure a faithful reconstruction, the original signal must be sampled at an appropriate rate, as described in the sampling theorem.

Theorem 10. Sampling theory: *A real-valued band-limited signal having no spectral components above a frequency of B Hz is determined uniquely by its values at uniform intervals spaced no greater than* $\frac{1}{2B}$ *seconds apart.*

In mathematics, Let $g_\delta(t)$ denote the ideal sampled signal as

$$g_\delta(t) = \sum_{n=-\infty}^{\infty} g(nT_s)\delta(t - nT_s), \tag{5.1}$$

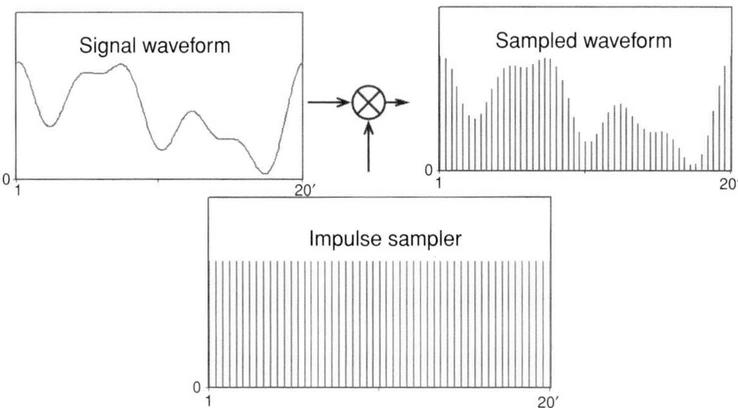

Figure 5.1 Sampling of analog signal.

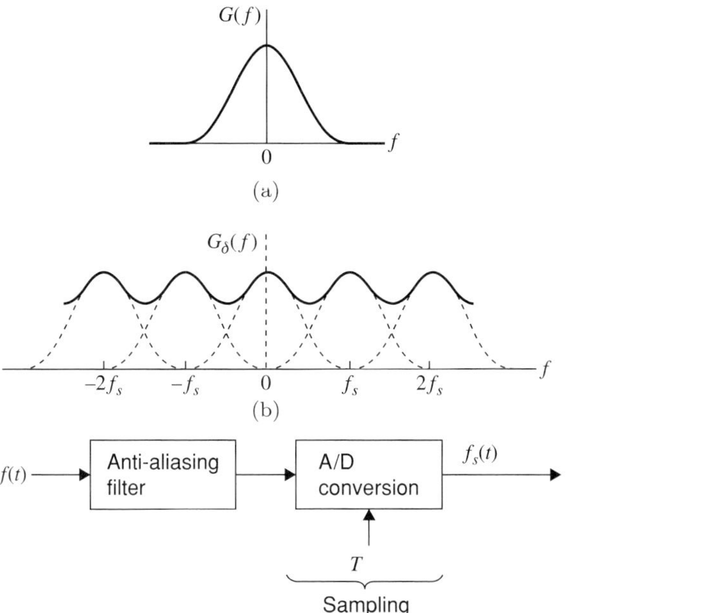

Figure 5.2 Anti-aliasing filter for ADC.

where T_s is the sampling period and $f_s = 1/T_s$ is the sampling rate. If the sampling is at exactly the Nyquist rate, then the original signal can be reconstructed without any error since

$$g(t) = \sum_{t=-\infty}^{\infty} g(nT_s)\text{sinc}\left(\frac{t - nT_s}{T_s}\right). \tag{5.2}$$

If the sampling rate is not sufficient, aliasing effects will occur as shown in Figure 5.2, where the original signal will be overlapped together and cannot be recovered without distortion. To prevent aliasing, before any ADC, there should be an analog anti-aliasing filter to ensure that the input signal to ADC is band-limited.

For the bandpass signal with bandwidth B from f_1 to f_2, there is no need to sample at $2f_2$ speed. Instead, the Bandpass Sampling theorem can be applied with sampling rate f_s satisfying

$$f_s \geq 2(f_2 - f_1)\left(1 + \frac{M'}{N'}\right), \tag{5.3}$$

where $M' = \frac{f_2}{f_2 - f_1} - N'$ and $N' = \lfloor \frac{f_2}{f_2 - f_1} \rfloor$.

If a signal is sampled at a rate much higher than the Nyquist frequency and then digitally filtered to limit it to the signal bandwidth, there are the following advantages:

- Digital filters can have better properties (sharper rolloff, phase) than analog filters, so a sharper anti-aliasing filter can be realized and then the signal can be downsampled giving a better result.
- An N-bit ADC can be made to act as a $N + M$-bit ADC with 2^M times oversampling, since the sampled signals are correlated and sampled noises are independent.
- The signal-to-noise ratio due to quantization noise will be higher than if the whole available band had been used. With this technique, it is possible to obtain an effective resolution larger than that provided by the converter alone.
- The improvement in SNR is 3 dB (equivalent to 0.5 bits) per octave of oversampling, which is not sufficient for many applications. Therefore, oversampling is usually coupled with noise shaping (see sigma-delta modulators). With noise shaping, the improvement is $6Q + 3$ dB per octave where Q is the order of the loop filter used for noise shaping, e.g., a second-order loop filter will provide an improvement of 15 dB/octave.

5.1.2 Quantization

The sampled signals still have real values in amplitude. In order to present the signal in the digital domain, we need quantization as shown in Figure 5.3. The input signal m is quantified to v_k if $m_k < m \leq m_{k+1}$, where $k = 1, 2, \ldots, L$ and L is the number of quantization steps (for example for 8-bit ADC, $L = 256$). The quantization error occurs since $v_k \neq m$ in general. The error energy can be written for a uniform quantizer as

$$\delta_Q^2 = \frac{\Delta^2}{12}, \tag{5.4}$$

where Δ is the quantization step. Then the signal-to-noise ratio (SNR) of a quantizer is linearly proportion to L^2. In other words, the SNR increases exponentially with increase of the number of ADC bits. In practice, the SNR increase 6 dB per ADC bit.

For different distributions of input signals, there are two conditions for an optimal quantizer as follows:

1. The partition boundary is located at the middle of quantized points, e.g., $m_k = (v_{k-1} + v_k)/2$.
2. The quantize point is located at the weighted center within the boundaries.

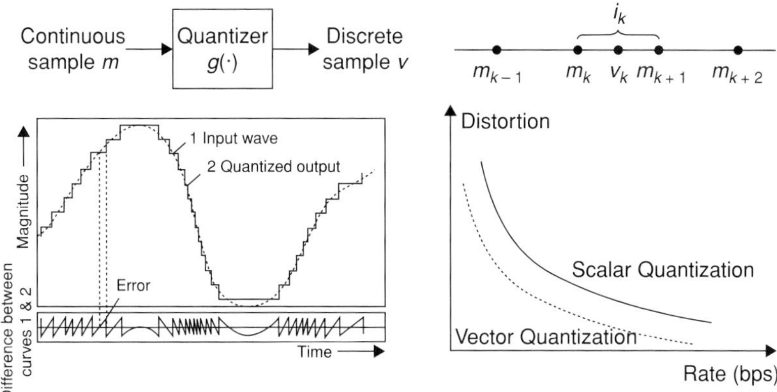

Figure 5.3 Quantization.

Example applications of the above optimal quantization laws are the μ law and the A law to enhance the SNR for small signals in voice communication. The μ-law algorithm is a companding algorithm, primarily used in the digital telecommunication systems of North America and Japan. Its purpose is to reduce the dynamic range of an audio signal. In the analog domain, this can increase the signal-to-noise ratio achieved during transmission, and in the digital domain, it can reduce the quantization error (hence increasing signal to quantization noise ratio). The A-law algorithm used in the rest of the world. The A-law algorithm provides a slightly larger dynamic range than the μ-law at the cost of worse proportional distortion for small signals. By convention, the A-law is used for an international connection if at least one country uses it.

If the quantization is done beyond the scalar, vector quantization is a classical quantization technique, which allows the modeling of probability density functions by the distribution of prototype vectors. It was originally used for data compression. It works by dividing a large set of points (vectors) into groups having approximately the same number of points closest to them. Each group is represented by its centroid point as shown in Figure 5.4. Typically, vector quantization shows superior performances to scalar quantization.

5.1.3 Practical implementation

Most commercial converters sample with 6 to 24 bits of resolution, and produce fewer than mega-samples per second. Thermal noise generated by passive components such as resistors masks the measurement when higher resolution is desired. For example, mega-sample converters are required in digital video cameras, video capture cards, and TV tuner cards to convert full-speed analog video to digital video files. In many cases, the most expensive part of an integrated circuit is the pins, because they make the package larger, and each pin has to be connected to the integrated circuit's silicon. To save pins, it is common for slow ADCs to send their data one bit at a time over a serial interface to the computer, with the next bit coming out when a clock signal changes state. This saves

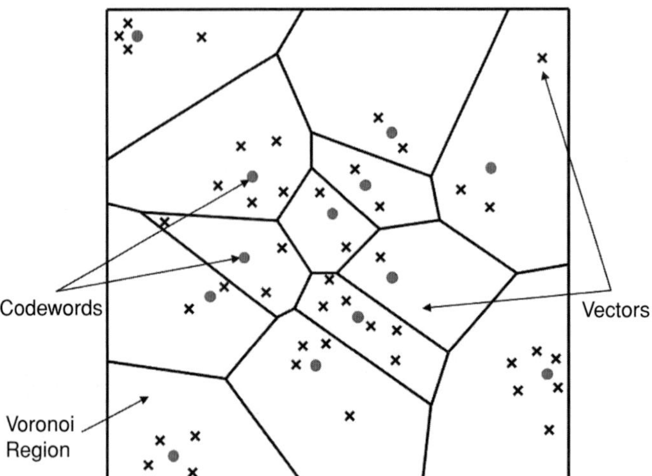

Figure 5.4 Vector quantization.

quite a few pins on the ADC package, and in many cases, does not make the overall design any more complex. Commercial ADCs often have several inputs that feed the same converter, usually through an analog multiplexer. Different models of ADC may include sample and hold circuits, instrumentation amplifiers or differential inputs, where the quantity measured is the difference between two voltages. In general, different types of ADC are listed as follows [243].

1. A direct-conversion ADC or flash ADC has a bank of comparators sampling the input signal in parallel, each firing for their decoded voltage range. The comparator bank feeds a logic circuit that generates a code for each voltage range. Direct conversion is very fast, capable of gigahertz sampling rates, but usually has only 8 bits of resolution or fewer, since the number of comparators needed, $2N - 1$, doubles with each additional bit, requiring a large, expensive circuit. ADCs of this type have a large die size, a high input capacitance, high power dissipation, and are prone to produce glitches at the output (by outputting an out-of-sequence code). They are often used for video, wideband communications or other fast signals in optical storage.
2. A successive-approximation ADC uses a comparator to successively narrow a range that contains the input voltage. At each successive step, the converter compares the input voltage to the output of an internal digital-to-analog converter that might represent the midpoint of a selected voltage range. At each step in this process, the approximation is stored in a successive approximation register (SAR).
3. A ramp-compare ADC produces a saw-tooth signal that ramps up or down then quickly returns to zero. When the ramp starts, a timer starts counting. When the ramp voltage matches the input, a comparator fires, and the timer's value is recorded. Timed ramp converters require the least number of transistors. The ramp time is sensitive to temperature because the circuit generating the ramp is often just some

simple oscillator. There are two solutions: use a clocked counter driving a DAC and then use the comparator to preserve the counter's value, or calibrate the timed ramp. A special advantage of the ramp-compare system is that comparing a second signal just requires another comparator, and another register to store the voltage value.

4. The Wilkinson ADC is based on the comparison of an input voltage with that produced by a charging capacitor. The capacitor is allowed to charge until its voltage is equal to the amplitude of the input pulse (a comparator determines when this condition has been reached). Then, the capacitor is allowed to discharge linearly, which produces a ramp voltage. At the point when the capacitor begins to discharge, a gate pulse is initiated. The gate pulse remains on until the capacitor is completely discharged. Thus, the duration of the gate pulse is directly proportional to the amplitude of the input pulse. This gate pulse operates a linear gate that receives pulses from a high-frequency oscillator clock. While the gate is open, a discrete number of clock pulses pass through the linear gate and are counted by the address register. The time the linear gate is open is proportional to the amplitude of the input pulse, thus the number of clock pulses recorded in the address register is proportional also. Alternatively, the charging of the capacitor could be monitored, rather than the discharge.
5. An integrating ADC (also dual-slope or multislope ADC) applies the unknown input voltage to the input of an integrator and allows the voltage to ramp for a fixed time period (the run-up period). Then a known reference voltage of opposite polarity is applied to the integrator and is allowed to ramp until the integrator output returns to zero (the run-down period). The input voltage is computed as a function of the reference voltage, the constant run-up time period, and the measured run-down time period. The run-down time measurement is usually made in units of the converter's clock, so longer integration times allow for higher resolutions. Likewise, the speed of the converter can be improved by sacrificing resolution. Converters of this type (or variations on the concept) are used in most digital voltmeters for their linearity and flexibility.
6. A delta-encoded ADC or counter-ramp has an up-down counter that feeds a digital-to-analog converter (DAC). The input signal and the DAC both go to a comparator. The comparator controls the counter. The circuit uses negative feedback from the comparator to adjust the counter until the DAC's output is close enough to the input signal. The number is read from the counter. Delta converters have very wide ranges and high resolution, but the conversion time is dependent on the input signal level, though it will always have a guaranteed worst case. Delta converters are often very good choices to read real-world signals. Most signals from physical systems do not change abruptly. Some converters combine the delta and successive approximation approaches; this works well when high frequencies are known to be small in magnitude.
7. A pipeline ADC (also called a subranging quantizer) uses two or more steps of subranging. First, a coarse conversion is done. In a second step, the difference from the input signal is determined with a digital-to-analog converter (DAC). This

difference is then converted finer, and the results are combined in a last step. This can be considered a refinement of the successive-approximation ADC wherein the feedback reference signal consists of the interim conversion of a whole range of bits (for example, four bits) rather than just the next-most-significant bit. By combining the merits of the successive approximation and flash ADCs this type is fast, has a high resolution, and only requires a small die size.

8. A sigma-delta ADC (also known as a delta-sigma ADC) oversamples the desired signal by a large factor and filters the desired signal band. Generally, a smaller number of bits than required are converted using a Flash ADC after the filter. The resulting signal, along with the error generated by the discrete levels of the Flash, is fed back and subtracted from the input to the filter. This negative feedback has the effect of noise shaping the error due to the Flash so that it does not appear in the desired signal frequencies. A digital filter (decimation filter) follows the ADC, which reduces the sampling rate, filters off unwanted noise signal and increases the resolution of the output (sigma-delta modulation, also called delta-sigma modulation).
9. A time-interleaved ADC uses M parallel ADCs where each ADC samples data every Mth cycle of the effective sample clock. The result is that the sample rate is increased M times compared to what each individual ADC can manage. In practice, the individual differences between the M ADCs degrade the overall performance reducing the SFDR. However, technologies exist to correct for these time-interleaving mismatch errors.
10. An ADC with intermediate FM stage first uses a voltage-to-frequency converter to convert the desired signal into an oscillating signal with a frequency proportional to the voltage of the desired signal, and then uses a frequency counter to convert that frequency into a digital count proportional to the desired signal voltage. Longer integration times allow for higher resolutions. Likewise, the speed of the converter can be improved by sacrificing resolution. The two parts of the ADC may be widely separated, with the frequency signal passed through an opto-isolator or transmitted wirelessly. Some such ADCs use sine-wave or square-wave frequency modulation; others use pulse-frequency modulation. Such ADCs were once the most popular way to show a digital display of the status of a remote analog sensor.
11. A time-stretch analog-to-digital converter (TS-ADC) digitizes a very wide bandwidth analog signal, that cannot be digitized by a conventional electronic ADC, by time stretching the signal prior to digitization. It commonly uses a photonic preprocessor frontend to time stretch the signal, which effectively slows the signal down in time and compresses its bandwidth. As a result, an electronic backend ADC, that would have been too slow to capture the original signal, can now capture this slowed-down signal. For continuous capture of the signal, the frontend also divides the signal into multiple segments in addition to time stretching. Each segment is individually digitized by a separate electronic ADC. Finally, a digital signal processor rearranges the samples and removes any distortions added by the frontend to yield the binary data that is the digital representation of the original analog signal.

All the above ADCs are limited by the Sampling Theorem. In the rest of this chapter, we discuss the CS-based ADCs, which explore the sparsity nature of CS so as not to be limited by the Sampling Theorem.

5.2 Random demodulator ADC

A data acquisition system, called a random demodulator, is mainly proposed by Rice University [244–246]. The input signal is assumed to be sparse in some domains such as the frequency domain. In a random demodulator system, the input signal is multiplied by a high-speed sequence from a pseudorandom number generator in the analog domain. As a result, the original signal is modulated and spread out like a CDMA system. Then after a low-pass filter, the modulated signal is sampled with a much slower rate than the Nyquist rate. The intuition behind this is that the sparse information can still be recovered from the piece of information obtained through the low-pass filter. In this section, we first explain the signal model, and then study the architecture of a random demodulator. Next, we study the reconstruction method. Finally, we show some theoretical results and discuss the implementation.

5.2.1 Signal model

The random demodulator fits the scenario in which the signal is band-limited, periodic and band-sparse. This fits many situations such as communication signals, acoustic signals, slowly varying chirps in radar and geophysics, smooth signals with only a few Fourier coefficients and piece-wise smooth signals. The signal can be written as

$$f(t) = \sum_{\omega \in \Omega} a_\omega e^{-j2\pi\omega\Delta t}, \tag{5.5}$$

where Δ is a constant representing the frequency resolution, Ω is a finite set of K integer-value harmonics $\Omega \subset \{0, \pm 1, \pm 2, \ldots, \pm(W/2-1), W/2\}$ and $\{a_\omega : \omega \in \Omega\}$ is a set of complex-valued amplitudes. Here, we have the sparsity assumption that $K \ll W$.

From the sampling theorem and information theorem point of view, it can be shown that the signal in (5.5) contains only $R = O(K \log(W/K))$ bits of information. So theoretically, it should be able to acquire these signals using only R digital samples.

5.2.2 Architecture

To achieve the low sampling rate, the random demodulation is proposed in [244–246] as shown in Figure 5.5. The random demodulator performs three basic actions: demodulation, low-pass filtering, and low-rate sampling. The intuition is that the random demodulator smears the sparse frequency signal across the entire spectrum so that after the low-pass filter it leaves a signature that can be detected by a low-rate sampler. The pseudorandom generator generates a (periodic) square wave that randomly alternates at or above the Nyquist rate. This random signal is an approximation to

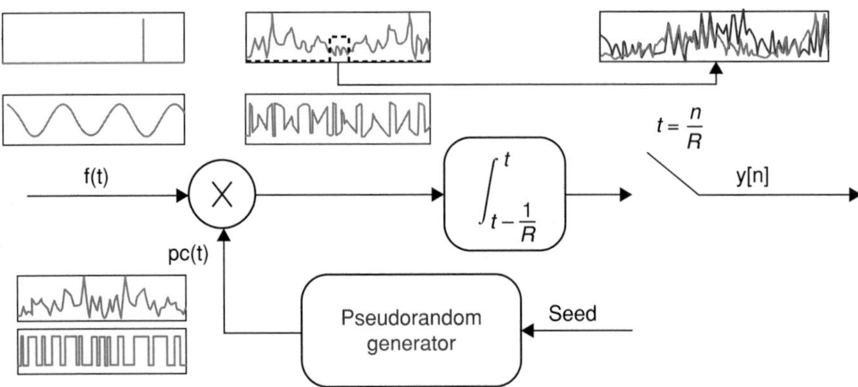

Figure 5.5 Random demodulator architecture.

white noise like a CDMA system. When the input signal is multiplied by this random signal, the narrowband sparse signal is spread out. The key point is that the input signals at different spectrums are transformed with different unique signatures even within the narrow low-pass band. A low-pass filter is used for anti-aliasing before the low-speed ADC.

To represent the random projection in a mathematical form, we first represent the continuous input signal to a discrete one by averaging over time as $\mathbf{x}$ whose component is

$$x_n = \int_{t_n}^{t_n+\Delta/W} f(t)\mathrm{d}t. \tag{5.6}$$

Define $\mathbf{F}$ as the permuted discrete Fourier transform (DFT) matrix. Then we can have $\mathbf{x} = \mathbf{Fs}$, where $\mathbf{s}$ is the sparse vector that we want to estimate. Secondly, the pseudorandom generator has the chipping sequence $\varepsilon_0, \varepsilon_1, \ldots, \varepsilon_{W-1}$, and we define the diagonal matrix $\mathbf{D} = \text{diag}(\varepsilon_0, \varepsilon_1, \ldots, \varepsilon_{W-1})$. Next, for the accumulate-and-dump sampler, we assume the sampling rate is R and R divides W. Therefore, the sampler can be written as an $R \times W$ matrix $\mathbf{H}$ whose rth row has W/R consecutive 1s starting in column $rW/R + 1$ for each $r = 0, 1, \ldots, R - 1$. In summary, the CS problem can be written as

$$\hat{\mathbf{s}} = \arg\min \|\mathbf{s}\|_0 \quad s.t.\ \mathbf{y} = \Phi\mathbf{s}, \tag{5.7}$$

where $\Phi = \mathbf{HDF}$. To solve the above problem, algorithms such as convex relaxation and greedy pursuits can be employed. The detail of the algorithm implementation is studied in Chapter 4.

From the numerical results [244–246], it is shown that the sampling rate necessary to recover random sparse signals is

$$R \approx 1.7K \log(W/K + 1). \tag{5.8}$$

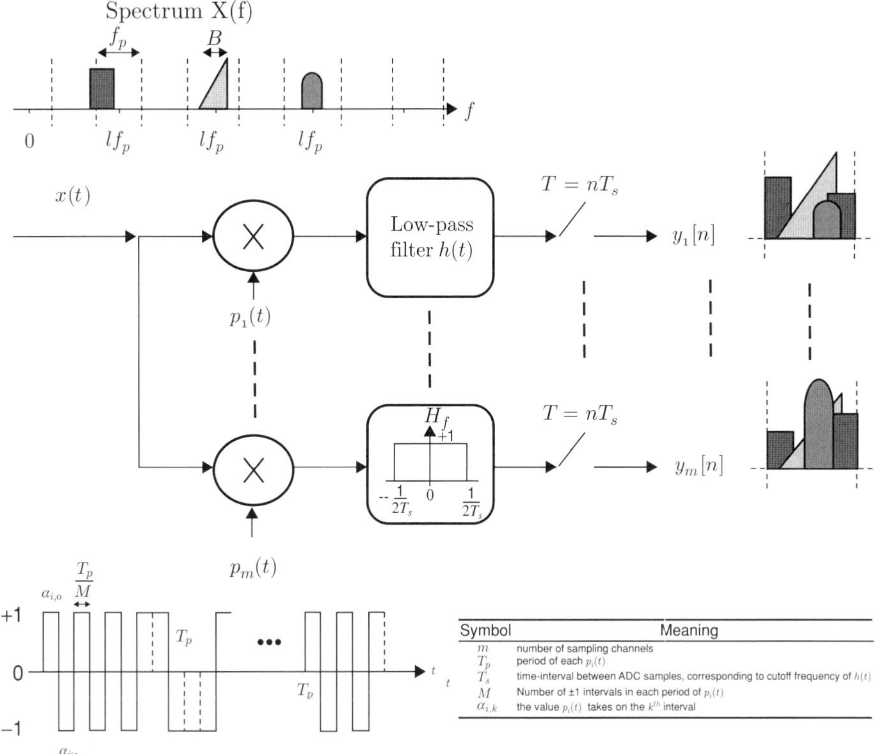

Figure 5.6 Modulated wideband converter architecture.

Some theoretical analysis also show that [248–250]

$$R \geq CK \log^{\epsilon}(W), \tag{5.9}$$

where C and ϵ are constants.

5.3 Modulated wideband converter ADC

In some sense, the random demodulator is implemented from the time domain. In this section, we study an alternative approach, a modulated wideband converter [247–249], which achieves the compression through the frequency domain. We study the general architecture of such an approach first, followed by the comparison with a random demodulator.

5.3.1 Architecture

The architecture for a modulated wideband converter is shown in Figure 5.6. A multi-band signal $x(t)$ has a sparse spectrum over N frequency bands (each band with a maximum bandwidth of B Hz). The overall spectrum ranges from 0 Hz to f_{max} Hz. The input $x(t)$ splits to $m \geq 4N$ different channels with identical structures. In the ith channel, $x(t)$ is multiplied by a periodic waveform $p_i(t)$ with period $T_p = 1/f_p$. Then a low-pass filter $h(t)$ with bandwidth of f_s is used as the anti-aliasing filter before ADC at sampling rate $f_s = 1/T_s \geq B$. The intuition behind a modulated wideband converter is similar to that behind a random demodulator, a mixture with unique signature, but in a different approach. The spectrum slices of $x(t)$ are overlaid in the spectrum of the output sequences $y_i[n]$ in different but unique ways, as shown in Figure 5.6. This generates the condition with which the CS technique can be employed.

Notice that N might be a large number. So the total number of channels m might also be large, which causes the hardware implementation challenge. To overcome this problem, it is possible to collapse the number of m by a factor of q at the expense of increasing the sampling rate of each channel by the same factor q; i.e., $f_s = q/T_s$. In the extreme case, a modulated wideband converter can be collapsed to a single sampling branch using $q = m$. For the purpose of analysis, we use $q = 1$ in the following.

The signal representation of a modulated wideband converter is illustrated as follows:

1. *Periodic waveform* $p_i(t)$ is a piece-wise deterministic function that alternates between ± 1 with M equal time intervals; i.e.,

$$p_i(t) = \alpha_{ik}, \quad k\frac{T_p}{M} \leq t \leq (k+1)\frac{T_p}{M}, \quad 0 \leq k \leq M-1, \tag{5.10}$$

   where $\alpha_{ik} \in \{+1, -1\}$. Binary patterns, such as the Gold or Kasami sequences, are suitable to $p_i(t)$ for the CS purpose. The Fourier expansion of $p_i(t)$ is

$$p_i(t) = \sum_{l=-\infty}^{\infty} c_{il} e^{j\frac{2\pi}{T_p}lt}. \tag{5.11}$$

2. *Input signal* $x(t)$ can be presented by sequence $z_l[n]$, which is obtained if $x(t)$ multiplies by $e^{j\frac{2\pi}{T_p}lt}$ and then passes through a low-pass filter. In other words, $z_l[n]$ are samples of frequency contents around lf_p Hz. The input $x(t)$ is determined by $z_l[n]$, $-L \leq l \leq L$, where $L = \lceil \frac{f_{max}}{f_p} \rceil$. Define $M = 2L + 1$ and $\mathbf{z}[n] = [z_{-L}[n], \ldots, z_L[n]]^T$.
3. *Sampled vector* $\mathbf{y}[n] = [y_1[n], \ldots, y_m[n]]^T$ obtained at $t = nT_s$ can be written as

$$\mathbf{y}[n] = \mathbf{C}\mathbf{z}[n], \tag{5.12}$$

   where $\mathbf{C}$ is an $m \times M$ matrix whose entries are c_{il}.

The general algorithms in Chapter 4 for reconstruction of $\mathbf{z}[n]$ in (5.12) can be employed. In [247], the continuous to finite (CTF) algorithm is proposed considering $\mathbf{z}[n]$ is jointly sparse in the time domain. Define the index set $\lambda = \{l | z_l[n] \neq 0\}$ is constant over consecutive time instances n. A matrix $\mathbf{V}$ is constructed from several (typically $2N$) consecutive samples $\mathbf{y}[n]$, either by directly stacking $\mathbf{y}[n]$ into the columns of $\mathbf{V}$, or via other computations for reducing noise. Then we can solve $\mathbf{V} = \mathbf{CU}$, where $\mathbf{U}$ has non-zero rows in the locations that coincide with the indices in λ. Then (5.12) reduces to

$\mathbf{y}[n] = \mathbf{C}_\lambda \mathbf{z}[n]$, where $\mathbf{C}_\lambda$ is the appropriate column subset of $\mathbf{C}$. Finally, we can write

$$\mathbf{z}_\lambda[n] = (\mathbf{C}_\lambda^H \mathbf{C}_\lambda)^{-1} \mathbf{C}_\lambda^H \mathbf{y}[n], \tag{5.13}$$

where $\mathbf{z}_\lambda[n]$ is the entry of $\mathbf{z}[n]$ indicated by λ. The required sampling rate is approximated by [247]

$$mf_s \approx 4NB \log(M/2N + 1). \tag{5.14}$$

5.3.2 Comparison with random demodulator

In [250], the comparison of a modulated wideband converter and a random demodulator is provided. In the following, we try to summarize the several main points.

- Robustness to model mismatch
  The random demodulator is sensitive to inputs with tones slightly displaced from the theoretical grid [250], while the modulated wideband converter is less sensitive to model mismatches.
- Hardware complexity
  The sampling stages of both approaches are similar. So different analog properties of the hardware to ensure accurate mapping to CS cause different design complexities. For a random demodulator, the time-domain properties of the hardware dictate the necessary accuracy. For a modulated wideband converter, the periodicity of the waveforms and low-pass filter need accurate frequency-domain properties. This is especially critical around the cutoff frequency of the low-pass filter.
- Computation load
  For a random demodulator, computational loads and memory requirements in the digital domain are the bottleneck. For a modulated wideband converter, it is limited for generating the periodic waveforms, which depends on the specific choice of waveform.

Overall, to select the best choice of CS ADC depends on the analog preprocessing complexity and the input signal properties. In [250], several operative conclusions are drawn for the choice.

1. Set system parameters with safeguards to accommodate possible model mismatches.
2. Incorporate design constraints that suit the technology generating the source signals.
3. Balance between non-linear and linear reconstruction complexities.

5.4 Xampling

The Xampling architecture was introduced in [250–253]. Xampling targets on low-rate sampling and processing of signals lying in a union of subspaces. Xampling consists of the following two blocks:

1. Analog compression: narrows down the input bandwidth prior to sampling with commercially available devices.
2. Non-linear algorithm: detects the input subspace prior to conventional signal processing.

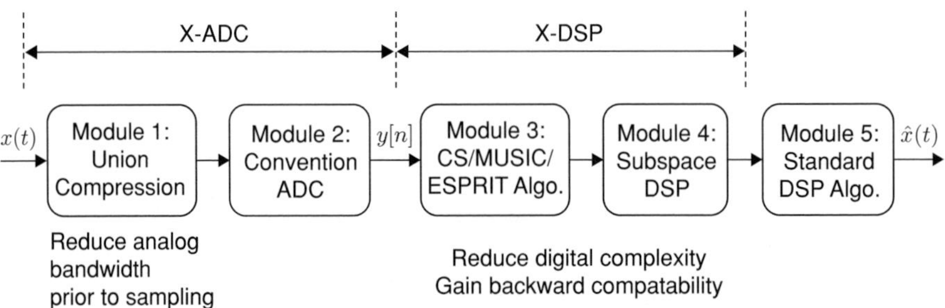

Figure 5.7 Xampling architecture.

The major difference is that the core DSP algorithm is the subspace-based algorithm (in addition to l_1 reconstruction for traditional CS algorithms). Based on this framework, there are many analog CS applications such as a general filter-bank scheme for sparse shift-invariant spaces, periodic non-uniform sampling and modulated wideband conversion for multiband communications with unknown carrier frequencies, acquisition techniques for a finite rate of innovation signals with applications to medical and radar imaging, and random demodulation of sparse harmonic tones [254]. Detailed hardware implementation is also discussed for the practical constraints. In the following, we discuss the single model, architecture, hardware implementation and subspace algorithms for Xampling.

5.4.1 Union of subspaces

The input analog signal $x(t)$ in the Hilbert space can be modeled by a union $\mathcal{U}$ of subspace as

$$x(t) \in \mathcal{U} \doteq \bigcup_{\lambda \in \Lambda} \mathcal{A}_\lambda, \tag{5.15}$$

where Λ is an index set and $\mathcal{A}_\lambda$ is a subspace of the Hilbert space. The size of Λ can be finite or infinite, and the dimension of the subspace $\mathcal{A}_\lambda$ can also be finite or infinite. Depending on the dimensions, there are three different approaches for X-ADC as

1. finite unions of infinite-dimensional spaces;
2. infinite unions of finite/infinite-dimensional spaces;
3. finite unions of finite-dimensional spaces.

5.4.2 Architecture

The high-level architecture of an Xampling system is shown in Figure 5.7 and consists of five modules.

- Analog part, which is called X-ADC, performs conversion of analog signal $x(t)$ to digital signal $y[n]$.

1. An operator compresses the high-bandwidth input $x(t)$ into a signal with lower bandwidth, effectively capturing the entire union by a subspace with substantially lower sampling requirements. The design of operator needs to exploit the union structure, in order not to lose any essential information while reducing the bandwidth. Moreover, the operator is done in the analog domain. So the operator has to be done by analog circuits, which have many limitations and are the major challenges for Xampling.
2. A conventional ADC device takes samples of the compressed signal, resulting in the sequence of low sampling rate samples $y[n]$. The ADC can be easily obtained in commercial markets.

- Digital part, which is called X-DSP, consists of three computational blocks that operate in a lower data rate than the Nyquist rate.

3. Non-linear detection detects the signal subspace from the low-rate samples. CS algorithms as well as comparable methods for subspace identification, e.g., MUSIC or ESPRIT, can be used.
4. Subspace DSP gains backward compatibility by reconstructing the signals, meaning standard DSP methods apply and commercial DAC devices can be used for signal reconstruction.
5. Standard DSP methods for signal processing.

In the following two subsections, we will discuss X-ADC and X-DSP in more detail, respectively.

5.4.3 X-ADC and hardware implementation

In [255], a circuit-level hardware prototype is illustrated based on a modulated wideband converter. There are two challenges for such a sub-Nyquist sampling system:

1. The analog preprocessing stage of the MWC needs to mix a signal with a set of multiple sinusoids. A standard switching-type mixer is used. In addition, a wideband equalization preceding the mixer, and a tunable power-control circuitry are added.
2. It is a challenge to generate highly transient periodic waveforms that are needed for the mixing. The high-speed transients, on the order of the 2 GHz Nyquist rate, need to meet strict timing constraints.

In Figure 5.8, the analog board consists of three consecutive stages of the analog path end-to-end: splitting the input into four channels, mixing with the sign patterns pi(t) and low-pass filtering. First, the input signal passes through a 100-MHz high-pass filter in order to reject the signals typically not of interest. A breakdown diode is employed to protect the circuit in case of instantaneous high input power. As is common in RF paths, a low-noise amplifier (LNA) leads a chain of analog amplifications at the very much frontend. Several attenuators are placed along the path, whose power loss can be digitally controlled between 0 dB to –15.5 dB. These devices allow widening of the dynamic range of the system. The actual split to four channels is carried out by splitting

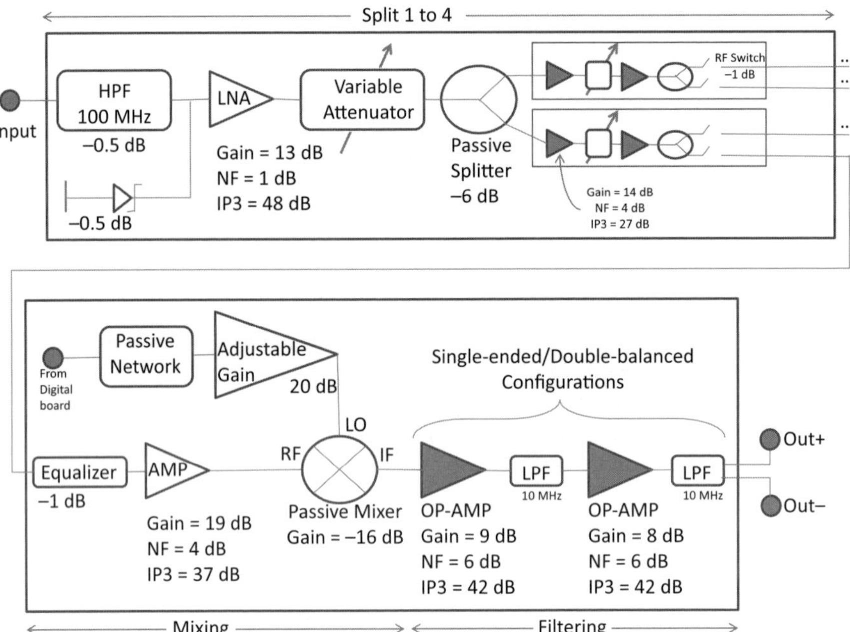

Figure 5.8 A block diagram of the Xampling analog board.

the signal twice. RF switches are the last components in the splitting stage. The switches allow shutting down any of the four channels, while isolating it from the active branches, so as to avoid possible RF reflections. Secondly, four identical blocks follow. In each block, the signal is equalized and then mixed with the corresponding waveform, which is provided by the digital board. The mixing stage is the key of the MWC system. An amplifier mitigates possible drops of the periodic waveform into the rest of the signal path. A two-stage low-pass filtering concludes the analog path before the signal is sent to the sub-Nyquist ADC.

5.4.4 X-DSP and subspace algorithms

As the signal comes out from convention ADC in the Xampling, it cannot be directly used by the conventional DSP algorithms, since the signal contains a mixture of information and is sampled at the sub-Nyquist rate. To reconstruct the signal under the Nyquist rate for further DSP purposes, the reconstruction algorithm is needed especially from subspace perspectives, since the original signal can be represented as the union of subspaces. The key is to estimate the carrier frequencies. In the following, we first discuss two traditional subspace estimation algorithms. Then we explain the refinement algorithms proposed in the literature.

Subspace algorithms: MUSIC and ESPRIT

A subspace estimator generates frequency component estimation based on the eigen-analysis or eigendecomposition of the correlation matrix, which is very effective for detection of sinusoids buried in noise, especially when the SNR is low. In the following, we discuss two popular subspace estimators.

MUltiple SIgnal Classification (MUSIC) is a subspace algorithm widely used for frequency estimation. MUSIC estimates the frequency content of a signal or autocorrelation matrix using an eigenspace method. This method assumes that an $N \times 1$ signal $\mathbf{x}$, consists of M complex exponentials in the presence of Gaussian white noise; i.e.,

$$\mathbf{x} = \mathbf{A}\mathbf{s} + \mathbf{n}, \tag{5.16}$$

where $\mathbf{A}$ is the $N \times M$ steering matrix, $\mathbf{s}$ is the $M \times 1$ source signal vector, and $\mathbf{n}$ is the $N \times 1$ noise vector. Given an $N \times N$ autocorrelation matrix, $\mathbf{R}_\mathbf{x} = E(\mathbf{x}\mathbf{x}^H)$, if the eigenvalues are sorted in decreasing order, the M eigenvectors $\mathbf{v}_1, \ldots, \mathbf{v}_M$ corresponding to the M largest eigenvalues span the signal subspace. Note that $N > M$, so that there are $N - M$ null spaces for noise and the noise eigenvectors are $\mathbf{v}_{M+1}, \ldots, \mathbf{v}_N$. The frequency estimation function for MUSIC is

$$\hat{w}(e^{jw}) = \frac{1}{\sum_{i=M+1}^{N} |\mathbf{a}^H \mathbf{v}_i|^2}, \tag{5.17}$$

where steering vector $\mathbf{a} = [1\ e^{jw}\ e^{2jw} \ldots e^{j(N-1)w}]^T$. In other words, when w is the same as the original frequency in $\mathbf{x}$, the denominator of (5.17) is approaching zero due to the orthogonality of the signal space $(\mathbf{v}_1, \ldots, \mathbf{v}_M)$ and noise space $(\mathbf{v}_{M+1}, \ldots, \mathbf{v}_N)$. As a result, $\hat{w}(e^{jw})$ has a peak at the corresponding frequency.

The MUSIC algorithm works for other array shapes, and needs to know sensor positions. It is very sensitive to sensor position, gain, and phase errors, and needs careful calibration to make it work well. Also it needs to search through all parameter space and could be computationally expensive.

A modification of the MUSIC algorithm without scanning, the Root MUSIC algorithm, is employed when the MUSIC spectrum is an all-pole function. Let the noise space $\mathbf{U}_n = [\mathbf{v}_{M+1}, \ldots, \mathbf{v}_N]$. The MUSIC problem in (5.17) can be written as

$$J(z) = \mathbf{z}^H \mathbf{U}_n \mathbf{U}_n^H \mathbf{z} = 0, \tag{5.18}$$

where the steering vector $\mathbf{z} = [1, z^{-1}, \ldots, z^{-(N-1)}]^T$. The estimation can be obtained by finding the M roots closest to the unit circle of the above equation, where M is the number of signals.

Estimation of Signal Parameter via Rotational Invariance Technique (ESPRIT) is robust with respect to array imperfections than MUSIC. So ESPRIT relaxes the calibration task somewhat. Moreover, computation complexity and storage requirements are lower than MUSIC as it does not involve an extensive search throughout all possible steering vectors. But ESPRIT takes twice as many sensors as MUSIC. If sensors are expensive and few, and if computation is not of concern, MUSIC is suitable. If there are plenty of sensors compared with the number of sources to detect, and if computational power is limited, ESPRIT is suitable.

The ESPRIT algorithm is based on "doublets" of sensors; i.e., in each pair of sensors the two should be identical, and all doubles should line up completely in the same direction with a displacement vector Δ. This results in the relation between doubles' steering matrix having the following relation $\mathbf{A}_1 = \mathbf{A}_2\mathbf{\Phi}$. The positions of the doublets are also arbitrary, which makes calibration a little easier. We assume there are N sets of doublets; i.e., $2N$ sensors, and $N > M$ where M is the number of sources. The correlation matrix for $2N \times 2N$ sensors is $\mathbf{R}_z$. Since there are M sources, the M eigenvectors of $\mathbf{R}_z$ corresponding to the M largest eigenvalues form the signal space $\mathbf{U}_s$. Therefore, there exists a unique non-singular $M \times M$ matrix $\mathbf{T}$ satisfying

$$\mathbf{U}_s = \begin{bmatrix} \mathbf{A} \\ \mathbf{A}\Phi \end{bmatrix} \mathbf{T} = \begin{bmatrix} \mathbf{U}_x \\ \mathbf{U}_y \end{bmatrix}. \tag{5.19}$$

Here $[\mathbf{U}_x\mathbf{U}_y]$ is an $N \times 2M$ matrix with rank M, and has a null space of dimension M. So there exists $\mathbf{F}_x$ and $\mathbf{F}_y$ such that $\mathbf{U}_x\mathbf{F}_x + \mathbf{U}_y\mathbf{F}_y = \mathbf{0}$. From (5.19), the final algorithm is performed by solving the following equation:

$$\mathbf{\Phi} = \mathbf{T}\mathbf{F}_x\mathbf{F}_y^{-1}\mathbf{T}^{-1}. \tag{5.20}$$

The estimation is "one shot," since even with matrix inversions the computation is much less than the MUSIC search.

Refinement subspace estimation

For multiband, the original signal can be described in quadrature representation as

$$x(t) = \sum_{i=1}^{N/2} I_i(t)\cos(2\pi f_i t) + Q_i(t)\sin(2\pi f_i t), \tag{5.21}$$

where $I_i(t)$ and $Q_i(t)$ are the real-valued narrowband signals for I and Q phases, respectively. f_i is the unknown carrier frequency. The subspace algorithm needs to estimate f_i, $I_i(t)$ and $Q_i(t)$.

In [250], a refinement subspace detection algorithm, Back-DSP, is proposed by assuming zero cross-correlation between $I_i(t)$ and $Q_i(t)$. The Back-DSP algorithm consists of four steps:

1. Refining the coarse support estimate f_i to the actual band edges using prior information of the minimal width of a single band and the smallest spacing between bands.
2. Obtain a sequence $s_i[n]$ for each $1 \leq i \leq N/2$, such that $s_i[n]$ contains the entire contribution of exactly one band.
3. Estimating f_i using a digital version of the balanced quadricorrelator (BQ) from $s_i[n]$.
4. The information signals $I_i(t)$ and $Q_i(t)$ are obtained based on f_i.

Another related work is called compressive signal processing (CSP) [256]. CSP is a framework in which signal processing problems are solved directly in the compressive measurement domain without first resorting to a full-scale signal reconstruction. Four fundamental signal-processing problems: detection, classification, estimation, and filtering are investigated. The first three enable the extraction of information from the

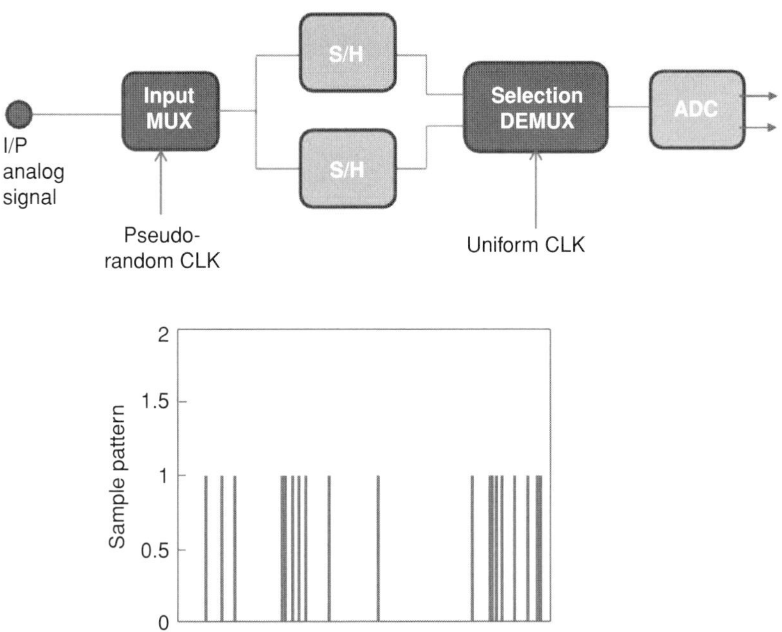

Figure 5.9 Block diagram of random sampling ADC.

samples, while the last enables the removal of irrelevant information and separation of signals into distinct components.

5.5 Other architecture

In this section, we briefly discuss several other architectures, followed by miscellaneous literature.

5.5.1 Random sampling

In [257, 258], a framework for analog-to-information conversion is proposed based on the theory of information recovery from random samples. The framework enables sub-Nyquist acquisition and processing of wideband signals that are sparse in a local Fourier representation. The random sampling theory associated with an efficient information recovery algorithm is presented to compute the spectrogram of the signal. A hardware design for the random sampling system demonstrates a consistent reconstruction fidelity in the presence of sampling jitter, which forms the main source of non-ideality in a practical system implementation.

The block diagram of random sampling-based CS ADC is shown in Figure 5.9 [257]. The input analog signal is fed to the input multiplexor. The MUX selects signal values with non-uniform timing and pushes them onto the analog queue, which stores the values

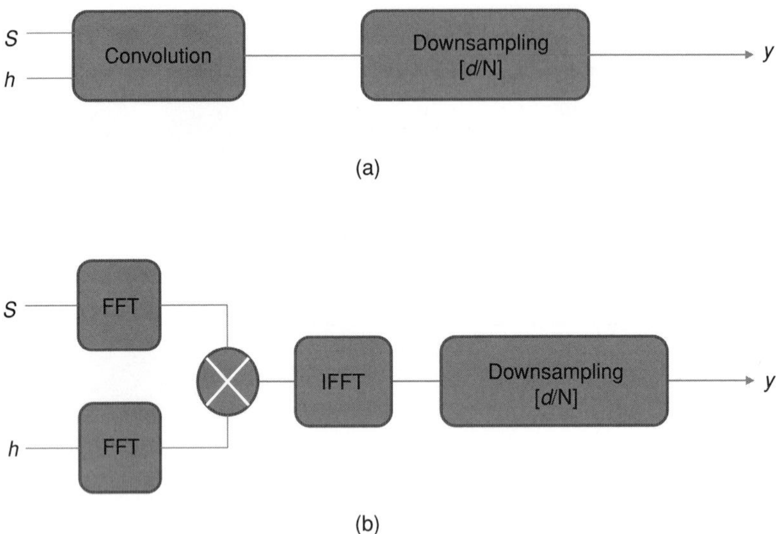

Figure 5.10 Block diagrams for signal acquisition through random filtering: (a) using convolution; (b) using FFT/IFFT.

via sample and hold (S/H). Finally, these values are read with uniform timing from the selection DEMUX and sampled with a sub-Nyquist rate ADC.

5.5.2 Random filtering

In [259], the random filter-based CS ADC is proposed as shown in Figure 5.10. This approach captures a signal s by convolving it with a random-tap FIR filter h and then downsampling the filtered signal to obtain a compressed representation y. The authors claim that this approach has the following benefits: measurements are time invariant and non-adaptive; the measurement operator is stored and applied efficiently; longer filters can be traded for fewer measurements; it is easily implementable in software or hardware; and it generalizes to streaming or continuous-time signals.

5.5.3 Random delay line

In [260], another CS ADC is proposed based on random delay lines as shown in Figure 5.11. It is somewhat like a dual of a modulated wideband converter ADC in the time domain. The differences are to implement the random delay line instead of filters. The random projection can be implemented by randomizing the delays. Moreover, for fiber communication, the delay can be implemented by random lengths of fiber links.

5.5.4 Miscellaneous literature

Finally, we discuss some miscellaneous literature for the CS-based ADC in this subsection.

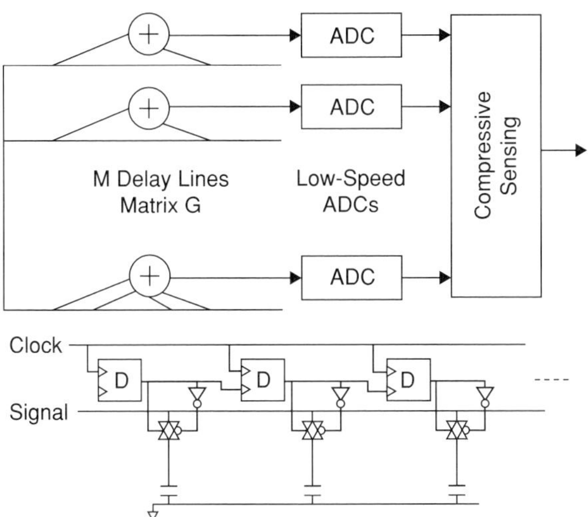

Figure 5.11 Block diagram for Random Delay Line Based CS ADC and Delay Line Implementation [260].

In [261], blind CS is defined for the problem of reconstructing a multiband signal from its sub-Nyquist pointwise samples, when the band locations are unknown. Here, only the number of bands and their widths are assumed without any other limitations on the support. The question is answered for how to choose the parameters of the multicoset sampling so that a unique multiband signal matches the given samples. To recover the signal, the continuous reconstruction is replaced by a single finite-dimensional problem without the need for discretization.

In [262], a new method is developed for tracking narrowband signals acquired via CS. The CS phase-locked loop (CS-PLL) enables one to track oscillating signals in very large bandwidths using sub-Nyquist sampling. A key feature of the approach is the fact that the frequency tracking is performed directly on the compressive measurements without ever recovering the signal. The CS-PLL has a wide variety of potential applications, including communications, phase tracking, and robust control. In [263], the CS-PLL is employed in FM receivers.

In [264, 265], it is demonstrated that since CS samples at a lower rate, CS-based systems can potentially attain a significantly larger dynamic range. The ability to control these aspects of performance provides an engineer new degrees of freedom in the design of CS wideband acquisition systems.

In [266], it is shown that the random demodulator and the MWC are both based on the general concept of random filtering, but employ significantly different sampling functions. The system sensitivities (or robustness) to sparse signal model assumptions are investigated. Lastly, "block convolution" is a fundamental aspect of the MWC, allowing it to successfully sample and reconstruct block-sparse (multiband) signals. Based on this concept, a new acquisition system is proposed for continuous-time signals whose amplitudes are block sparse.

In [267], two generic VLSI architectures implement the approximate message-passing (AMP) algorithm for sparse signal recovery. The first architecture, referred to as AMP-M, employs parallel multiply-accumulate units, which are suitable for recovery problems based on unstructured (or random) matrices. The second architecture, referred to as AMP-T, takes advantage of fast linear transforms, which arise in many real-world applications. To demonstrate the effectiveness of both architectures, corresponding ASIC and FPGA implementation results are presented for an audio restoration application. AMP-T is superior to AMP-M with respect to the silicon area, throughput, and power consumption, whereas AMP-M offers more flexibility and can be reconfigured at run time. In [268], a parallel-path structure that employs current mode sampling techniques is studied using a 90-nm CMOS process.

In [269], Discrete Prolate Spheroidal Sequences (DPSSs) are studied on the effects of time-limiting and band-limiting operations. DPSSs form a highly efficient basis for sampled band-limited functions; by modulating and merging DPSS bases, a dictionary is obtained that offers high-quality sparse approximations for most sampled multiband signals. This multiband modulated DPSS dictionary can be readily incorporated into the CS framework.

5.6 Summary

In digital signal processing, the Nyquist sampling theorem states that in order to reconstruct a signal without aliasing, the signal should be sampled at least two times faster than the signal bandwidth. However, modern sensors and applications produce a large amount of data, resulting in enormous numbers of samples and requiring high speed. To overcome this problem, the CS-based ADC is studied in this chapter. Two major types, a random demodulator and a modulated wideband converter, are explained in detail. Then a unified subspace view is studied for Xampling. Other architectures and miscellaneous literature are covered at the end. A variety of applications can be developed based on the CS ADC. In the next part of this book, we concentrate on how to use CS in the wireless networks.

Part II

CS-Based Wireless Communication

6 Compressed channel estimation

6.1 Introduction and motivation

The wireless channels place fundamental limitations on the performance of many wireless communication systems. A transmitted radio signal propagating from a transmitter to a receiver is generally reflected, scattered, and attenuated by the obstacles in the field. At the receiver, two or more slightly delayed versions of the transmitted signal will superpose together. This so-called fading phenomena in wireless communication channels is a double-edged sword to wireless communication systems [1, 270]. On the one hand, this multipath propagation phenomenon will result in severe fluctuations of the amplitudes, phases and delays of a radio signal over a short period of time or distance. Interpreting the resultant received signals, which vary widely in amplitude and phase, is a critical design challenge to the receiver. On the other hand, multipaths will enhance the time, spatial and frequency diversity of the channel available for communication, which will lead to gains in data transmission and channel reliability. To obtain the diversity gain and alleviate the side effect of the channel fading, knowledge of channel state information (CSI) is necessary. Thus, channel estimation is of significant importance for the transceiver design.

The CS theory [69, 70, 271] is a new technology that has emerged in the area of signal processing, statistics and wireless communication. By utilizing the fact that a signal is sparse or compressible in some transform domain, CS can powerfully acquire a signal from a small set of randomly projected measurements with a sampling rate much lower than the Nyquist sampling rate. The word "sparse" means that a signal vector only contains very few non-zero entries. As more and more experimental evidences suggest that broadband or multiantenna communication channels have an inherent sparse multipath structure, CS has recently become a promising tool to deal with the channel estimation problem [272–277].

In this chapter, we present several new approaches to estimating sparse (or effectively sparse) multipath channels that are based on CS for different scenarios. The traditional channel estimation method is also mentioned and compared. We highlight channel estimation via CS and explain the problem formulations for different kinds of wireless channels. Specifically, this chapter is organized as follows:

- In Section 6.2, based on [272], we study how to formulate channel estimation problems as the CS problems, under different types of multipath channels. Specifically, we discuss the following channels: sparse frequency-selective channel, sparse doubly selective channel, sparse non-selective MIMO channel, and sparse frequency-selective MIMO channel. For comparison reason, we also explain the traditional least-squares algorithm for training-based methods.

- In Section 6.3, based on [278], we formulate OFDM channel estimation as a CS problem, which takes advantage of the sparsity of the channel impulse response and reduces the number of probing measurements, which in turn reduces the ADC speed needed for channel estimation. Specifically, we propose sending out pilots with random phases in order to spread out the sparse taps in the impulse response over into the receiver's measurements, so that the impulse response can be faithfully recovered by sparse optimization. This contribution leads to high-resolution channel estimation with low-speed ADCs. We also propose a novel estimator that performs better than the commonly used ℓ_1 minimization. Specifically, it significantly reduces estimation error by combing ℓ_1 minimization with iterative support detection and support-restricted least-squares. While letting the receiver ADC run at a much lower speed of the transmitter DAC, we simulated various numbers of multipaths and different measurement SNRs and obtained great estimation performance, compared to the Cramér–Rao lower bound.
- In Section 6.4, we investigate another class of the channel, underwater acoustic channel. First, we describe the channel model for this specific situation, and explain the differences to the RF channels. Then based on [274], we study the basic pursuit and orthogonal matching pursuit methods as the CS approaches for such channels. Next, based on [279], we investigate the pilot design for cyclic-prefix OFDM. Finally, based on [280], we study the iterative channel estimation and decoding.
- In Section 6.5, detailed measurement campaigns are needed to collect the spectrum occupancy data to obtain a deeper understanding of the spectrum-usage characteristics especially in cognitive radio networks. This approach, however, is usually inefficient due to the ignorance of the spatial, temporal and spectral correlations of spectrum occupancies, and impractical because of the geographical and hardware limitations of the sensing nodes. In this section, we apply the theory of random fields to model the spatial-temporal correlated spectrum usage data, using the two-dimensional Ising model and the Metropolis–Hastings algorithm, respectively. To efficiently obtain the spectrum occupancy, we adopt the matrix completion technique that leverages the low-rank structure of the data matrix to recover the original data from limited measurements. Simulation results validate the effectiveness of the proposed algorithm.
- In Section 6.6, we investigate some other CS methods for channel estimation. For blind channel estimation [281, 282], we study the case in which the sparsity bases are not unknown. For adaptive algorithms [283], the slow-changing algorithm estimation is investigated. Finally, if the channel has the bursty responses, we study the group sparsity method as in [97, 284, 285].
- Finally, we provide a summary of this chapter in Section 6.7.

6.2 Multipath channel estimation

In this section, based on [272], we discuss the multipath channel model first, and then we illustrate the traditional least-squares method for channel estimation. Finally, we discuss the CS-based channel sensing and study four different scenarios: sparse frequency-selective channel, sparse doubly selective channel, sparse non-selective MIMO channel, and sparse frequency-selective channel.

6.2.1 Channel model and training-based method

For the MIMO multipath channel, the channel response can be written as

$$\mathbf{H}(t, f) = \sum_{i=1}^{N_R} \sum_{k=1}^{N_T} \sum_{l=0}^{L-1} \sum_{m=-M}^{M} H_v(i, k, l, m) \alpha_R \left(\frac{i}{N_R}\right) \alpha_T \left(\frac{k}{N_T}\right) e^{-j2\pi \frac{l}{W} f} e^{j2\pi \frac{m}{T} t}, \tag{6.1}$$

where N_R is the dimension of the channel output, N_T is the dimension of the channel input, L is the maximal number of resolvable delays, M is the maximal number of resolvable one-side doppler shifts, $H_v(i, k, l, m)$ is the sampled channel response, α_R is the receiver array response vector, α_T is the transmitter array steering vector, W is the bandwidth, and T is the duration.

In the traditional training-based solution, the transmitter sends a known sequence $\mathbf{X}$ to the receiver. The received signal can be written as

$$\mathbf{Y} = \sqrt{\frac{\varepsilon}{N_T}} \mathbf{X}\mathbf{H}_v + \mathbf{z}, \tag{6.2}$$

where ε is the transmit power, $\mathbf{H}_v$ is the channel response comprising of the channel coefficient $H_v(i, k, l, m)$, and $\mathbf{z}$ is the thermal noise. The goal is to estimate the least-squares channel response as the following problem:

$$\mathbf{H}_v^{LS} = \arg\min_{\mathbf{H}} \|\mathbf{Y} - \sqrt{\frac{\varepsilon}{N_T}} \mathbf{X}\mathbf{H}\|^2. \tag{6.3}$$

The well-known solution is given by

$$\mathbf{H}_v^{LS} = \sqrt{\frac{N_T}{\varepsilon}} (\mathbf{X}^H \mathbf{X})^{-1} \mathbf{X}^H \mathbf{Y}. \tag{6.4}$$

The reconstruction error can be expressed and bounded by

$$E[\Delta(\mathbf{H}_v^{LS})] = \frac{trace((\mathbf{X}^H \mathbf{X})^{-1}) N_R N_T}{\varepsilon} \geq \frac{N_R N_T^2 L(2M+1)}{\varepsilon}. \tag{6.5}$$

6.2.2 Compressed channel sensing

The maximal total number of coefficients to estimate is given by

$$D = N_R N_T L(2M+1). \tag{6.6}$$

However, most of the coefficients are below the noise level, and there are only very few components that are large. This sparsity motivates the CS-based approaches. From the perspective of CS, in (6.2) $\mathbf{Y}$ is the compressed vector, $\mathbf{X}$ is the random projection matrix, and $\mathbf{H}_v$ is the unknown sparse signal. In this subsection, we discuss four special cases and study how to write the CS problem formulations.

Sparse frequency-selective channel

For the single-antenna case, the frequency-selective channel in (6.1) is reduced to

$$H(f) = \sum_{l=0}^{L-1} H_v(l) e^{-j2\pi \frac{l}{W} f}. \tag{6.7}$$

The question is how to construct the random projection matrix $\mathbf{X}$ so that the CS can be employed. We discuss two cases for CDMA and OFDM, respectively.

- In the CDMA case, the training sequence can be written as

$$x(t) = \sqrt{\varepsilon} \sum_{n=0}^{N_0-1} x_n g(t - nT_c), \ 0 \le t \le T, \tag{6.8}$$

  where $g(t)$ is the unit energy chop waveform and x_n is the N_0-dimensional spreading code with unit energy $\sum_n |x_n|^2 = 1$. The received signal can be expressed by

$$\mathbf{y} = \sqrt{\varepsilon} \mathbf{X} \mathbf{h}_v + \mathbf{z}, \tag{6.9}$$

  where X is an $N_0 \times L$ Toeplitz matrix[1] whose first row and first column can be written as $[x_0 \mathbf{0}_{L-1}^T]$ and $\mathbf{x}^T \mathbf{0}_{L-1}^T]^T$, respectively.
- For the OFDM case, the training sequence takes the form as

$$x(t) = \sqrt{\frac{\varepsilon}{N_{tr}}} \sum_{n \in S_{tr}} g(t) e^{j2\pi \frac{n}{T} t}, \ 0 \le t \le T, \tag{6.10}$$

  where $g(t)$ is a prototype pulse having unit energy, S_{tr} is the set of indices of pilot tones used for training, and N_{tr} is the number of receive training dimensions or the number of pilot tones ($N_{tr} = |S_{tr}|$). The received signal can be expressed similarly as $\mathbf{y} = \sqrt{\varepsilon} \mathbf{X} \mathbf{h}_v + \mathbf{z}$. But here $\mathbf{X}$ is the $N_{tr} \times L$ sensing matrix that comprises $\{(1/\sqrt{N_{tr}})[1\ w_{N_0}^n \dots w_{N_0}^{n(L-1)}] : n \in S_{tr}\}$ as its row with $w_{N_0} = e^{-j2\pi/N_0}$.

Sparse doubly selective channel

For the single-antenna doubly selective case [287–289], the channel in (6.1) is reduced to

$$\mathbf{H}(t, f) = \sum_{l=0}^{L-1} \sum_{m=-M}^{M} H_v(l, m) e^{-j2\pi \frac{l}{W} f} e^{j2\pi \frac{m}{T} t}. \tag{6.11}$$

[1] Some work on the Toeplitz matrix in CS can be found in [286].

For the CDMA case, $\mathbf{X}$ is an $N_0 \times L(2M+1)$ block matrix of the form $\mathbf{X} = [\mathbf{X}_{-M} \ldots \mathbf{X}_0 \ldots \mathbf{X}_M]$. Here, $\mathbf{X}_m$ is an $(N_0 \times L)$-dimension matrix as

$$\mathbf{X}_m = \mathbf{W}_m \mathbf{T}, \tag{6.12}$$

where $\mathbf{W}_m = \mathrm{diag}(w_{N_0}^{-m0}, w_{N_0}^{-m1}, \ldots, w_{N_0}^{-m(N_0-1)})$ and $\mathbf{T}$ is an $N_0 \times L$ Toeplitz matrix whose first row and first column are given by $[x_0 \mathbf{0}_{L-1}^T]$ and $\mathbf{x}^T \mathbf{0}_{L-1}^T]^T$, respectively.

The multicarrier short-time Fourier (STF) waveforms are a generalization of OFDM waveforms for doubly selective channels. The training signal can be written as

$$x(t) = \sqrt{\frac{\mathcal{E}}{N_{tr}}} \sum_{(n,m)\in S_{tr}} g(t - nT_0) e^{j2\pi m W_0 t}, \ 0 \le t \le T, \tag{6.13}$$

where $g(t)$ is a prototype pulse having unit energy, $S_{tr} \subset \{0, \ldots N_t - 1\} \times \{0, \ldots N_f - 1\}$, $N_f = N_0/N_t$, T_0 is the time separation of STF basis waveform $g(t - nT_0)e^{j2\pi m W_0 t}$, W_0 is the frequency separation of the STF basis waveform, $T_0 W_0 = 1$, $N_t = T/T_0$, and $N_f = W/W_0$. Now the $N_{tr} \times L(2M+1)$ random projection matrix (or training sequence) matrix $\mathbf{X}$ has the following as its row

$$\left\{ \frac{1}{\sqrt{N_{tr}}} \left[w_{N_t}^{nM} \ w_{N_t}^{n(M-1)} \ w_{N_t}^{-nM} \right] \bigotimes \left[1 \ w_{N_f}^{1} \ldots w_{N_f}^{m(L-1)} \right] : (n, m) \in S_{tr} \right\}.$$

Sparse non-selective MIMO channel

For a non-selective MIMO channel, the channel in (6.1) is reduced to

$$\mathbf{H} = \sum_{i=1}^{N_R} \sum_{k=1}^{N_T} H_v(i,k) \alpha_R \left(\frac{i}{N_R} \right) \alpha_T \left(\frac{k}{N_T} \right) = \mathbf{A}_R \mathbf{H}_v^T \mathbf{A}_T^H. \tag{6.14}$$

The training sequence for MIMO can be written as

$$x(t) = \sqrt{\frac{\mathcal{E}}{N_T}} \sum_{n=0}^{M_{tr}-1} x_n g(t - \frac{n}{W}), \ 0 \le t \le \frac{M_{tr}}{W}, \tag{6.15}$$

where $g(t)$ is a prototype pulse having unit energy, $x_n \in \mathcal{C}^{N_T}$ is the training sequence with energy $\sum_n \|x_n\|^2 = N_T$, and M_{tr} is the number of temporal training dimensions (the total number of time-slots dedicated to training). For MIMO, the received signal can be written as

$$\mathbf{y} = \sqrt{\frac{\mathcal{E}}{N_T}} \mathbf{X} \mathbf{H}_v + \mathbf{Z}, \tag{6.16}$$

where $\mathbf{X}$ is an $M_{tr} \times N_T$ matrix of form $\mathbf{X} = [\mathbf{x}_0 \ \ldots \mathbf{x}_{M_{tr}-1}]^T \mathbf{A}_T^*$.

Sparse frequency-selective MIMO channel

For a frequency-selective MIMO channel, we have

$$\mathbf{H}(f) = \sum_{l=0}^{L-1} \mathbf{A}_R \mathbf{H}_v^T(l) \mathbf{A}_T^H e^{-j2\pi \frac{l}{W} f}. \tag{6.17}$$

The training sequence for this case can be written as

$$x(t) = \sqrt{\frac{\varepsilon}{N_T}} \sum_{(n,m)\in S_{tr}} x_{n,m} g(t - nT_0) e^{j2\pi m W_0 t}, \tag{6.18}$$

where $S_{tr} \subset \{0, 1, \ldots, N_t - 1\} \times \{0, 1, \ldots, N_f - 1\}$, and $x_{n,m} \in C^{N_t}$ is the training sequence with energy $\sum_{S_{tr}} \|x_{n,m}\|^2 = N_T$. The random projection matrix $\mathbf{X}$ is an $M_{tr} \times N_T L$ matrix with the following as its row

$$\{[1\ w_{N_f}^m \ldots w_{N_f}^{m(L-1)}] \bigotimes \mathbf{x}_{n,m}^T \mathbf{A}_T^* : (n, m) \in S_{tr}\}.$$

In summary, four different types of wireless channels are formulated as the CS problem in this section.

6.3 OFDM channel estimation

Orthogonal frequency division multiplexing (OFDM) is a technique that has been widely applied in wireless communication systems and will prevail in the next-generation wireless communication, because it transmits at a high rate, achieves high bandwidth efficiency, and is robust to multipath fading and delay [290]. OFDM applications can be found in digital television and audio broadcasting, wireless networking, and broadband internet access. Current OFDM-based WLAN standards (such as IEEE802.11a/g) use variations of QAM schemes for subcarrier modulations, which require a coherent detection at the OFDM receiver and consequently requires an accurate (or near accurate) estimation of Channel State Information (CSI). The structure of the OFDM signal makes it difficult to balance complexity and performance in channel estimation. The design principles for channel estimators are to reduce the computational complexity and bandwidth overhead while maintaining sufficient estimation accuracy.

Some channel estimation schemes proposed in the literature are based on pilots, which form the reference signal used by both the transmitter and the receiver. This approach has two main challenges: (i) the design of pilots; and (ii) the design of an efficient estimation algorithm (i.e., the estimator). There is a tradeoff between the spectrum efficiency and the channel estimation accuracy. Most of the existing pilot-assisted OFDM channel estimation schemes rely on the use of a large number of pilots to increase the estimation accuracy; the spectral efficiency is therefore reduced. For example, there are approaches based on a time-multiplexed pilot, frequency-multiplexed pilot, and scattered pilot [291], all achieving higher estimation accuracy at the price of using more pilots. There have been attempts to reduce the number of pilots; i.e., Byun and Natarajan, in [292]. The solutions generally require extra "test signals" for channel pre-estimation. By sending out a "test signal," they try to find out how many pilots are needed by first inserting a relatively small number of pilots and then, based on the results of the "test," the number of pilots is decided. Therefore, there is no guaranteed overall reduction of pilot insertion.

CS has been applied to channel estimation in [287, 293–295], which are reviewed in Section 6.3.2. For OFDM channel estimation, there are papers [296–302], to which our work differs in various ways as follows.

As a sensing problem, OFDM channel estimation can benefit from the emerging technique of CS, which acquires and reconstructs a signal from fewer samples than what is dictated by the Nyquist–Shannon sampling theorem, mainly by utilizing the signal's sparse or compressible property. The field has exploded since the pioneering work by Donoho [70] and Candes, Romberg and Tao [68]. The main idea is to encode a sparse signal by taking its "incoherent" linear projections and recovering the signal through algorithms, such as ℓ_1 minimization. To maximize the benefits of CS for OFDM channel estimation, one must skillfully perform the CS encoding and decoding steps, which are precisely the two focuses of this chapter: the designs of pilots and an estimator, respectively.

Based on [278], we skillfully designed CS encoding and decoding strategies for OFDM channel estimation. Compared to existing work, we are able to obtain channel response in much higher resolutions and from far fewer pilots (thus taking much shorter times). This is achieved by designing pilots with uniform random phases and using a novel estimator. The pilot design preserved the information of high-resolution channel responses during aggressive uniform downsampling, which means that receiver ADC can run at a much lower speed. The estimator was tailored for OFDM channel response; in particular, instead of the generic ℓ_1 minimization, iterative support detection (ISD) [238] and limited-support least-squares were adopted in order to take advantage of the characteristics of channel response. The resulting algorithm are very simple and performs better.

6.3.1 System model

A baseband OFDM system is shown in Figure 6.1. In this system, the modulated signal in the frequency domain, represented by $\mathbf{X}(k)$, $k \in [1, N]$, is inserted with the pilot signal and guard band, and then an N-point IDFT transforms the signal into the time domain, denoted by $x(n)$, $n \in [1, N]$, where a cyclic extension of time length T_G is added to avoid intersymbol and inter-subcarrier interferences. The resulting time-series data are converted by a digital-to-analog converter (DAC) with a clock speed of $1/T_S$ Hz into an analog signal for transmission. We assume that the channel response comprises P propagation paths, which can be modeled by a time-domain complex baseband vector with P taps, written as

$$h(n) = \sum_{p=1}^{P} \alpha_p \delta(n - \tau_p T_S),\ n = 1, \ldots, N, \tag{6.19}$$

where α_p is a complex multipath component and τ_p is the multipath delay ($0 \leq \tau_p T_S \leq T_G$). Since T_G is less than the OFDM symbol durations, the non-zero channel response concentrates at the beginning, which translates to $h = [h_1, h_2, \ldots, h_{\tilde{N}}, 0, \ldots, 0]$; i.e., only the first $\tilde{N}$ components of h can possibly take non-zero values and $\tilde{N} < N$.

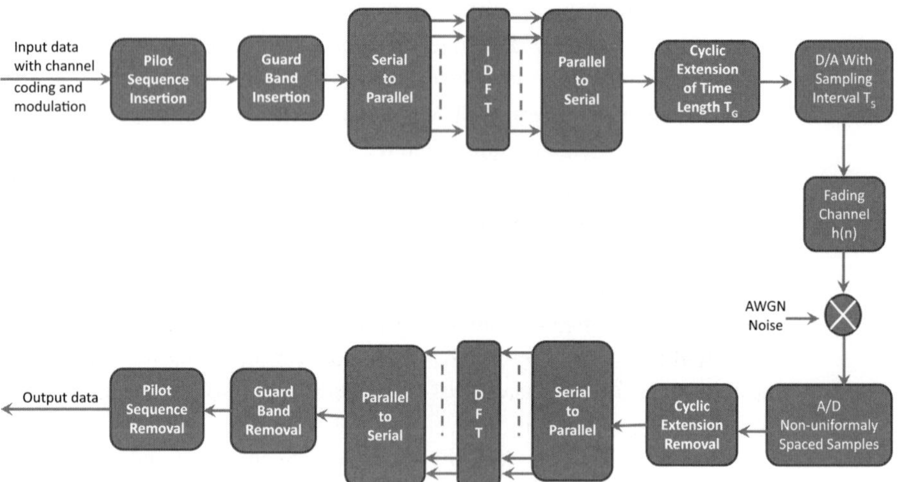

Figure 6.1 Baseband OFDM system.

Assuming that interferences are eliminated, what arrives at the receiver is the convolution of the transmitted signal and the channel response plus noise, denoted by $z(n)$ given by

$$z = x \otimes h + \xi, \tag{6.20}$$

where $\otimes$ denotes convolution and $\xi(n), n \in [1, N]$ denote the sampled AWGN noise. Passing through the analog-to-digital converter (ADC), $z(n), n \in [1, N]$ is sampled as $y(m), m \in [1, M]$, and the cyclic prefix (CP) is removed. Traditional OFDM channel estimation schemes assume $M = N$. If $M < N$, then y is a downsample of z. An M-point DFT converts y to $\mathbf{Y}(k), k \in [1, M]$, where the guard band and pilot signal will be removed.

For pilot-assisted OFDM channel estimation, we shall design the pilots $\mathbf{X}$ (and thus x) and recover h from the measurements $\mathbf{Y}$ (or, equivalently y).

6.3.2 Compressive sensing OFDM channel estimator

Motivations

CS, which will be reviewed in the next subsubsection, allows sparse signals to be recovered from very few measurements, which translates to slower sampling rates and shorter sensing times. Because the channel impulse response h is very sparse, we are motivated to apply CS to recover h by using a reduced number of pilots, so that the estimation becomes much easier. Furthermore, in sharp contrast to conventional OFDM channel estimation in which ADC and DAC run at the same sampling rate, we can obtain a higher-resolution h by increasing the resolution of only the transmitter DAC, or we can reduce the receiver ADC speed, which often defines the system cost. In other words, we have $N > M$. The rest of this subsection reviews CS and introduces our proposed approach for OFDM channel estimation.

CS background

CS theories [67, 69, 70] state that a S-sparse signal[2] h can be stably recovered from linear measurements $y = \Phi h + \xi$, where Φ is a certain matrix with M rows and N columns, $M < N$, by minimizing the ℓ_1-norm of h. Classic CS often assumes that Φ, after scaling, satisfies the restricted isometry property (RIP) $(1-\delta)\|h\|_2^2 \leq \|\Phi h\|_2^2 \leq (1+\delta)\|h\|_2^2$ for all S-sparse h, where $\delta > 0$ is the RIP parameter. The RIP is satisfied with a high probability by a large class of random matrices, e.g., those with entries independently sampled from a sub-Gaussian distribution. By minimizing the ℓ_1-norm, one can stably recover h, as long as $M \geq O(S \log N)$.

The classic random sensing matrices are not admissible in OFDM channel estimation, because the channel response h is *not* directly multiplied by a random matrix; instead, as was described in Section 6.3.1, h is first convoluted with x. The noise is added, and then the received signal z is uniformly down-sampled to y. Because convolution is a circulant linear operator, we can present this process by

$$y = \downarrow_\Omega z = \downarrow_\Omega (Ch + \xi), \tag{6.21}$$

where the sensing matrix C is a full circulant (convolution) matrix determined by x, and $\downarrow_\Omega$ denotes the uniform downsampling action at points in $\Omega = [1, 1 + N/M, \ldots, N - N/M + 1]$. As is widely perceived, CS favors fully random matrices, which admit stable recovery from the fewest measurements (in terms of order of magnitude), but C in our case is structured and thus much less "random." This factor seemingly suggests that C would unlikely to be favored by CS. Nevertheless, carefully designed circulant matrices can deliver the same optimal CS performance.

Pilot with random phases

To design the sensing matrix C, we propose to generate pilots $\mathbf{X}$ in either one of the following two ways: (i) the real and imaginary parts of $\mathbf{X}(k)$ are sampled independently from a Gaussian distribution, $k = 1, \ldots, N$; (ii) (the same as [294]) $\mathbf{X}(k)$, $k = 1, \ldots, N$, have independent random phases but a uniform amplitude. Note that $\mathbf{X}(k)$ of type (i) also have independent random phases. Let F denote the discrete Fourier transform. Following from the convolution theorem $x \otimes h = F^{-1}(F(x) \cdot F(h))$ and $x = F^{-1}(\mathbf{X})$, we have $x \otimes h = F^{-1}\text{diag}(\mathbf{X})Fh$, so the measurements y can be written as

$$y = \downarrow_\Omega (Ch + \xi) = \downarrow_\Omega \left(F^{-1}\text{diag}(\mathbf{X})Fh + \xi\right). \tag{6.22}$$

Let us explain intuitively why (6.22) is an effective encoding scheme for a sparse vector h. First, it is commonly known that Fh is non-sparse, and its mass is somewhat evenly spread over all its components. The random phases of $\mathbf{X}$ by design are of critical importance. They "scramble" the components of Fh wisely and break the delicate relationships among Fh's components; as a result, in contrast to the sparse $F^{-1}Fh = h$, $F^{-1}\text{diag}(\mathbf{X})Fh$ is not sparse at all. Furthermore, $\mathbf{X}$ has a random-spreading effect. Because of a phenomenon called the concentration of measures [303], the mass of Ch

[2] In our case, S is equal to P, the number of non-zero taps in (6.19).

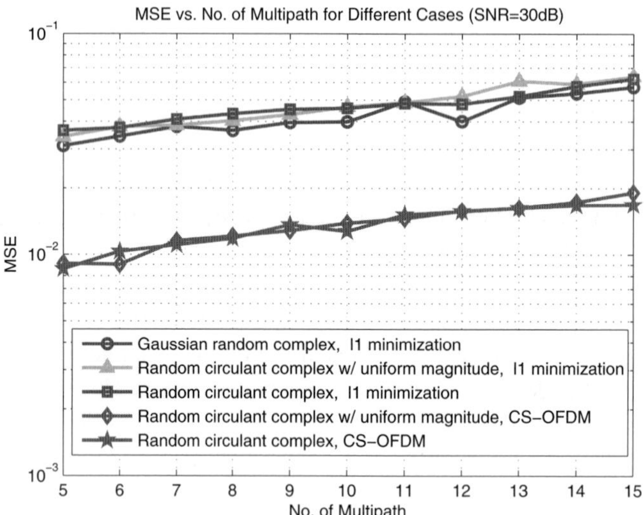

Figure 6.2 MSE vs. number of multipaths for different cases (SNR $=30$ dB).

spreads over its components in a way that, with a high probability, the information of h is preserved by the downsampling of a size essentially linear in P – the sparsity of h (whether or not the downsampling is equally spaced; i.e., uniform). Up to a log factor, the down-sampled measurements permit stable ℓ_1 recovery. Both types (i) and (ii) of $\mathbf{X}$ have similar encoding strength, but $\mathbf{X}$ of type (ii) gives an orthogonal C; i.e., $C^*C = I$, so $x \otimes h$ transforms h into a random ortho-basis. Such orthogonality results in multiple benefits, such as faster convergence of our recovery algorithm. Because of the page limitation, we omit rigorous mathematical analysis of (6.22) and its guaranteed recovery.

Note that the proposed sampling $\downarrow_\Omega (F^{-1}\text{diag}(\mathbf{X})F)$ is very different from partial Fourier sampling $\downarrow_\Omega F$. The latter requires a random Ω to avoid the aliasing artifacts in the recovery, but the former, with random-phased $\mathbf{X}$, permits both random and uniform Ω. Below we numerically demonstrate its encoding efficiency.

Numerical evidence of effective random convolution

CS performance is measured by the number of measurements required for stable recovery. To compare the proposed sensing schemes with the well-established Gaussian random sensing, we conduct numerical simulations and show their results in Figure 6.2. We compare three types of CS encoding matrices: the i.i.d. Gaussian random complex matrix, and the circulant random complex matrices corresponding to $\mathbf{X}$ of types (i) and (ii) above, respectively. In addition, ℓ_1 minimization is compared to our proposed algorithm CS-OFDM, which is detailed in the next subsection. The simulation results show that the random convolutions of both types perform as well as the Gaussian random sensing matrix under ℓ_1 minimization, and our algorithm CS-OFDM further improves the performance by half an order of magnitude.

Relationship to existing results

Random convolution CS

Random convolution has been used and proved to be an effective way of taking CS measurements that allow the signal to be recovered using ℓ_1 minimization. In [293], Toeplitz[3] measurement matrices are constructed with i.i.d. random row 1 (the same as type (i)) but with only ± 1 or $\{-1, 0, 1\}$; their downsampling effectively takes the first M rows; and the number of measurements needed for stable ℓ_1 recovery is shown as $M \geq O(S^3 \cdot \log N/S)$. Reference [287] uses a "partial" Toeplitz matrix, with i.i.d. Bernoulli or Gaussian row one, for sparse channel estimation, where the downsampling effectively also takes the first M rows. Their scheme requires $M \geq O(S^2 \cdot \log N)$ for stable ℓ_1 recovery. In [294], random convolution of type (ii) above with either random downsampling or random demodulation is proposed and studied. It is shown that the resulting measurement matrix is incoherent with any given sparse basis with a high probability and ℓ_1 recovery is stable given $M \geq O(S \cdot \log N + \log^3 N)$. Our proposed type (ii) is motivated by [294]. On the other hand, no existing work proposes uniform downsampling or shows its recovery guarantees. In addition, most existing analysis is limited to real-valued matrices and signals.

CS-based channel estimation

Our work is closely related to [287] and [295]. In [287], i.i.d. Bernoulli or Gaussian vector is used as a training sequence, and downsampling is carried out by taking only the first M rows. While channel estimation is obtained as a solution to the Dantzig selector. In [295], MIMO channels are estimated by activating all sources simultaneously. The receivers measure the cumulative response, which consists of random convolutions between multiple pairs of source signals and channel responses. Their goal is to reduce the channel estimation time, and ℓ_1 minimization is used to recover channel response.

Our current work is limited to estimating a signal h-vector. While our work is based on similar random convolution techniques, we have proposed to use a pair of source and receiver, which consists of a high-speed source and a low-speed receiver for the novel goal of high-resolution channel estimation. Furthermore, we apply a novel algorithm for the channel response recovery based on iterative support detection and limited-support least-squares, which is described in detail in Section 6.3.3.

6.3.3 Numerical algorithm

Problem formulation

As a result of rapid decaying of wireless channels, P – the number of significant multipaths – is small, so the channel response h is a highly sparse signal. Recall that the non-zero components of h only appear in the first $\tilde{N}$ components. We shall recover a sparse high-resolution signal h with a constraint from the measurements y at a lower resolution of M. We define operation $|\cdot|$ as the amplitude of a complex number,

[3] which is slightly more general than a circulant.

$\|h\|_0$ as the total number of non-zeros of $|h|$ and $\|h\|_1 = \sum_i |h_i|$. The corresponding model is

$$\min_{h \in \mathbb{R}^N} \|h\|_0, \text{ s.t. } \phi h = y \text{ and } h_i = 0 \; \forall i > \tilde{N}, \tag{6.23}$$

where ϕ denotes $\downarrow_\Omega C$ in (6.22), the submatrix of C formed by its rows corresponding to the downsampling points in Ω. Generally speaking, problem (6.23) is NP-hard and is impossible to solve, even for moderate N. A common alternative is its ℓ_1 relaxation model with the same constraints.

$$\min_{h \in \mathbb{R}^N} \|h\|_1, \text{ s.t. } \phi h = y \text{ and } h_i = 0 \; \forall i > \tilde{N}, \tag{6.24}$$

which is convex and has polynomial time algorithms. If y has no noise, both (6.23) and (6.24) can recover h exactly given enough measurements, but (6.24) requires more measurements than (6.23).

Algorithm

Instead of using a generic algorithm for (6.24), we design an algorithm to exploit the OFDM system features, including the special structure of h and noisy measurements y. At the same time, we maintain its simplicity to achieve low complexity and match it with easy hardware implementation.

First, we can simply collaborate two constraints into one by letting the variables be $\tilde{h} = [h_1, h_2, \ldots, h_{\tilde{N}}]$ and dropping the rest of the components of h. Let $\tilde{\phi}$ be the matrix formed by first $\tilde{N}$ columns of ϕ. Hence, the only constraints are $\tilde{\phi}\tilde{h} = y$, which reduces the size of our problem.

We also develop our algorithm CS-OFDM for the purpose of handling noisy measurements. The iterative support detection (ISD) scheme proposed in [238] has a very good performance for solving (6.24) even with noisy measurements. Our algorithm uses the ISD, as well as a final denoising step. In the main iterative loop, it estimates a support set I from the current reconstruction and reconstructs a new candidate solution by solving the minimization problem $\min\{\sum_{i \in I^c} |\tilde{h}_i| : \tilde{\phi}\tilde{h} = y\}$, and it iterates these two steps for a small number of iterations. The idea of iteratively updating the index set I helps to catch missing spikes and erase fake spikes. This is an ℓ_1-based method but outperforms ℓ_1. The analysis and numerical performance of ISD can be found in [304]. Because the measurements have noise, reconstruction is never exact. Our algorithm uses a final denoising step, which solves least-squares over the final support T, to eliminate tiny spikes likely due to noise. Our pseudocode is listed in Algorithm 1.

In Algorithm 1, at each iteration j, (6.25) solves a weighted ℓ_1 problem, and the solution h^j is used for support detection to generate a new I^{j+1}. After the main loop is done, a support T is estimated above a threshold, which is selected based on empirical experiences. If the support detection is executed successfully, T would be the set of all channel multipath delays. Finally, $\tilde{h}$ is constructed by solving a small least-squares problem, and $\tilde{h}_i, \forall i \notin T$ falls to zero.

Algorithm 1 CS-OFDM

Input: ϕ, y;

Initalize:

$\tilde{\phi}$ as the first $\tilde{N}$ columns of ϕ.

$I^0 \leftarrow \emptyset$ and $w_i^0 = 1, \forall i \in \{1, 2, \ldots, \tilde{N}\}$

while the stopping condition is not met, **do**

Subproblem:

$$\tilde{h} \leftarrow \arg\min \sum_{i \notin I^j} |\tilde{h}_i|, \text{ s.t. } \tilde{\phi}\tilde{h} = y. \tag{6.25}$$

Support detection: $I^{j+1} \leftarrow \{i : |\tilde{h}_i^j| \geq 2^{-j} \|\tilde{h}^j\|_\infty\}$, where $\|\tilde{h}^j\|_\infty = \max_i\{|\tilde{h}_i^j|\}$.

Weights update:

$w_i^{j+1} \leftarrow 0, \forall i \in I^{j+1}$; otherwise $w_i^{j+1} \leftarrow 1$.

$j \leftarrow j + 1$

end while

Final least-squares:

let $T = \{i : |\tilde{h}_i| > \text{threshold}\}$, then:

$$\tilde{h}_T \leftarrow \arg\min_{\tilde{h}} \|\tilde{\phi}_T \tilde{h} - y\|_2^2, \text{ and } \tilde{h}_{T^c} \leftarrow 0. \tag{6.26}$$

Return $\tilde{h}$

Complexity analysis

This algorithm is efficient since every step is simple and the total number of iterations needed is small. The subproblem is a standard weighted ℓ_1 minimization problem, which can be solved by various ℓ_1 solvers. Since ϕ is a convolution operator, we choose YALL1 [187] since (i) it allows us to customize the operators involving $\tilde{\phi}$ and its adjoint to take advantages of DFTs, making it easier to implement the algorithm in hardware, (ii) YALL1 is asymptotically, geometrically convergent and efficient even when the measurements are noisy. With our customization, all YALL1 operations are either an DFT/IDFT or one-dimensional vector operations, so the overall complexity is $O(N \log N)$. Moreover, for support detection, we run YALL1 with a more forgiving stopping tolerance and always restart it from the last step solution. Furthermore, YALL1 converges faster as the index I^j gets closer to the true support. The total number of YALL1 calls is also small, since the detect support threshold decays exponentially and is bounded below by a positive number. Numerical experience shows that the total number of YALL1 calls never exceeds P.

The computational cost of the final least-squares step is negligible, because the associated matrix $\tilde{\phi}_T$ has its number of columns approximately equal to the number of spikes in h, which is far fewer than its rows. For example, if the system has P multipaths, the associated matrix for least-squares has size $M \times P$. Generally speaking, the complexity for this least-squares is $O(MP + P^3)$. Since P and M are much smaller than N, the complexity of the entire algorithm is dominated by that of YALL1, which is $O(N \log N)$.

Cramér–Rao lower bound

The Cramér–Rao Lower Bound (CRLB) is an indicator of how good an unbiased estimator performs. In this subsection, we derive the CRLB for pilot-assisted OFDM channel estimation. The CRLB for each entry of y is

$$\mathrm{CRLB}(h_i) = (\mathbf{I}^{-1}(h))_{ii},\ 0 \leq i \leq M-1, \tag{6.27}$$

where $\mathbf{I}(h)$ is the Fisher information matrix, written as

$$\mathbf{I}(h) = -E\{\frac{\partial}{\partial h}\log f(y|h)(\frac{\partial}{\partial h}\log f(y|h))^*\}. \tag{6.28}$$

In (6.28), E denotes the expectation, and $f(y|h)$ is the conditional PDF of y given h. Recall that the OFDM system model can be written as

$$y = \phi h + \xi, \tag{6.29}$$

where ϕ denotes $\downarrow_\Omega C$ in (6.22) and ξ is the AWGN noise with σ variance and zero mean. Following (6.29), we can derive the conditional PDF of y given h

$$f(y|h) = \frac{1}{(\pi\sigma^2)}\exp\{-\frac{1}{\sigma^2}\|y-\phi h\|^2\}. \tag{6.30}$$

Then the partial derivative of the logarithm of (6.30) with respect to h is

$$\frac{\partial}{\partial h}\log f(y|h)) = \frac{1}{\sigma^2}[y^* - h^*\phi^*]^T. \tag{6.31}$$

Substitute (6.31) into (6.28), and the Fisher information matrix is

$$\mathbf{I}(h) = \frac{1}{\sigma^2}\phi^*\phi. \tag{6.32}$$

Since the uniform downsample operation $\downarrow_\Omega C$ does not affect the orthogonality, and $C^*C = I$, we have $\phi^*\phi = I$, where I is an $M \times M$ identity matrix. Then the Fisher information matrix in (6.32) can be written as

$$\mathbf{I}(h) = \frac{1}{\sigma^2}I. \tag{6.33}$$

The CRLB for the ith entry of y is $\mathbf{I}^{-1}(h)_{ii}$. Then the overall CRLB is

$$\mathrm{CRLB}(h) = \sum_{i=1}^{M}\mathrm{CRLB}(h_i) = \mathrm{trace}(\mathbf{I}^{-1}(h)). \tag{6.34}$$

6.3.4 Numerical simulations

In this subsection, we performed numerical simulations to illustrate the performance of the proposed CS-OFDM algorithm for high-resolution OFDM channel estimation. We focused on the mean square error (MSE) of channel estimation, as well as the multipath delay detection, when the channel profile and signal-to-noise ratio (SNR) changes.

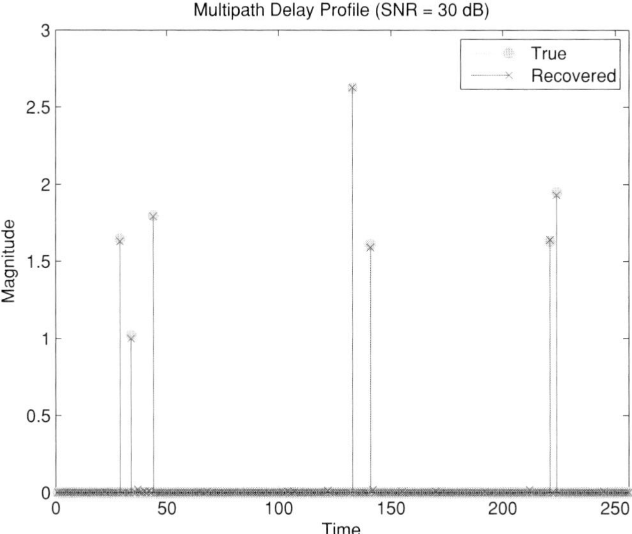

Figure 6.3 Multipath delay profile.

Simulation Settings

We considered an OFDM system with $1k$-point IDFT ($N = 1024$) at the transmitter and 64-point DFT ($M = 64$) at the receiver, where the compression ratio is 16. The number of silent subcarriers, which acts as a guard band is 256 among 1024 subcarriers. The channel is estimated based on 768 pilot tones with uniformly random phases and a unit amplitude, with the Gaussian noise level ranging from 10 dB to 30 dB. We assume the usage of cyclic prefix and the impulse response of the channel is shorter than cyclic prefix, which means there is no intersymbol interference. For all simulations, we test the total numbers of multipaths from five to fifteen. Moreover, we use only one OFDM symbol; i.e., use all non-silent subcarriers only once to carry pilot signals. All reported performances will substantially improve if more pilots are inserted.

MSE performance

Figure 6.3 is a snapshot of one-channel estimation simulation. It suggests that the proposed pilot arrangement and CS-OFDM successfully detect an OFDM channel with 7 multipaths when the signal-to-noise ratio is 30 dB. Our method not only exactly estimates the multipath delays, but also correctly estimates the values of the corresponding multipath components. Figure 6.4 depicts the MSE performance on OFDM channels with the number of multipaths ranging from five to fifteen and the noise level ranging from 10 dB to 30 dB. When there are only a moderate number of multipaths on the OFDM channel, e.g., 10 multipaths, even when the SNR is 20 dB, MSE is as low as -17 dB. Figure 6.5 shows the reconstructed SNR vs. the number of multipaths when the input SNR changes. We can see that CS-OFDM achieves a gain in SNR. For example, when the input SNR is 10 dB, we obtain a reconstructed SNR higher than 20 dB when there are five multipaths. As the number of multipaths increases, the SNR gain from the

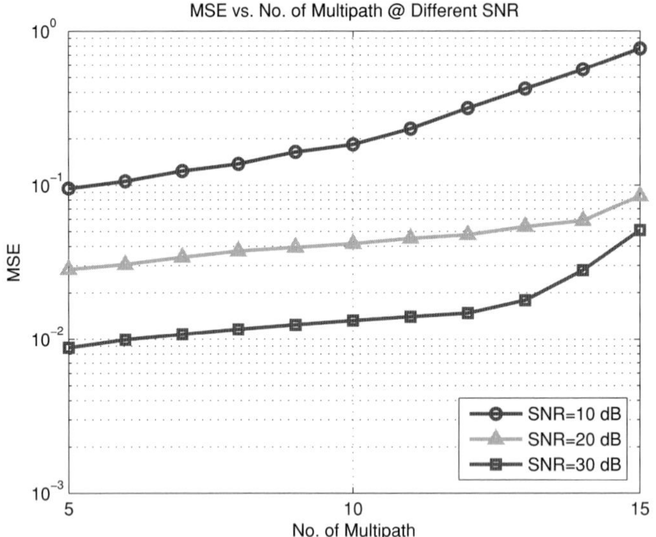

Figure 6.4 MSE performance.

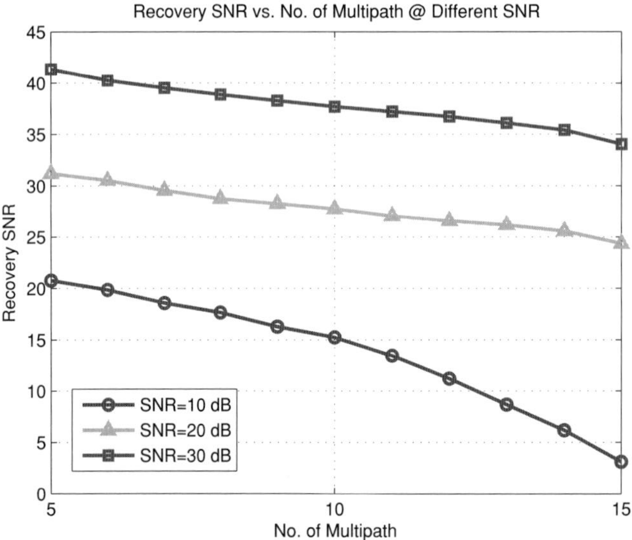

Figure 6.5 Reconstructed SNR.

reconstructed signal to the input signal decreases. However, even when the number of multipaths is 10, we still have a 5 dB gain, e.g., reconstructed SNR is 15 dB when the input signal SNR is 10 dB. The similar SNR gain appears for input SNR = 20 dB and SNR = 30 dB cases. From the entire input SNR and the number of multipath ranges we have tested, there is an average gain of 6 dB from the input SNR to the recovered SNR.

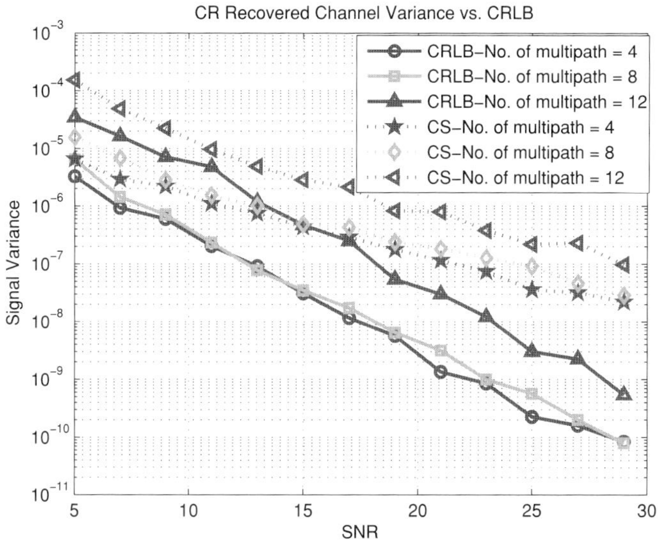

Figure 6.6 CS recovered channel variance vs. CRLB.

CRLB performance

Figure 6.6 shows the CS estimated channel variance vs. CRLB with the change of SNR and number of multipaths. The gap between the CS estimated channel variance and the CRLB is relatively small at low SNR and small number of multipaths. For example, when SNR = 5 dB, and the number of multipaths equals four, there is only a 4 dB performance gap. While as SNR increases at the same number of multipaths when SNR increases to 29 dB, performance loss increases to 22 dB. Besides, the performance gap increases with the increase in the number of multipaths. For instance, when SNR = 5 dB, as the number of multipaths increases from 4 to 12, the performance gap increases from 4 dB to 9 dB.

Multipath delay detection performance

Figure 6.7 and Figure 6.8 depict the probability of correct detection (POD) of the multipath delay and the false detection rate (FAR) while we change the SNR and the number of multipaths. When the SNR is above 10 dB, simulation shows almost 100% POD when the number of multipaths changes from 5 to 12. When there are a relatively large number of multipaths, e.g., 15, the probability of correct multipath delay detection is higher than 95% as SNR $\geq$ 10 dB. Even when SNR is low, as long as the number of multipaths does not exceed 10, we still have a POD of greater than 95%. The FAR performance shows similar results, as the SNR decreases and the number of multipaths increases, performance decreases. But, in a large range, e.g., SNR $\geq$10 dB, the number of multipaths $\leq$10, we have almost zero FAR.

In summary, efficient OFDM channel estimation will drive OFDM to carry the future of wireless networking. A great opportunity for high-efficiency OFDM channel estimation is lent by the sparse nature of channel response. Riding on the recent development

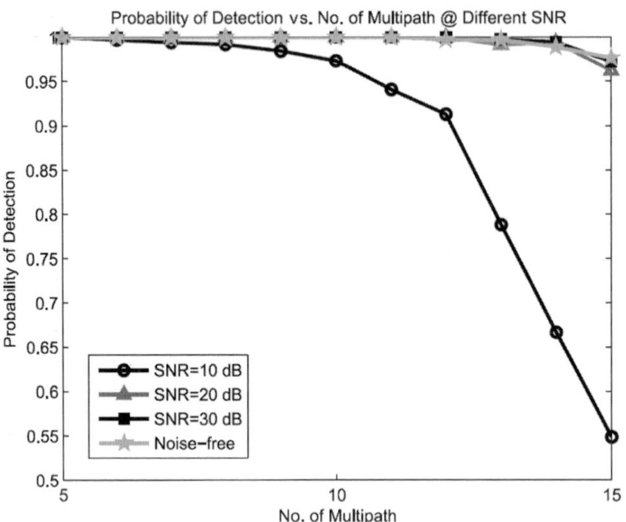

Figure 6.7 Probability of support detection.

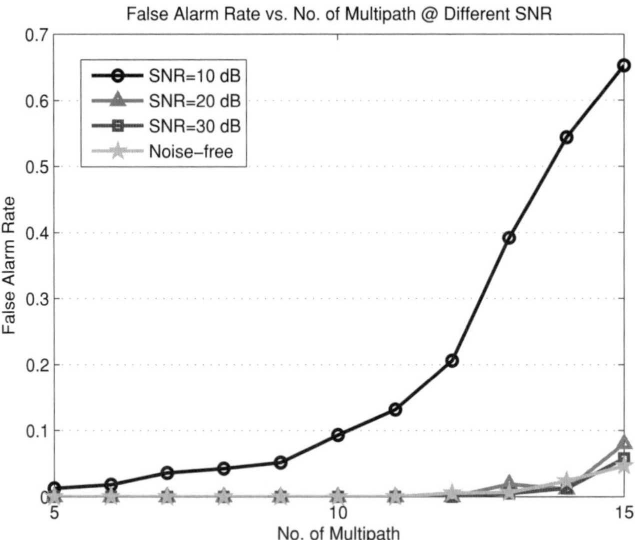

Figure 6.8 Probability of false support detection.

of CS, we propose, in this chapter, a design of probing pilots with random phases, which preserves the information of channel response during the convolution and downsampling processes, and a sparse recovery algorithm, which returns the channel response in high SNR. These benefits translate to the high resolution of channel estimation, the lower speed of the receiver ADC, as well as shorter probing times. In this section, the presentation is limited to an idealized OFDM model, intuitive explanations, and simulated experiments. The results presented here hint at a high-efficiency improvement for OFMD in practice.

6.4 Underwater acoustic channel estimation

In this section, we discuss the CS channel estimation for the underwater acoustic scenario. We first explain the channel model, and then discuss several existing approaches in the literature.

6.4.1 Channel model

Generally speaking, energy loss during underwater sound propagation is caused by geometric spreading, attenuation, and other path-related losses. The attenuation sources can be further categorized into medium absorption and scattering. Among these factors, the attenuation caused by absorption and spreading is relatively stable compared to other factors in a given area.

Depending on the spatial boundary of the water, the *spreading loss* can be modeled as either spherical for deep water or cylindrical for shallow water, corresponding to an inverse-square or inverse first-power relation with the propagation range. The spherical model and cylindrical model can also be combined [305]. *Attenuation* is normally caused by medium absorption and scattering. The absorption loss can be described as

$$I_2 - I_1 = -\alpha d, \tag{6.35}$$

where I_2 and I_1 are the intensities in dB at two locations, d is the distance in kilometers between these locations, and α is the *absorption coefficient* in dB per kilometer. The absorption coefficient increases with sound frequency. It is also related to the density of water. Thorp's formula [306–309] approximates this relation as:

$$\alpha = \frac{0.11 f^2}{1 + f^2} + \frac{44 f^2}{4100 + f^2} + 2.75 \times 10^{-4} f^2 + 0.003, \tag{6.36}$$

where f is the sound frequency in kilohertz.

The speed of sound depends on water salinity, temperature, and density. Briefly speaking, the speed of sound increases with the increment of any one of the above three factors. Several empirical formulae have been proposed by oceanographists. Here we present the Coppens equation as an example (refer to [310] for coefficient values).

$$\begin{aligned} c_{O,S,T} &= c_0 + c_1 T + c_2 T^2 + c_3 T^3 + (c_4 + c_5 T + c_6 T^2)(S - 35), \qquad (6.37)\\ c_{D,S,T} &= c_{O,S,T} + (c_7 + c_8 T)D + (c_9 + c_{10} T)D^2 + [c_{11} + c_{12}(S - 35)](S - 35)TD, \end{aligned}$$

where D is the depth, S is salinity, T is water temperature, $c_{O,S,T}$ is the speed without considering the depth, and $c_{D,S,T}$ is the speed considering the depth. The distribution of sound speed in the ocean normally conforms to a general profile. However, near shores and estuaries, high gradients of sound speed can occur frequently [311], which results in irregular acoustic channels.

Since the speed of a wave depends on many factors as shown in (6.37), the arrival time for the ith path is given by

$$t_i = \int_{\mathcal{L}_i} \frac{1}{c_{D,S,T}(L)} dL \,, \tag{6.38}$$

where $\mathcal{L}_i$ is the ith propagation path. If we define $|\mathcal{L}_i|$ as the path length, the propagation loss for each path is given by

$$l_i = (\nu_s)^{n_s} (\nu_b)^{n_b} 10^{-\alpha |\mathcal{L}_i|}, \tag{6.39}$$

where n_s is the number of times the path hits the surface, n_b is the number of times the path hits the sea bottom, and ν_s and ν_b are the loss of reflections at sea surface and bottom, respectively. $\nu_s \approx 1$ and

$$\nu_b = \frac{\rho_2 k_{z,1} - \rho_1 k_{z,2}}{\rho_2 k_{z,1} + \rho_1 k_{z,2}}, \tag{6.40}$$

where ρ_1 and ρ_2 are the material density for water and sea bottom, respectively, and $k_{z,1}$ and $k_{z,2}$ are vertical wave numbers for water and sea bottom, respectively.

Using the above model, the channel impulse response can be written as

$$h(t) = \sum_i l_i \delta(t - t_i). \tag{6.41}$$

Here we define $G = |h|^2$ as the general term for the channel gain.

With the knowledge of sound-speed distribution and boundary conditions, the transmission loss of sound can be computed by solving the wave equation of the acoustic field. Most numerical solutions are based on ray theory, normal-mode theory, or the parabolic equation method [308, 309].

6.4.2 Compressive sensing algorithms

Based on the underwater acoustic channel described in the previous subsection, the received signal can be written as $\mathbf{y} = \mathbf{H}\mathbf{x} + \mathbf{z}$ for CS. In the following, we first discuss two basic algorithms proposed in the literature. Secondly, we discuss the pilot design for OFDM approach. Finally, we study an iterative channel estimation and decoding algorithm.

Basic pursuit and orthogonal matching pursuit

In [274], the authors focus on two popular algorithms in CS. First for basic pursuit, the constrained l_1 compressive optimization problem is transformed to the following unconstrained optimization problem $\min_{\mathbf{x}} \frac{1}{2}\|\mathbf{H}\mathbf{x} - \mathbf{z}\|^2 + \tau\|\mathbf{x}\|_1$. The efficient algorithms to solve the problem can be easily found [146], and the parameter τ is quite robust against the choice of values.

Secondly, the orthogonal matching pursuit method is a greedy algorithm to iteratively identify one multipath component at a time and solve a constrained least-squares problem at each iteration to measure the fitting error. Since the number of multipaths is unknown, a certain stopping criterion is required. A simple method is to compare the residual errors in the consecutive iterations.

In both algorithms, it is important to consider that multiplying of channel matrix **H** can be done efficiently. The authors in [274] reduce the complexity by considering interchannel interference from only a limited number of neighboring subcarriers. As a result, the matrix computation complexity can be greatly reduced.

Pilot design for cyclic-prefix OFDM

In [279], the authors discuss several important questions that remain unanswered in the context of multicarrier transmission such as OFDM:

1. should pilots be placed in clusters as for time-varying, non-sparse channels, or randomly dispersed between the data as common in the CS literature;
2. as pilot and data subcarriers cannot be perfectly separated at the receiver due to intercarrier interference (ICI) should channel estimation be based on observations corresponding to pilot subcarriers only, or can the data subcarriers be used to practically extract additional information about the channel;
3. how does the performance vary with the number of pilot symbols?

The authors use data recorded at the practical experiment to investigate these questions. Because of the time variation the orthogonality between subcarriers is lost, making it costly in terms of spectral efficiency to use pilot designs that keep pilots and data symbols fully orthogonal at the receiver. Observations on adjacent subcarriers are required to estimate the time-varying channel, which matches existing results for estimation of non-sparse time-varying channels. This contradicts some early concepts that for sparse channel estimation, observations should simply be placed randomly among the data. Moreover, when pilots and data symbols are not separated at the receiver, subcarriers corresponding to data symbols can be used to observe the ICI caused by the neighboring pilot subcarriers. In this case, the effect of the unknown data symbols should be modeled as additional colored noise, which is especially needed for the basic pursuit method, while the orthogonal matching pursuit method seems to be less affected by an unknown noise spectrum. Finally, varying the number of pilots is considered to improve the performance at the cost of reduced data rate.

Iterative channel estimation and decoding

In [280], a block-by-block iterative receiver is designed for underwater MIMO-OFDM, which couples channel estimation with MIMO detection and low-density parity-check channel decoding. The channel estimator is based on a CS technique to exploit the channel sparsity, the MIMO detector consists of a hybrid use of successive interference cancelation and soft minimum mean-square error equalization, and channel coding uses non-binary LDPC codes. Some feedback strategies from the channel decoder to the channel estimator are studied. The experiment was held off the coast of Martha's Vineyard, MA, from October 14 to November 1, 2008, which shows that iterative receiver processing, including sparse channel estimation, leads to impressive performance gains.

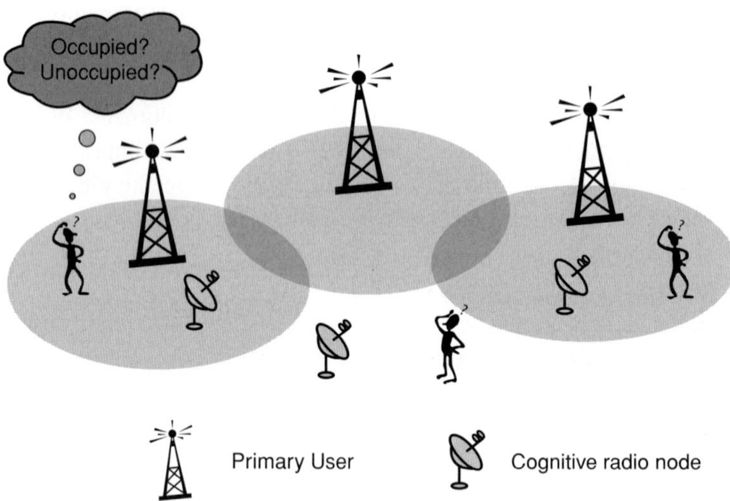

Figure 6.9 Illustration of the spectrum occupancy data sampling. Only limited measurements can be sampled in cognitive radio networks, while spectrum occupancy at the spots without cognitive radio nodes are unknown.

These gains are more pronounced when the number of available pilots to estimate the channel is decreased. For the feedback strategies for iterative channel estimation, soft decision feedback slightly outperforms hard decision feedback.

6.5 Random field estimation

The task of identifying the unoccupied spectrum is crucial for the efficiency and reliability of future wireless networks, especially for cognitive radio networks. A quantitative and precise understanding of current spectrum occupancy can help the secondary users to make more effective dynamic spectrum access decisions. Moreover, knowledge of the spectrum occupancy can help us analyze connectivity, capacity and other properties of cognitive radio networks. In order to obtain the spectrum usage data extensive measurement campaigns are needed. An illustration of the spectrum occupancy sampling is shown in Figure 6.9. Spectrum occupancy data can be obtained at the spots where cognitive radio nodes are present. Because of the hardware and geographical limitations, however, limited samples only can be collected at specific locations during a certain period of time, which is impractical and inefficient for the longtime measurements collection. Other approaches [312, 313] of spectrum occupancy prediction are to apply the model of primary users' activity or the experience of the secondary users, and few works take the correlations of the spectrum usage data into account.

Existing works [314] show that the spectrum occupancy is usually correlated in space, time and frequency. For example, adjacent secondary users may share very similar spectrum occupancy, and the spectrum usage is usually not independent of time, thus incurring spatial-temporal correlation. Many existing works [315–319] have exploited mathematically tractable models to describe the correlations. Note that the spectrum occupancy data reconstruction problem is very similar to the matrix completion [69, 320]: only a portion of cognitive radio nodes report the spectrum usage samplings to the fusion center, where the spectrum occupancy data for the whole network are correlated in nature and can be reconstructed. By leveraging the low-rank nature of the spectrum usage data matrix, we can reconstruct the original spectrum occupancy from a limited number of measurements. Some other existing work for CS-based random fields can be found in [321] for wireless sensor networks and in [322] for image processing.

In this section, we model the spatial-temporal correlation of the spectrum occupancy using the theory of random fields. The spatial correlation is described by the two-dimensional Ising model, and the Metropolis–Hastings algorithm is used to model the correlated spectrum occupancy over time. To effectively and efficiently obtain the spectrum usage in the network, we adopt the matrix completion technique to recover the spectrum occupancy data with a limited number of measurements. We study the performance of the proposed algorithm at different sampling rates and then consider the scenarios for different number of primary users and different communication interruption ranges. The performance of the proposed mechanism versus the dynamics of the spectrum usage data is also evaluated. The simulation results show the effectiveness of the proposed algorithm.

6.5.1 Random field model

In this subsection, we describe the modeling of cognitive radio networks using random fields. The features of cognitive radio networks can be realistically captured by many random field models, and we will focus on the classical Ising model [323]. We use this model to consider spatial correlation of the spectrum occupancy in the cognitive radio network, and then use the Metropolis–Hastings algorithm to model the spectrum-usage situation during a period of time.

Topology of cognitive radio network

We adopt the plenary grid topology for the cognitive radio networks. The geographic area is divided into many grids, and for simplicity, we assume each gird only has two states: busy or idle. An illustration of the topology is shown in Figure 6.10, where colors of black and white represent the busy and idle state, respectively. Each plane stands for the spectrum occupancy at a specific time, and thus a series of the planes detail the spatial-temporal spectrum usage in cognitive radio networks.

For facilitating the modeling using the random fields, we define the neighborhood and clique in the network topology. For any grid that is not on the boundary of the network, the four immediately adjacent grids are considered as its neighbors. Also, a clique is defined as a set of points that are all neighbors of each other. Examples of the plenary grid topology, neighbors and two-point cliques are shown in Figure 6.10.

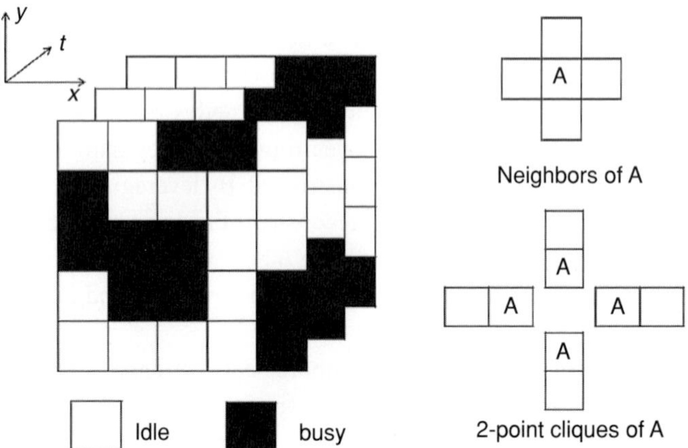

Figure 6.10 Illustration of network topology, neighbors and cliques.

Ising model

The Ising model defines a probability measure on the set of all possible configurations. We define the sample space Ω as a collection of all possible realizations of random variables. Here we assume the random variable X stands for the spectrum occupancy in the cognitive radio networks. The energy function corresponding to a specific spectrum occupancy $X = \{x_0, \dots, x_i, \dots, x_n\} \in \Omega$, is defined as

$$U(X) = \alpha \sum_i x_i + \beta \sum_{ij} x_i x_j. \tag{6.42}$$

Here, the first term describes the effect of the exogenous inputs and the second sum represents the endogenous effect. Each random variable x_i represents a grid in the topology and has two possible values, –1 and 1. $x_i = -1$ means busy and $x_i = 1$ means idle. α and β are two parameters of the Ising model. Intuitively, the Ising model can address the fact that the spectrum occupancy in the cognitive radio network depends on the natural (exogenous) and manufactured (endogenous) inputs.

Note that the second sum is taken over all pairs i, j of points that are neighbors. The Ising model makes the simplifying assumption that only interactions between neighboring points need to be taken into account. This is called a first-order Markovian property. The case $\beta > 0$ is called the attractive case, and the $\beta < 0$ is called the repulsive case.

In cognitive radio networks, the first-order Markovian property holds rigorously when the interruption ranges of primary users are small. As the interruption range increases, this may no longer be valid. However, some useful properties can still be quantified by the Ising model [317].

In the Ising model, specific spectrum occupancy $X \in \Omega$ is assigned to a probability according to the probability distribution:

$$P(X) = Z^{-1} e^{-\frac{1}{KT} U(X)}, \tag{6.43}$$

where T is the temperature and K is a universal constant. For simplicity of analysis, we assume $T = 1$ and $K = 1$. The normalizing constant Z is called the partition function, which is given by:

$$Z = \sum_{X} e^{-\frac{1}{KT}U(X)}. \tag{6.44}$$

Formula (6.43) is called the Gibbs distribution and it defines the probability distribution of the random variables X over the sample space Ω.

Metropolis–Hastings algorithm

The Metropolis–Hastings algorithm can be described as follows: suppose the initial state we sampled is X^t, we draw a new proposal state X' with probability $Q(X'|X^t)$. Then we calculate a ratio γ:

$$\gamma = \frac{P(X')Q(X^t|X')}{P(X^t)Q(X'|X^t)}, \tag{6.45}$$

where $P(X')/P(X^t)$ is the likelihood between the proposed sample X' and the previous sample X^t, $Q(X^t|X')/Q(X'|X^t)$ is the ratio of the proposal density in two directions. Each time the new state X^{t+1} is chosen according to the following rules: when $\gamma \geq 1$, we assign X' to X^{t+1}; when $\gamma < 1$, $X^{t+1} = X'$ with probability γ, and $X^{t+1} = X^t$ with probability $1 - \gamma$.

Thus, once the α and β for the Ising model are specified, we can obtain a sequence of random samples from the given Gibbs distribution, which represent the spectrum occupancy during a period of time.

Related work

Modeling and analyzing the geometrical configuration of wireless communication networks has been the subject of extensive study recently. Nowadays, the theory of random field is an emerging mathematical model used to describe the spectrum occupancy in cognitive radio networks. Currently, the research of the cognitive radio would benefit greatly from the possibility to generate synthetic data. Reference [315] provides a partial answer to the spectrum modeling problem by characterizing the spectrum maps with spatial statistics and random fields. Reference [316] uses random fields and the semi-variogram to describe the spatial correlation of spectrum usage. The proposed model can accurately reproduce the power spectral density as measured at the sampling locations. Reference [317] applies the famous Ising model to analyze the connectivity of cognitive radio network. The average delay and jitter of an arbitrary packet in a multi-channel cognitive radio network is also analyzed. Reference [318] has shown that the power spectral density measured over space can be interpreted as a random field whose properties can be analyzed in detail. Two approaches, deterministic and empirical, are proposed for the scenario evaluations. The impact of different network properties on the models is elaborated and realistic parameter sets for generation of simulation scenarios are provided. Reference [319] presents a similar work in the spectrum occupancy reconstruction. The author models the geographical-spectral pattern of interruptions from primary users within an area using the Gibbs random field. Then, Bayesian compressed sensing is used to reconstruct the spectrum occupancy patterns from a reduced number of measurements.

6.5.2 Matrix completion algorithm

In this subsection, we discuss the reconstruction of the spectrum occupancy in cognitive radio networks from a limited number of measurements. Assume $\boldsymbol{X} = \{X^{t_0}, \ldots, X^{t_n}\}$ represents the spectrum occupancy of the whole network from t_0 state to t_n state. Each $X^{t_i} \in \boldsymbol{X}$ is the spectrum occupancy matrix corresponds to the network topology at that time. For simplicity, we just consider the spectrum usage in a single channel. Usually, $\boldsymbol{X}$ is a 3D data matrix that is inconvenient for algebraic operations and optimization. Without loss of generality, for every matrix $X^{t_i} \in \boldsymbol{X}$, we can stack the columns of X^{t_i} to form a vector that stands for the spectrum utilization of t_i state. Then, we can combine these vectors into the columns of a larger matrix that represent the spectrum occupancy over a period of time. For consistency, here we still use the same notation $\boldsymbol{X}$ to denote the overall larger matrix.

At each time instant, only a limited number of cognitive radio nodes in the network measure the spectrum occupancy data and report their measurements to the fusion center. The sampled measurements can be expressed as:

$$\boldsymbol{Y} = \boldsymbol{M} \circ \boldsymbol{X}, \tag{6.46}$$

where

$$\boldsymbol{M}(i, j) = \begin{cases} 1, & \text{if } \boldsymbol{X}(i, j) \text{ is collected,} \\ 0, & \text{otherwise,} \end{cases}$$

and $\circ$ is the notion of a Hadamard product operator. $\boldsymbol{Y} = \boldsymbol{M} \circ \boldsymbol{X}$ means $\boldsymbol{Y}(i, j) = \boldsymbol{M}(i, j)\boldsymbol{X}(i, j)$.

Our goal is to recover spectrum occupancy of cognitive radio networks from the partial measurements. It is usually difficult to unambiguously determine the original $\boldsymbol{X}$ for this kind of underdetermined linear inverse problems. However, because of the spatial-temporal correlations of the spectrum occupancy in the cognitive radio network, matrix $\boldsymbol{X}$ inherits the property of low rank. We can use the matrix completion techniques to recover it:

$$\min Rank(\boldsymbol{X}) \quad s.t. \quad \boldsymbol{M}(i, j) = \boldsymbol{X}(i, j). \tag{6.47}$$

Generally, (6.47) is NP-hard due to the combinational natural of the $rank$ function. Instead of solving (6.47), we can replace $Rank(\boldsymbol{X})$ by its convex envelop to get a convex and more computationally tractable approximation [150]:

$$\min \|\boldsymbol{X}\|_* \quad s.t. \quad \boldsymbol{M}(i, j) = \boldsymbol{X}(i, j), \tag{6.48}$$

where $\|\boldsymbol{X}\|_*$ is the nuclear-norm that is defined as the sum of singular values of $\boldsymbol{X}$,

$$\|\boldsymbol{X}\|_* = \sum_{i=1}^{r} \sigma_i(\boldsymbol{X}), \tag{6.49}$$

where $\sigma_i(\boldsymbol{X})$ are the positive singular values of matrix $\boldsymbol{X}$, and r is the rank of matrix $\boldsymbol{X}$.

Algorithm 2 Proposed spectrum occupancy data matrix completion algorithm

Require: The sampling process matrix $\boldsymbol{M}$
Estimated rank of original matrix k
Tuning parameter λ
Maximum number of iterations N.

Initialize $\boldsymbol{L}, \boldsymbol{R}$ using the size of the matrix $\boldsymbol{M}$ and k.
Initialize the outputs of the optimization algorithm.
$\boldsymbol{L}_{opt} = \boldsymbol{L}, \boldsymbol{R}_{opt} = \boldsymbol{L}$
Calculate the error of known entries between original data matrix and reconstructed data matrix.
$E = \|\boldsymbol{M} \circ (\boldsymbol{Y} - \boldsymbol{X})\|_F$

for $i = 1 : N$ **do**
  Fix $\boldsymbol{R}$, and then find $\boldsymbol{L}$ that minimizes
    $\|\boldsymbol{M} \circ (\boldsymbol{Y} - \boldsymbol{X})\|_F + \lambda\|\boldsymbol{L}\|_F$
  Fix $\boldsymbol{L}$, and then find $\boldsymbol{R}$ that minimizes
    $\|\boldsymbol{M} \circ (\boldsymbol{Y} - \boldsymbol{X})\|_F + \lambda\|\boldsymbol{R}\|_F$
  Normalize $\boldsymbol{L}$ and $\boldsymbol{R}$, make sure they have the same norm as well as minimize $(\|\boldsymbol{L}\|_F + \|\boldsymbol{R}\|_F)$.
  Calculate the error $e = \|\boldsymbol{M} \circ (\boldsymbol{Y} - \boldsymbol{X})\|_F$
  if $e < E$ **then**
    $\boldsymbol{L}_{opt} = \boldsymbol{L}, \boldsymbol{R}_{opt} = \boldsymbol{R}$
    $e = E$
  end if
end for
RETURN $\boldsymbol{L}_{opt}, \boldsymbol{R}_{opt}$
Output $\boldsymbol{X} = \boldsymbol{L}_{opt} \times \boldsymbol{R}^T_{opt}$

The original data matrix $\boldsymbol{X}$ represents the spectrum occupancy data that can be factorized by the singular value decomposition:

$$\boldsymbol{X} = \boldsymbol{U}\boldsymbol{\Sigma}\boldsymbol{V}^T = \boldsymbol{L}\boldsymbol{R}^T, \tag{6.50}$$

where $\boldsymbol{U}$ and $\boldsymbol{V}$ are unitary matrices and $\boldsymbol{\Sigma}$ is a diagonal matrix containing the singular values of $\boldsymbol{X}$. $\boldsymbol{V}^T$ is the transpose of $\boldsymbol{V}$. $\boldsymbol{L} = \boldsymbol{U}\boldsymbol{\Sigma}^{1/2}$ and $\boldsymbol{R} = \boldsymbol{V}\boldsymbol{\Sigma}^{1/2}$. Note that the singular value decomposition cannot reconstruct the original matrix, and typical operations for singular value decomposition of the matrix assume the matrix is completely known. Hence, we adopt an algorithm to first estimate the $\boldsymbol{L}$ and $\boldsymbol{R}$, and then we reconstruct the original data matrix by calculating $\boldsymbol{X} = \boldsymbol{L}\boldsymbol{R}^T$. The optimization problem (6.48) can thus be solved by the following optimization [324, 325]:

$$\min_{L,R} \|\boldsymbol{L}\|_F + \|\boldsymbol{R}\|_F \quad s.t. \quad \boldsymbol{M}(i,j) = \boldsymbol{X}(i,j), \tag{6.51}$$

where the $\|\boldsymbol{L}\|_F$ and $\|\boldsymbol{R}\|_F$ denote the Frobenius norm of the matrix $\boldsymbol{L}$, $\boldsymbol{R}$, respectively,

$$\|\boldsymbol{L}\|_F = (\sum_{i,j} \boldsymbol{L}_{i,j}^2)^{1/2}, \quad \|\boldsymbol{R}\|_F = (\sum_{i,j} \boldsymbol{R}_{i,j}^2)^{1/2}. \tag{6.52}$$

In order to reduce the recovery error as well as guarantee a low-rank solution of the optimization function, we rewrite the formula (6.51) as:

$$\min \|\boldsymbol{M} \circ (\boldsymbol{Y} - \boldsymbol{X})\|_F + \lambda(\|\boldsymbol{L}\|_F + \|\boldsymbol{R}\|_F), \tag{6.53}$$

where λ is a tunable parameter that balances between the precise fit of the measurement data and the goal of achieving low rank. The proposed algorithm to reconstruct the spectrum occupancy data matrix is shown in Algorithm 2.

6.5.3 Simulation results

The simulation setup is described as follows. We assume a 1500 m $\times$ 1500 m area and equally divide it into a 50 $\times$ 50 grid. A number of primary users are placed randomly within the squares formed by the grid. The corresponding parameters α and β in the Ising model are obtained by matching the first-order and second-order statistics of the spectrum occupancy of the disc model: interruption ranges of the primary users are fixed and identical, and the secondary users within the primary users' interruption range are unable to communicate. We generate the spectrum occupancy data using the Ising model as the initial state, and then the Metropolis–Hastings algorithm is used to generate a series of states that represent the spectrum usage data over a period of time.

Each state will last a period of Δt. Totally, there are N grids in the network, and we collect the spectrum occupancy data for a time T.

At each time instant, only a limited number of cognitive radio nodes are able to collect the spectrum occupancy data, and report their measurements to the fusion center. The size of the overall spectrum usage data matrix is supposed to be $N \times T$. However, we only know partial entries of the spectrum usage matrix due to the limited number of secondary users present in the cognitive radio network. In the simulation, $N = 2500$ and $T = 500$ min. The fusion center gathers the samples from the cognitive radio nodes once per minute. We define the sampling rate as the ratio of the collected measurements to the total number of grids. For example, suppose 500 cognitive radio nodes report their measurements, then the sampling rate is $500/2500 = 20\%$.

Performance vs. sampling rate

We assume 50 primary users in the network with an interruption range of 200 m. Every spectrum occupancy state lasts for 5 min. The number of the secondary users ranges from 250 to 750, so the sampling rate varies from 10% to 30%. The numerical simulation results are shown in Figure 6.11.

The proposed algorithm can reconstruct the spectrum occupancy matrix with high correctness rate at relatively low sampling rate. As to the missing detection rate and false alarm rate, they are around 5% at a relative low sampling rate of 10%. When the sampling rate increases, the miss detection rate and false alarm rate decrease. As the

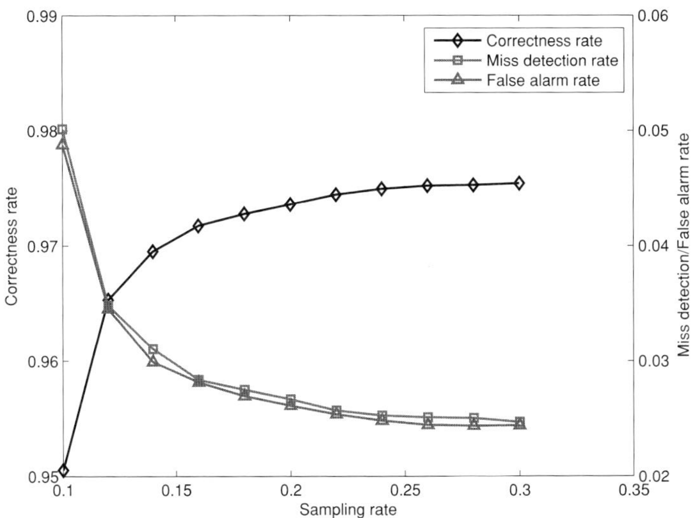

Figure 6.11 Performance vs. sampling rate.

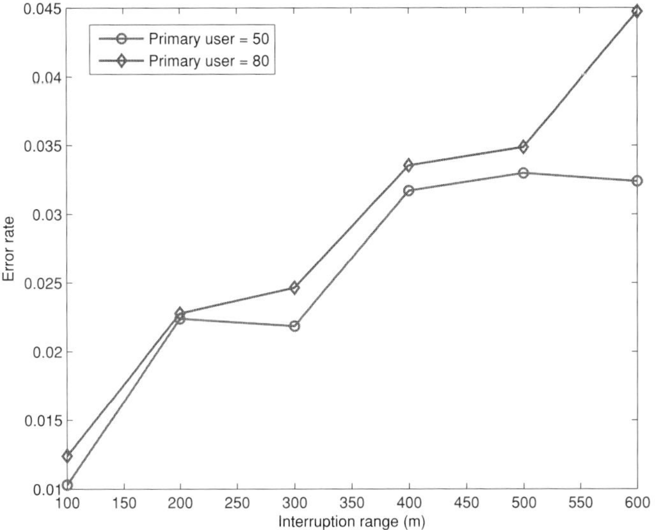

Figure 6.12 Performance vs. interruption range and number of primary users.

sampling rate reaches 30%, the algorithm can guarantee the miss detection rate and false alarm rate to be less than 3%.

Performance vs. interruption range and number of primary users

We fix the sampling rate to 20% and change the interruption range of the primary user from 100 to 600 m. We consider the cases for when the number of the primary users are 50 and 80, respectively. Figure 6.12 shows the error rate in different scenarios.

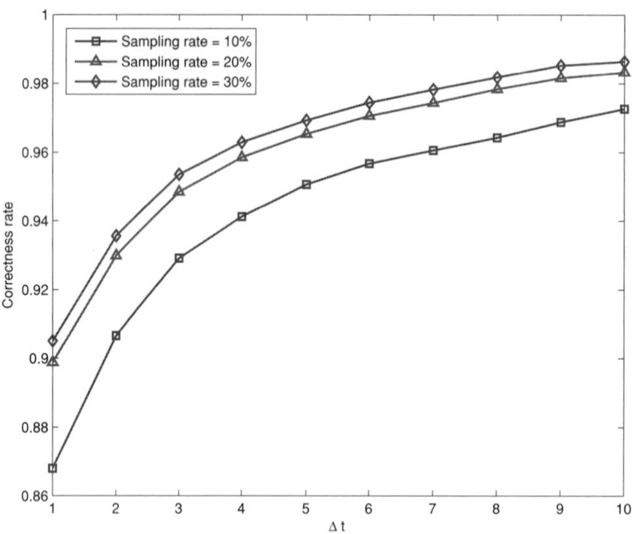

Figure 6.13 Performance vs. dynamics of the spectrum usage data.

When the interruption range is 100 m, the error rate is very low (around 1%) for both cases. As the interruption range increases, we observe that the reconstruction precision decreases. This is probably because a larger interruption range leads to a more complicated spectrum occupancy situation, which makes the reconstruction of the original data matrix more difficult. Also, as the number of primary users becomes larger, the error rate increases.

Performance vs. dynamics of the spectrum usage data

We set the number of primary users to 50 and the interruption range is 200 m. We range the Δt from 1 min to 10 min. The smaller value of Δt means a more rapidly changing of the spectrum occupancy status. We study the performance of the proposed mechanism at different sampling rates varying from 10% to 30%. The correctness rate is shown in Figure 6.13.

From the figure we can see that when Δt is small, the correctness rate is relative low at all the three sampling rates. As Δt increases, the performance is better. A sampling rate of 10% will obtain an acceptable correctness rate when the spectrum usage is changing drastically ($\Delta t = 1$ min). When the spectrum occupancy data are changing very slowly ($\Delta t = 10$ min), a sampling rate of 10% will result in a correctness rate of 96%. The simulation results demonstrate that to achieve a better performance, more measurements need to be collected in a violently changing environment.

In this section, we use the theory of random fields that characterizes the spatial-temporal correlation of spectrum usage data to model the spectrum occupancy in cognitive radio networks. The Ising model is used to model the spatial correlation of the spectrum usage data, and the correlated spectrum occupancy ia described by the Metropolis–Hastings algorithm. To effectively and efficiently obtain the spectrum usage data, the matrix completion technique that leverages the low-rank structure of the spectrum occupancy data are used to reconstruct the original data matrix from a limited number of measurements. The simulation results validate the effectiveness of our algorithm.

6.6 Other channel estimation methods

In this section, we study several other channel estimation methods in the following subsections, and show the broad scope of the CS channel estimation approaches.

6.6.1 Blind channel estimation

In [281, 282], the concept of blind CS is introduced so as to avoid the need to know the sparsity basis in both the sampling and the recovery process. In other words, the only prior is that there exists some basis in which the signal is sparse. This is very different from CS, in which the basis representation must be known. The goals are to investigate the basic conditions under which blind recovery from compressed measurements is possible theoretically, to propose concrete algorithms for this task, and to study the applications in the spectrum sensing.

Since the sparsity basis is unknown, the uncertainty about the signal is larger in the blind case than in the traditional CS case. A straightforward solution is to increase the number of measurements, which is shown to have multiple solutions. In order for the measurements to determine the signal uniquely, additional constraints are needed. In [281], three constraints are listed together with the condition for uniqueness and algorithms as follows.

1. Finite set: The basis is in a given finite set of possible bases.
2. Sparse basis: the columns of the basis are sparse under some known dictionary.
3. Structure: the basis is block diagonal and orthogonal.

In [282], the above ideas are employed in multichannel systems, in which the constraint is implied from the system structure. For a system such as microphone arrays or antenna arrays, the ensemble of signals can be constructed by dividing the signals from each channel into time intervals, and constructing each column of the signal as a concatenation of the signals from all the channels over the same time interval. In other words, we can show that the basis is block diagonal and orthogonal. Under certain conditions, it is shown that the blind CS algorithms can achieve results similar to those of standard CS, which rely on prior knowledge of the sparsity basis.

6.6.2 Adaptive algorithm

Many signal-processing applications require adaptive channel estimation with minimal complexity and small memory requirements. Most of the CS algorithms have to compute the results for each time instance, and do not consider the adaptive algorithms. In [283], a conversion procedure that turns greedy algorithms into adaptive schemes is established for sparse system identification. A Sparse Adaptive Orthogonal Matching

Pursuit (SpAdOMP) algorithm of linear complexity is developed to provide optimal performance guarantees. The steady-state mean square error (MSE) of the SpAdOMP algorithm is studied analytically. The developed algorithm is used to estimate Autoregressive moving-average (ARMA) and non-linear ARMA channels. In addition, in [326], sparse adaptive regularized algorithms based on Kalman filtering and expectation maximization are reported.

6.6.3 Group sparsity method

The methodology of group sparse CS (GSCS) [145] allows the efficient reconstruction of signals whose support is contained in the union of a small number of groups (sets) from a collection of predefined disjoint groups. Conventional CS is a special case of GSCS, with each group containing only a single element. Several CS recovery algorithms like basis pursuit denoising and orthogonal matching pursuit have been extended to the group sparse case [97, 284] and more details can be found in Chapter 4.

GSCS techniques are interesting in channel estimation, because the propagation paths of typical wireless channels are often structured in clusters. This is an additional physical reason why the channelòs delay-Doppler components occur in groups. The performance of compressive channel estimation tends to be limited by leakage effects that are caused by the finite bandwidth and block length, and that impair the effective delay-Doppler sparsity of doubly dispersive channels. In [285], the authors demonstrate that the leakage components in the channelòs delay-Doppler representation exhibit a group sparse structure, which can be exploited by the use of GSCS recovery techniques.

6.7 Summary

In summary, channel state information is of significant importance for high-speed wireless communication over a multipath channel both at transmitters and receivers. Traditional training-based methods usually probe the channel response in time, frequency and space domains using a sequence of known training symbols, and then the channel response is reconstructed from the received signals by linear reconstruction techniques. This method is known to be optimal for rich multipath channels. However, more and more physical measurements and experimental evidence indicate that many wireless channels encountered in practice tend to exhibit a sparse multipath structure. In this chapter, we present a new approach to estimating sparse (or effectively sparse) multipath channels that is based on recent advances in the theory of CS. The CS-based approaches can potentially achieve a target reconstruction error using far less energy and, in many instances, latency and bandwidth than that dictated by the traditional least-squares-based training methods.

7 Ultra-wideband systems

Ultra-wideband (UWB) is one of the major breakthroughs in the area of wireless communications, which is highly sparse in the time domain. Hence, it is natural to introduce CS in the design of UWB systems in order to improve the performance of UWB signal acquisition. In this chapter, we will discuss both the compression and reconstruction procedures in UWB systems, which serves as a good tutorial for how to apply CS in communication systems having sparsity. Note that channel estimation is also an important topic in UWB systems. However, since the CS-based channel estimation for general wireless communication systems has been discussed in Chapter 6, we omit the corresponding discussion in this chapter.

7.1 A brief introduction to UWB

7.1.1 History and applications

Although it has been studied since the late 1960s [327], the term "ultra-wideband" was not applied until around 1989. A major breakthrough of UWB was the invention of micropower impulse radar (MIR) that, for the first time, operated UWB signals at extremely low power and inexpensive hardware. According to the wide applications of UWB, the Federal Communications Commission (FCC) in the United States authorized the unlicensed use of UWB in the spectrum band of 3.1 GHz to 10.6 GHz in a report and order issued on February 14, 2002. The more detailed history of UWB can be found in [328].

Nowadays, UWB is widely applied in various scenarios. Two examples are as follows.

- Sensor networks [329]: It is well known that a bottleneck of wireless sensor networks is the limited power storage. Hence, it is very natural to use UWB devices having very low power consumption in sensor networks, particularly in scenarios requiring short-range communications, such as health care and factory equipment tracking. A typical UWB system specification is IEEE802.15.4.a published in 2004 whose physical layer can support transmission rates up to 27.24 Mb/s.
- Precise locationing [330]: Because of their extremely short pulse width, UWB signals can provide very precise timing information that is of key importance for precise locationing in applications such as robotic surgery. It is reported that the UWB-based locationing system in [330] can achieve the precision of centimeters.

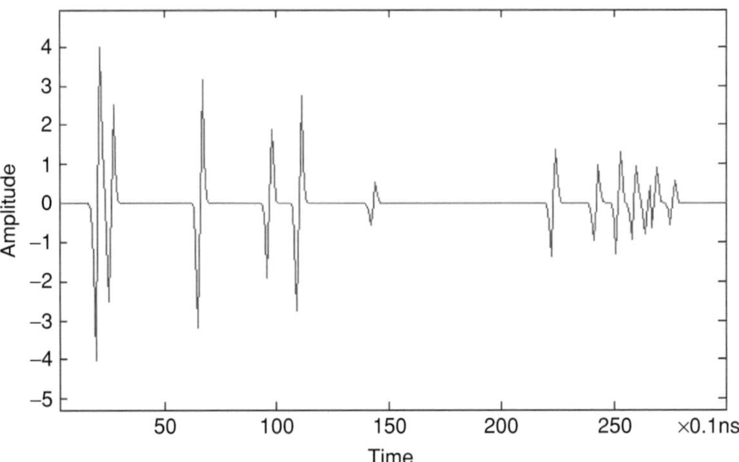

Figure 7.1 An illustration of UWB pulses.

7.1.2 Characteristics of UWB

UWB is characterized by its low power consumption, large bandwidth, short range and coexistence with licensed communication systems. From the viewpoint of receiver design, the major challenge of UWB is its narrow pulse width, which is illustrated in Figure 7.1. In typical UWB systems, a pulse could be of the order of nanoseconds; hence, the corresponding Nyquist frequency, which guarantees perfect reconstruction of the original signal from sampling, could be of the order of GHz. Analog-to-digital converters (ADCs) with such a high sampling frequency could be very expensive. Although it is possible to use an array of samplers with different time offsets to realize a serial sampling, the hardware cost increases with the number of samplers. For commercial applications, the UWB receiver may ignore the details of UWB signals, e.g., detecting only the signal energy and realizing the on-off keying (OOK) modulation. However, the missing information may cause significant performance loss. For example, the peak of an UWB pulse can be used to determine precise timing, thus obtaining precise locations. When energy detection is used, the information of the impulse peak is then lost, thus degrading the locationing precision. This also causes the loss of channel capacity in the scenario of communications. Hence, it is important to recover the fine structure of the UWB signal at the receiver, which motivates the application of CS.

7.1.3 Mathematical model of UWB

We denote by $h(t)$ the impulsive channel response and by $x(t)$ the transmit signal waveform. Then, the received signal at the receiver is given by

$$y(t) = x(t) \star h(t), \tag{7.1}$$

where $\star$ stands for convolution. Note that here we ignore the thermal noise and interference. If the channel is a multipath one; i.e., $h(t) = \sum_{l=0}^{L-1} h_l \delta(t - \tau_l)$, the received signal can be written as

$$y(t) = \sum_{l=1}^{L-1} h_l x(t - \tau_l), \tag{7.2}$$

where h_l and τ_l are the channel gain and delay of path l, respectively, and L is the total number of paths. A typical waveform for the transmitter is the Gaussian pulse, which is given by

$$x(t) = x_n(t)e^{-\frac{t^2}{2\sigma^2}}, \tag{7.3}$$

where $x_n(t)$ is a polynomial and σ^2 is a parameter controlling the pulse width.

7.2 Compression of UWB

In sharp contrast to signals such as images, the UWB signal is sparse in the time domain, while the sampling is also in the time domain. A low-rate sampler may miss the UWB signal in the time domain. Hence, it is necessary to compress the UWB signal first; i.e., "spreading" the information of the UWB signal over the time such that a few samples in the time domain can recover the original UWB signal.

Actually, the channel response can be considered as a compression. If we consider N time points in the transmit signal $x(t)$ and denote the corresponding samples by an N-vector $\mathbf{x}$, then the received signal can be written as

$$\mathbf{y} = \mathbf{H}\mathbf{x}, \tag{7.4}$$

where $\mathbf{y}$ is the vector of the received signal after sampling and the matrix $\mathbf{H}$ represents the channel response. We observe that (7.4) is a standard form of CS; i.e., reconstructing the sparse signal $\mathbf{x}$ from the measurement $\mathbf{y}$. However, the matrix $\mathbf{H}$ is completely determined by the channel and may not satisfy the conditions for signal reconstruction. In particular, when there are not many paths, $\mathbf{H}$ is also sparse. It is intuitive to see that CS does not work in this scenario. Hence, we need to add an artificial compression procedure for the UWB signal. It can be done on either the transmitter side or the receiver side, as we will explain in the following.

7.2.1 Transmitter side compression

One approach is to mix the UWB signal at the transmitter [331]. The advantage is that it can reduce the complexity of the receiver, if the receiver requires a simple design. An illustration of the UWB transmitter with the functionality of signal mixing is given in Figure 7.2. In this structure, the UWB impulses are generated according to the big sequence and then passed through a finite impulse response (FIR) filter. The filter output is sent through the wireless channel and then reaches the receiver. Suppose that the UWB pulse, together with the silent period, is denoted by an N-vector $\mathbf{x}$ in the discrete

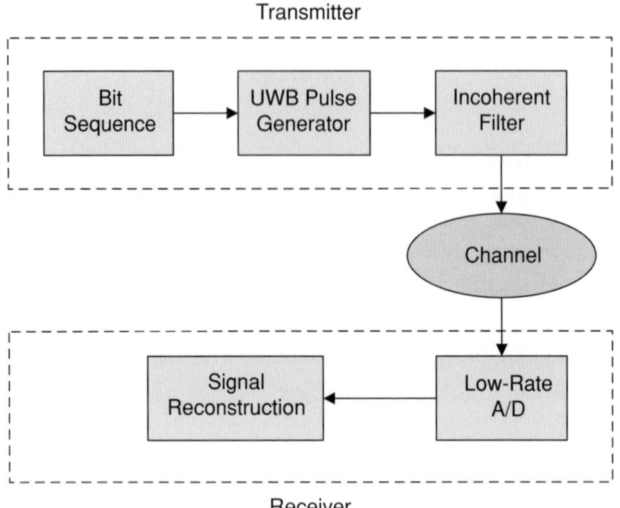

Figure 7.2 An illustration of the UWB transmitter structure with signal mixing.

time domain. Then, the received signal at the receiver is given by

$$\mathbf{y}_0 = \mathbf{f} \star \mathbf{h} \star \mathbf{x}, \tag{7.5}$$

where $\mathbf{f}$ is the discrete time impulse response of the FIR filter at the transmitter.

Because of the sampling-rate limitation, the vector $\mathbf{y}_0$ cannot be obtained directly. After downsampling to M uniform measurements, the signal after the sampler is given by

$$\mathbf{y} = \Phi \mathbf{x}, \tag{7.6}$$

where we ignore the quantization error, for simplicity of analysis. Note that the measurement matrix Φ is determined by the following detailed expression:

$$y(n) = \sum_{j=0}^{B-1} x\left(n \left\lfloor \frac{N}{M} \right\rfloor + j\right) (\mathbf{f} \star \mathbf{h})(B - j), \tag{7.7}$$

where B is the dispersion of the composite filter $\mathbf{f} \star \mathbf{h}$.

The relationship among the original signal $\mathbf{x}$, the measurement matrix Φ and the signal after downsampling $\mathbf{y}$ is illustrated in Figure 7.3. We observe that Φ is a quasi-Toeplitz matrix in which the non-zero elements are located within a strip and each column has B elements. Intuitively, we should design an FIR filter with a longer dispersion, namely, with a larger B, in order to make the measurement matrix more filled with non-zero elements and $\mathbf{y}$ more dispersed in the time domain.

As summarized in [331], the advantages of the proposed structure include:

- The mixing procedure can be implemented by a FIR filter.
- The dispersion B can be designed flexibly.
- The architecture can easily handle streaming signals.

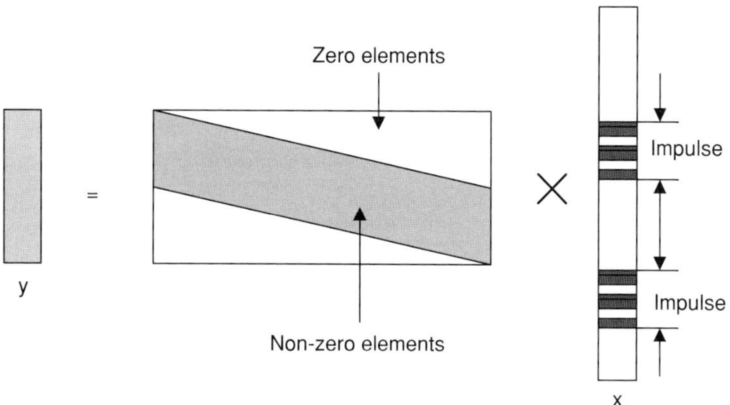

Figure 7.3 An illustration of the relationships between $\mathbf{x}$, Φ and $\mathbf{y}$.

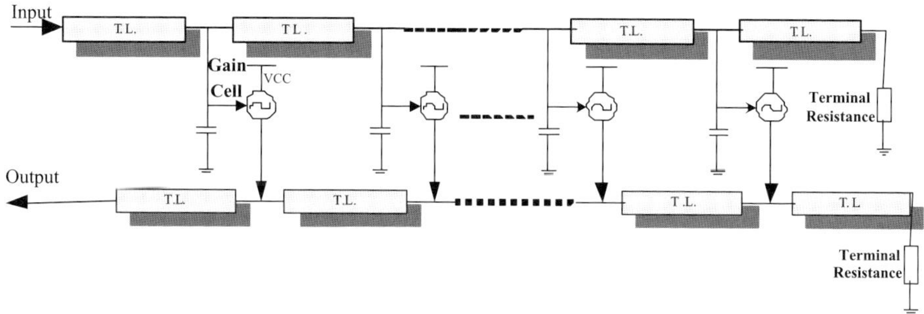

Figure 7.4 An illustration of the signal compression at the receiver.

7.2.2 Receiver side compression

The other approach for compressing the signal is to implement the compressor at the receiver as shown in Figure 7.4. Compared with the transmitter side signal compression, the advantage of receiver side compression is to reduce the complexity of the transmitter, which coincides with the purpose of simplifying the design and structure of transmitters. Hence, it is suitable for scenarios such as sensor networks, in which the transmitter requires a simple structure and low cost.

A possible receiver-side signal compression scheme is proposed in [332], which is based on distributed amplifiers (DAs). Essentially, the compression is accomplished by letting the UWB signal pass through a microstrip circuit, along which the signal is added with predetermined weights and then sampled.

Note that DA, also called a transversal filter, is a microwave circuit that has been used for many decades. A DA consists of multiple repeated taps, each containing a section of microstrip input and output transmission lines, and the gain cell. This periodic architecture forms a special transmission line [333, 334]. The output of a DA at time t

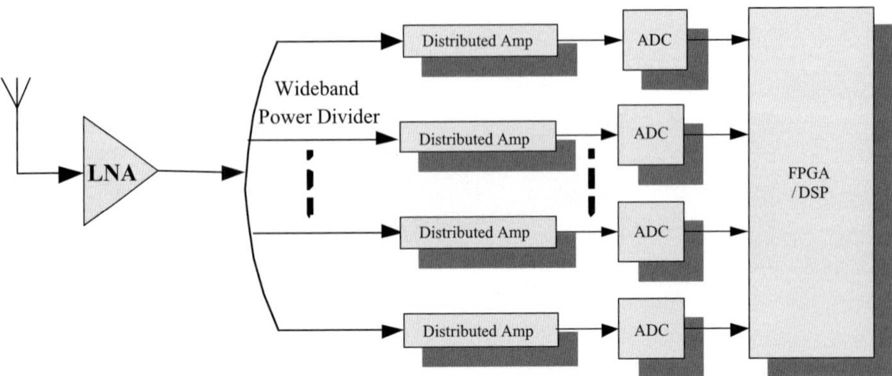

Figure 7.5 An illustration of the DA-based receiver structure.

is given by

$$y(t) = \sum_{j=1}^{m} a_j x(t - j\tau), \tag{7.8}$$

where m is the number of taps in a DA, $\{a_j\}_j$ are attenuation coefficients of different gain cells, which can be set randomly, x is the input signal and τ is the fixed time delay of each section of transmission line. Note that the attenuation coefficients and the time delay can be predetermined in the hardware design. For example, the coefficients can use random numbers, while the time delay can be uniform.

DA is suitable for compressing UWB signals due to the following reasons:

- The transmission line in DA supports UWB signal propagation with almost perfect impedance match. The characteristic impedance of a transmission line changes very little over several GHz, thus, the waveform of the propagating UWB signal is maintained.
- The time delays, determined by the length of the transmission line along which the signal propagates, can easily achieve a time scale of 50 ps, or less based on different substrates and technologies without changing the structure of DA [334].

The major issue of the proposed structure is the non-linearity of the gain cell in different frequency bands, which introduces polynomial multiplication terms. However, when considering small signals, the non-linear effect can be substantially alleviated.

Based on the proposed DA structure, we can obtain the linear relationship between the input, which is the original signal, and the output, which will be sampled with a lower sampling rate. As shown in Figure 7.5, the received UWB signal is put into the compressor with M DAs each having N gain cells, and then sampled by M ADCs. The output of the ith ADC is given by (here the noise is ignored)

$$y_i = \sum_{j=1}^{N} h_{ij} x_j, \tag{7.9}$$

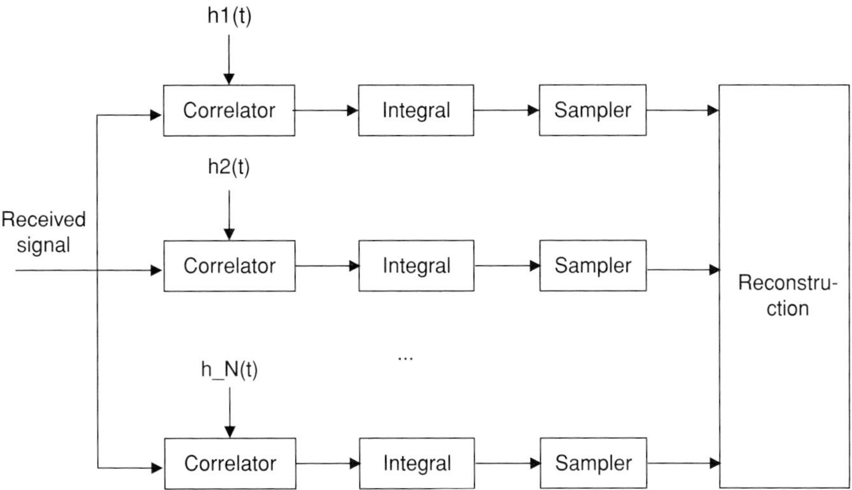

Figure 7.6 An illustration of the correlator-based receiver structure.

where $x_j = x(lT_s - j\tau)$, T_s is the sampling period of ADC and h_{ij} is the coefficient of the jth gain cell of the ith AD.

Defining $\mathbf{x} = (x_1, \ldots, x_N)$ and $\mathbf{y} = (y_1, \ldots, y_M)$, we obtain the linear compressing equation in (7.4), where the measurement matrix is given by

$$\mathbf{H} = \begin{pmatrix} h_{11} & h_{12} & \cdots & h_{1N} \\ h_{21} & h_{22} & \cdots & h_{2N} \\ \vdots & \vdots & \ddots & \vdots \\ h_{M1} & h_{M2} & \cdots & h_{MN} \end{pmatrix}. \tag{7.10}$$

The compression can also be realized by using an array of correlators shown in Figure 7.6, as proposed in [332]. In such a structure, the received signal $x(t)$ in the continuous time domain is passed through M analog correlators. The output of the mth correlator is given by

$$y_m = \int_0^T x(t)h_m(t)dt, \tag{7.11}$$

where T is the time duration of the received signal and $h_m(t)$ is the correlation function of the mth correlator.

Then, we can approximate the continuous time function using a series of characteristic functions that discretize the continuous time signal; i.e.,

$$x(t) \approx \sum_{n=1}^{N} x_n I\left(\frac{(n-1)T}{N} \leq t \leq \frac{nT}{N}\right). \tag{7.12}$$

We do the same approximation for $h_m(t)$; i.e.,

$$h_m(t) \approx \sum_{n=1}^{N} h_{mn} I\left(\frac{(n-1)T}{N} \leq t \leq \frac{nT}{N}\right). \tag{7.13}$$

Then, the integral can be approximated by summation, which is given by

$$y_m \approx \sum_{n=1}^{N} x_n h_{mn}. \tag{7.14}$$

When N is sufficiently large, the approximation can be very precise. Stacking the M outputs together, we have $\mathbf{y} = \mathbf{Hx}$, where $\mathbf{x} = (x_1, \ldots, x_N)$, $\mathbf{y} = (y_1, \ldots, y_N)$ and $(\mathbf{H})_{mn} = h_{mn}$.

7.3 Reconstruction of UWB

In this section, we discuss how to reconstruct the original UWB signal from the compressed version of signal obtained in the previous section. Many standard CS algorithms can be applied, such as the BP and OMP algorithms. However, these standard algorithms are designed for general systems. We can tailor the reconstruction algorithm according to the following features of UWB signals:

- Block structure: The UWB impulses arrive in a block manner. Hence, in the unknown vector $\mathbf{x}$, the non-zero elements are grouped in blocks. It is impossible to have the non-zero elements distributed in arbitrary locations of $\mathbf{x}$. Note that the traditional algorithms such as BP or OMP do not consider this a priori information. Hence, it can improve the reconstruction performance to take this feature into account.
- Noisy observations: Many traditional signal reconstruction algorithms do not assume noise or consider only weak noise in the compressed signal. In the context of UWB systems, the noise could be significant, which could come from thermal noise or co-channel interference. Without considering the disturbance incurred by noise, the signal reconstruction could fail.
- Possible collaborations: In many cases, it is possible for several receivers to obtain multiple replicas of the same UWB pulse. For example, in the case of UWB-based locationing, one UWB pulse will be received by several base stations that can communicate with each other. Hence, it may improve the performance of signal reconstruction via the collaboration of the receivers.

In the subsequent discussion, we will address the above features, either challenge or advantage, in UWB systems. All the algorithms could be integrated into a uniform protocol.

7.3.1 Block reconstruction

We first exploit the block property of UWB signals. For simplicity, we ignore the noise in the discussion. To interpret this property into mathematical language, we first define the block sparsity by following the definition in [284]. For simplicity, we rewrite the unknown vector $\mathbf{x}$ as

$$\mathbf{x} = \left(\mathbf{x}_1, \mathbf{x}_2, \ldots, \mathbf{x}_{\lceil \frac{N}{d} \rceil}\right), \tag{7.15}$$

where each $\mathbf{x}_k$, except the last one, has d elements. Each of the subvectors is called a block. Then, if at most k blocks have non-zero elements, we say that the vector $\mathbf{x}$ is block k-sparse.

For UWB signals, we can fix the block length d as the width of received UWB pulses (measured in discrete time), which could incorporate the multipath effect. Hence, the block sparsity of the received signal $\mathbf{x}$ is upper bounded by

$$K \triangleq \left\lceil \frac{N}{d} \right\rceil \leq 2 \left\lceil \frac{N}{d+g} \right\rceil, \tag{7.16}$$

where d is the width of received UWB pulses, g is the minimum gap between two pulses, and the scalar 2 is due to the possibility that one pulse crosses the boundary of two blocks. We define the norm of $\mathbf{x}$ as

$$\|\mathbf{x}\|_{2,0} = \sum_{k=1}^{K} I\left(\|\mathbf{x}_k\|_2 > 0\right), \tag{7.17}$$

where I is the indicator function, namely, the number of non-zero blocks. Note that we can also rewrite the measurement matrix $\mathbf{H}$ as

$$\mathbf{H} = \left(\mathbf{H}_1, \ldots, \mathbf{H}_K\right), \tag{7.18}$$

where the submatrix $\mathbf{H}_k$ corresponds to the elements in the block $\mathbf{x}_k$.

Two approaches have been proposed in [284] to solve the reconstruction of block sparse signals, namely, the l_2/l_1-optimization program (L-OPT) and the Block OMP/MP algorithm. We provide detailed explanations below.

L-OPT

The L-OPT algorithm minimizes a combination of l_1 and l_2 norms of the unknown vector $\mathbf{x}$, namely, the following optimization problem:

$$\begin{aligned} & \min_{\mathbf{x}} \sum_{k=1}^{K} |\mathbf{x}_k|_2 \\ s.t. \quad & \mathbf{y} = \mathbf{H}\mathbf{x}. \end{aligned} \tag{7.19}$$

Obviously, when $d = 1$, is the objective function the l_1-norm of $\mathbf{x}$; when $d = N$, the objective function is the l_2-norm of $\mathbf{x}$. For a general d, the objective function is something between the l_1-norm and l_2-norm of $\mathbf{x}$. On the one hand, the l_1-norm part utilizes the block sparsity property of $\mathbf{x}$; on the other hand, using the l_2-norm for each block $\mathbf{x}_k$ is because $\mathbf{x}_k$ itself is not necessarily sparse.

Note that the optimization problem in (7.19) can be rewritten as

$$\begin{aligned} &\min_{\{\mathbf{x}_k, t_k\}_{k=1,\ldots,K}} \sum_{k=1}^{K} t_k \\ s.t.\quad & \mathbf{y} = \mathbf{H}\mathbf{x}, \\ & t_i \geq \|\mathbf{x}_k\|_2, \qquad k = 1, \ldots, K, \\ & t_i \geq 0, \qquad k = 1, \ldots, K. \end{aligned} \tag{7.20}$$

This optimization problem can be solved by employing standard optimization packages.

Next, we discuss two properties of the L-OPT algorithm, namely, the condition for perfect reconstruction and the robustness to perturbations.

- Condition of perfect reconstruction: The authors in [284] proposed the Block-RIP condition; i.e., the matrix $\mathbf{H}$ is said to satisfy the Block-RIP condition if there exist parameters δ_k such that, for every $\mathbf{c}$ that is block k-sparse, we have

$$(1 - \delta_k)\|\mathbf{c}\|_2^2 \leq \|\mathbf{H}\mathbf{c}\|_2^2 \leq (1 + \delta_k)\|\mathbf{c}\|_2^2. \tag{7.21}$$

  Then, [284] showed that the optimization problem in (7.19) results in the unique solution $\mathbf{x}$ if the matrix $\mathbf{H}$ satisfies the Block-RIP condition with $\delta_{2k} < \sqrt{2} - 1$.
- Perturbations: We assume that the received signal is perturbed by some random factors such as noise or interference and then is given by

$$\mathbf{y} = \mathbf{H}\mathbf{x} + \mathbf{w}, \tag{7.22}$$

  where $\mathbf{w}$ is the random factor. We assume that the random perturbation is bounded; i.e., $\|\mathbf{w}\| \leq \epsilon$.

  Taking the unknown perturbation into account, the L-OPT algorithm can be revised into the following form:

$$\begin{aligned} &\min_{\mathbf{x}} \sum_{k=1}^{K} |\mathbf{x}_k|_2 \\ s.t.\quad & \|\mathbf{y} - \mathbf{H}\mathbf{x}\|_2 \leq \epsilon. \end{aligned} \tag{7.23}$$

  The impact of the perturbation $\mathbf{w}$ has been analyzed in [284]. Suppose that the original signal is $\mathbf{x}_0$. We denote by $\mathbf{x}^k$ the best block k-sparse approximation of $\mathbf{x}_0$. Here the "best" means that $\mathbf{x}^k$ minimizes $\|\mathbf{x}_0 - \mathbf{x}\|_{2,0}$ for all block k-sparse vector $\mathbf{x}$. We denote the solution to (7.23) by $\mathbf{x}^*$. If the matrix $\mathbf{H}$ satisfies the Block-RIP condition in (7.21) with $\delta_k < \sqrt{2} - 1$, then the following inequality has been shown in [284], which is given by

$$\|\mathbf{x}_0 - \mathbf{x}^*\|_2 \leq \frac{2(1 - \delta_{2k})}{1 - (1 + \sqrt{2})\delta_k} \frac{1}{\sqrt{k}} \|\mathbf{x}_0 - \mathbf{x}^k\|_{2,0} + \frac{4\sqrt{1 + \delta_k}}{1 - (1 + \sqrt{2})\delta_k}. \tag{7.24}$$

Block OMP and block MP

The key idea of block OMP is very similar to that of OMP. The only difference is to incorporate the feature of block sparsity in the UWB signal. It follows the following steps to minimize the residual:

1. Initialization: We initialize the residual as $\mathbf{r}_0 = \mathbf{y}$ and the block set $\mathcal{I}$ as the empty set, which is the same as that in OMP.
2. The lth iteration ($l \geq 1$):
   - Block selection: We choose the block that is closest to the residual $\mathbf{r}_{l-1}$; i.e.,

$$k_l = \arg\max_k \|\mathbf{H}_k^H \mathbf{r}_{l-1}\|_2. \tag{7.25}$$

   Include k_l in the set $\mathcal{I}$. Note that, when $d = 1$, this selection degenerates to the traditional OMP case.
   - Residual minimization: We choose $\{\mathbf{x}_k\}_{k\in\mathcal{I}}$ to minimize the residual, namely,

$$\mathbf{x}_k = \arg\min_{\{\mathbf{x}_k\}_{k\in\mathcal{I}}} \left\| \mathbf{y} - \sum_{k\in\mathcal{I}} \mathbf{H}_k \mathbf{x}_k \right\|. \tag{7.26}$$

   - Residual update: Then, we update the residual by subtracting the contribution from block $\mathbf{x}_k$, namely,

$$\mathbf{r}_l = \mathbf{y} - \sum_{k\in\mathcal{I}} \mathbf{H}_k \mathbf{x}_k. \tag{7.27}$$

The above approach can also be applied to extend from the MP algorithm, when the column vectors of $\mathbf{H}_k$ are orthogonal to each other. The block MP algorithm is different from the block OMP algorithm in the step of least-squares minimization in (7.26). It updates the residual vector directly, namely,

$$\mathbf{r}_l = \mathbf{r}_{l-1} - \mathbf{H}_{k_l} \mathbf{H}_{k_l}^H \mathbf{r}_{l-1}. \tag{7.28}$$

In [284], a sufficient condition to perfectly reconstructing the signal using Block-OMP has been found: if the matrix $\mathbf{H}$ satisfies

$$\rho_c \left(\mathbf{H}_0^T \bar{\mathbf{H}}_0 \right) < 1, \tag{7.29}$$

where $\mathbf{H}_0$ is the matrix containing the blocks in $\mathbf{H}$ corresponding to the non-zero blocks of $\mathbf{x}$, $\bar{\mathbf{H}}_0$ is the matrix containing the remaining blocks, and ρ_c is defined as

$$\rho_c(\mathbf{A}) = \max_r \sum_l \rho\left(\mathbf{A}_{l,r}\right), \tag{7.30}$$

where $\mathbf{A}_{l,r}$ is the (l, r)th block of matrix $\mathbf{A}$ and ρ is the spectral norm, namely, $\rho(\mathbf{A}) = \sqrt{\lambda_{\max} \mathbf{A}^H \mathbf{A}}$. Meanwhile, the block OMP algorithm finds a correct new block in each iteration. The iteration converges to the correct solution in k steps if $\mathbf{x}$ is block k-sparse.

It is not easy to directly verify the sufficient condition in (7.29). There are some other easier conditions proposed in [284]. To understand these conditions, we need to define the following concepts:

- Block-coherence: The block-coherence of $\mathbf{H}$ is defined as

$$\mu_B = \max_{l,r\neq l} \frac{1}{d}\rho\left(\mathbf{H}_l^T\mathbf{H}_r\right). \tag{7.31}$$

- Subcoherence: The subcoherence of $\mathbf{H}$ is defined as

$$\nu = \max_l \max_{i,j\neq i} \left|\mathbf{d}_i^H\mathbf{d}_j\right|, \qquad \mathbf{d}_i, \mathbf{d}_j \in \mathbf{H}_l. \tag{7.32}$$

Based on the above definitions, [284] has proved that (7.29) is satisfied if

$$kd < \frac{1}{2}\left(\mu_B^{-1} + d - (d-1)\frac{\nu}{\mu_B}\right). \tag{7.33}$$

In the special case in which the column vectors in $\mathbf{H}_k$ are orthogonal to each other, it is easy to verify that $\nu = 0$. Then, the condition in (7.33) is simplified to

$$kd < \frac{1}{2}\left(\mu_B^{-1} + d\right). \tag{7.34}$$

When the above orthogonal condition and the inequality in (7.34) satisfy, [284] has shown that the norm of the residual vector $\mathbf{r}_l$ decays exponentially; i.e.,

$$\|\mathbf{r}_l\|_2^2 \leq \beta^l \|\mathbf{r}_0\|_2^2, \tag{7.35}$$

where

$$\beta = 1 - \frac{1-(k-1)d\mu_B}{k}. \tag{7.36}$$

7.3.2 Bayesian reconstruction

In the previous discussion, we have assumed that the receiver is noise free or has bounded noise. This is reasonable for the case of high signal-to-noise ratio (SNR) or when the only source of noise is the quantization noise (note that quantization noise is bounded). However, in practical receivers, the noise could assume large values, even though this probability is small. Usually, thermal noise is modeled as Gaussian random variables. Assuming that the UWB signal has been compressed at the transmitter and ignoring the quantization noise, the received signal is given by

$$\mathbf{y} = \mathbf{H}\mathbf{x} + \mathbf{w}, \tag{7.37}$$

where $\mathbf{w}$ is a Gaussian random variable having expectation 0 and variance σ_n^2. We assume that the noise variance σ_n^2 is known in advance. The framework can also be extended to the case of unknown noise variance.

Again, $\mathbf{x}$ is sparse due to the characteristic of UWB signals.[1] Since the distribution of noise is assumed to be known, we will put the reconstruction in the framework of Bayesian CS [335, 336], which is significantly different from the approaches such as BP or OMP, due to its probabilistic essence. An illustration is given in Figure 7.7.

[1] It is interesting to integrate the block sparsity of UWB signals and Bayesian reconstruction. However, there has not been any study on this.

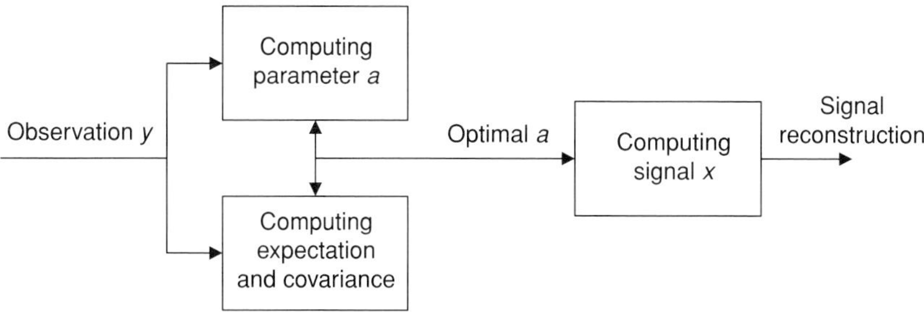

Figure 7.7 An illustration of Bayesian CS.

Now, we formulate the signal reconstruction in a probabilistic framework. We assume that each element in $\mathbf{x}$, e.g., x_n, satisfies a Gaussian distribution with expectation zero and variance a_n^{-1}. Furthermore, we assume that the elements in $\mathbf{x}$ are mutually independent; then, we have

$$P(\mathbf{x}|\mathbf{a}) = \prod_{n=1}^{N} \mathcal{N}\left(x_n|0, a_n^{-1}\right), \tag{7.38}$$

where $\mathcal{N}\left(x_n|0, \alpha_n^{-1}\right)$ means a Gaussian distribution with expectation zero and variance a_n^{-1}, and $\mathbf{a}$ is the set of all a_n. Note that the parameter α_n^{-1} is directly related to the sparsity of $\mathbf{x}$:

- When a_n is very large, the variance of x_n is very small. Since the expectation of x_n is 0, we can consider x_n as a zero element.
- When a_n is small, the variance x_n is large. Hence, x_n is highly probably a non-zero element.

By applying the Bayes's rule, we can compute the a posteriori probability of $\mathbf{x}$, which is given by

$$P(\mathbf{x}|\mathbf{y}, \mathbf{a}, \sigma_n^2) \sim \mathcal{N}(\mathbf{x}|\mu, \mathbf{\Sigma}), \tag{7.39}$$

where

$$\mu = \frac{1}{\sigma_n^2}\mathbf{\Sigma}\mathbf{H}^T\mathbf{y}, \tag{7.40}$$

and

$$\mathbf{\Sigma} = \left(\frac{1}{\sigma_n^2}\mathbf{H}^T\mathbf{H} + \mathbf{A}\right)^{-1}, \tag{7.41}$$

where $\mathbf{A} = \text{diag}(\mathbf{a})$. Once the expectation μ and covariance matrix $\mathbf{\Sigma}$ are known, we have various approaches to estimate $\mathbf{x}$. For example, we can adopt the maximum a posteriori (MAP) estimation and obtain the estimation given by

$$\hat{\mathbf{x}} = \mu. \tag{7.42}$$

However, the parameters in $\mathbf{a}$, or equivalently in the matrix $\mathbf{A}$, are unknown. Hence, the challenge of reconstructing $\mathbf{x}$ is converted to estimating the parameters in $\mathbf{a}$. Actually, when $\mathbf{a}$ is obtained, we can roughly know where the non-zero elements are and thus obtain $\mathbf{x}$ easily.

To obtain an estimation of $\mathbf{a}$, we can use the maximum likelihood (ML) estimation, which is given by

$$\begin{aligned}\hat{\mathbf{a}} &= \arg\max_{\mathbf{a}} P(\mathbf{y}|\mathbf{a}, \sigma_n^2) \\ &= \arg\max_{\mathbf{a}} \int P(\mathbf{y}|\mathbf{x}, \sigma_n^2) P(\mathbf{x}|\mathbf{a}) d\mathbf{x}. \end{aligned} \tag{7.43}$$

With some algebra, we can prove that the likelihood $P(\mathbf{y}|\mathbf{a}, \sigma_n^2)$ is given by

$$\begin{aligned}\int P(\mathbf{y}|\mathbf{x}, \sigma_n^2) P(\mathbf{x}|\mathbf{a}) d\mathbf{x} &= (2\pi)^{-N/2} \left|\sigma_n^2 \mathbf{I} + \mathbf{H}\mathbf{A}\mathbf{H}^T\right|^{-1/2} \\ &\times \exp\left\{-\frac{1}{2}\mathbf{y}^T \left(\sigma_n^2 \mathbf{I} + \mathbf{H}\mathbf{A}^{-1}\mathbf{H}\right)^{-1} \mathbf{y}\right\}. \end{aligned} \tag{7.44}$$

It is fortunate to have an explicit expression for the likelihood $P(\mathbf{y}|\mathbf{a}, \sigma_n^2)$. Unfortunately, there is no closed form for the optimal $\mathbf{a}$ that maximizes (7.44). In [335], an iterative algorithm is used to compute the optimal $\mathbf{a}$. By differentiating (7.44) with respect to $\mathbf{a}$ and equating to zero, we can obtain

$$a_n = \frac{\gamma_n}{\mu_n^2}, \tag{7.45}$$

where

$$\gamma_n = 1 - a_n \Sigma_{nn}. \tag{7.46}$$

Recall that μ_n and Σ_{nn} are the nth element in the expectation and covariance matrix in (7.40) and (7.41), respectively. It has been shown in [335] that, by iterating the equation in (7.45), the estimation of $\mathbf{a}$ converges to the optimal value.

In [330], the framework of Bayesian CS is applied in acquiring UWB signals used in precise positioning systems. This framework can be extended to incorporate the collaboration of multiple receivers receiving similar UWB pulses.

7.3.3 Computational issue

Since there is a delay requirement for the UWB signal acquirement, the signal reconstruction must be real time. Hence, the computation is of key importance to guarantee an online processing. In [330], the computational issue in the UWB signal reconstruction has been studied. A key difficulty for implementing the signal reconstruction is the least-squares problem (e.g., in the OMP algorithm or the Bayesian reconstruction), which is equivalent to solving the following problem:

$$\left(\mathbf{H}^T\mathbf{H}\right)\mathbf{x}_t = \mathbf{H}\mathbf{y}. \tag{7.47}$$

There could be many approaches to solve the above equation, e.g., the singular value decomposition (SVD) and Cholesky decomposition. In this book, we consider the Cholesky decomposition, which decomposes the symmetric matrix $\mathbf{H}^T\mathbf{H}$ into

$$\mathbf{H}^T\mathbf{H} = \mathbf{L}_t\mathbf{L}_t^T, \tag{7.48}$$

where $\mathbf{L}_t$ is a lower triangular matrix.

The advantage of choosing the Cholesky decomposition is that we can update the matrix $\mathbf{L}_t$ incrementally and reuse many intermediate computation results. To see this, we have

$$\mathbf{H}_t^T\mathbf{H}_t = \begin{pmatrix} \mathbf{H}_{t-1}^T\mathbf{H}_{t-1} & \mathbf{h}_t \\ \mathbf{h}_t^T & g \end{pmatrix}, \tag{7.49}$$

where t is the iteration index and $\left(\mathbf{h}_t^T, g\right)^T$ is the new column in matrix $\mathbf{H}_t^T\mathbf{H}_t$, compared with the smaller matrix $\mathbf{H}_{t-1}^T\mathbf{H}_{t-1}$. It is easy to verify

$$\mathbf{L}_t\mathbf{L}_t^T = \begin{pmatrix} \mathbf{L}_{t-1} & 0 \\ \mathbf{w}_t^T & v \end{pmatrix} \begin{pmatrix} \mathbf{L}_{t-1}^T & \mathbf{w}_t \\ 0 & v \end{pmatrix}. \tag{7.50}$$

Obviously, we need only compute the new vector $\mathbf{w}$ by solving the equation given by

$$\mathbf{L}_{t-1}\mathbf{w}_t = \mathbf{h}_t, \tag{7.51}$$

where the scalar $v = \sqrt{g}$.

To avoid too many operations of divisions and to utilize the advantage of pipelining in FPGA structure, we modify the Cholesky decomposition to an LDL decomposition, i.e.,

$$\mathbf{H}_t^T\mathbf{H}_t = \mathbf{L}_t\mathbf{\Sigma}_t\mathbf{L}_t^T, \tag{7.52}$$

where $\mathbf{\Sigma}_t$ is a diagonal matrix and $\mathbf{L}_t$ is a lower triangular matrix with unit diagonal elements. The updates of $\boldsymbol{L}_t$ and $\mathbf{\Sigma}_t$ are similar to the above discussions.

A pipelining structure as shown in Figure 7.8 can be used for the least-squares problem in the UWB signal reconstruction based on the above discussion. Take a four-dimensional equation for instance; the equation is equivalent to

$$\begin{pmatrix} 1 & 0 & 0 & 0 \\ h_{21} & 1 & 0 & 0 \\ h_{31} & h_{32} & 1 & 0 \\ h_{41} & h_{42} & h_{43} & 1 \end{pmatrix} \mathbf{\Sigma} \begin{pmatrix} x_1 \\ x_2 \\ x_3 \\ x_4 \end{pmatrix} = \begin{pmatrix} y_1 \\ y_2 \\ y_3 \\ y_4 \end{pmatrix}. \tag{7.53}$$

We can solve the equation using pipelining within the following four clocks.

- **Clock 1**: Get $r_1 = y_1$ from the RAM; compute

$$r_2 = y_2 - a_{21}r_1$$

$$r_3 = y_3 - a_{31}r_1$$

$$r_4 = y_4 - a_{41}r_1,$$

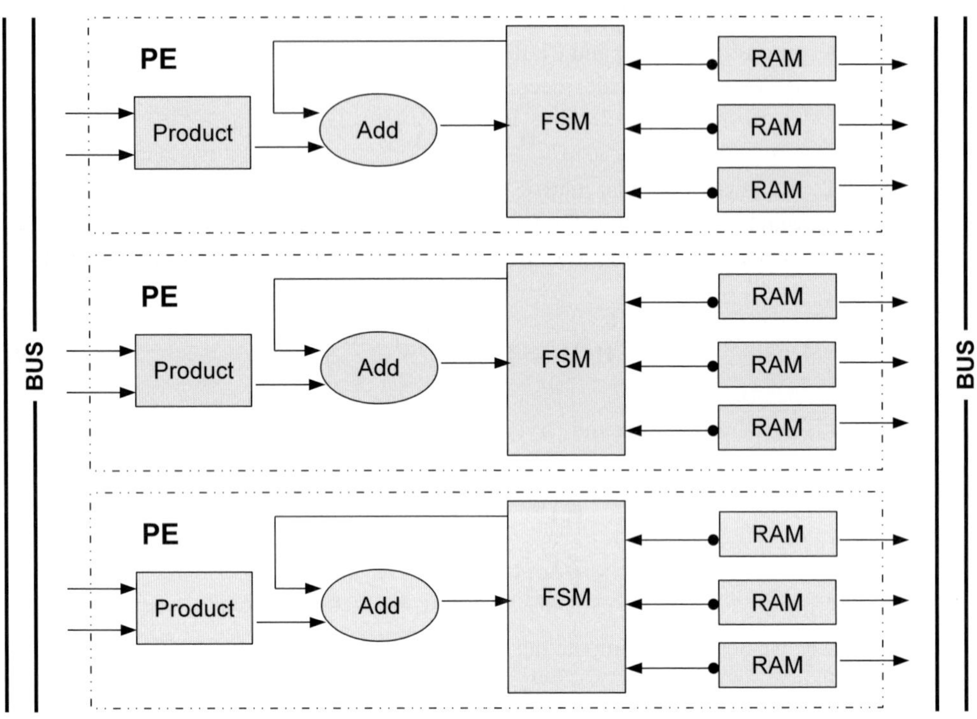

Figure 7.8 Pipelining structure of the UWB signal reconstruction.

and

$$x_1 = r_1/\Sigma_{11}.$$

- **Clock 2**: Get r_2 from the RAM; compute

$$r_3 = r_3 - a_{32}r_2,$$
$$r_4 = r_4 - a_{42}r_2$$

and

$$x_2 = r_2/\Sigma_{22}.$$

- **Clock 3**: Get r_3 from the RAM; compute

$$r_4 = r_4 - a_{43}r_3$$

and

$$x_3 = r_3/\Sigma_{33}.$$

- **Clock 4**: Get r_4 from the RAM; compute

$$x_4 = r_4/\Sigma_{44}.$$

7.4 Direct demodulation in UWB communications

In the previous sections, we considered general UWB signals and used CS to reconstruct the original signals. However, in UWB communication systems, the purpose of the receiver is not to reconstruct the received waveform, since the eventual goal is to recover the transmitted information bits. Hence, an efficient way is to carry out demodulation directly, without obtaining the original waveform. In this section, we introduce the demodulation algorithm for CS-based UWB communication systems proposed in [337]. Note that it does not apply in UWB positioning systems since the original waveform is of key importance for locationing.

7.4.1 Transceiver structures

We first introduce the structures of transmitter and receiver in the UWB communication system considered in [337]. Suppose the transmitter adopts the binary phase shift keying (BPSK) in which the information symbols are either +1 or –1. Note that it is easy to extend to general modulations such as M-PSK or QAM. The principle of the bit reconstruction, which will be discussed later, can also be extended to the case in which the information bits arrive in random blocks. Consider K bits that are conveyed by K UWB pulses. The waveform of each UWB pulse is denoted by $\phi(t)$. Then the transmitted signal is given by

$$S(t) = \sum_{k=0}^{K-1} b_k \phi(t - kT), \tag{7.54}$$

where b_k is the kth symbol (+1 or –1 when there is an information bit, or 0 when there is no information bit), T is the time interval between two pulses.

Then, the receiver receives the signal given by

$$R(t) = \sum_{k=0}^{K-1} b_k g(t - kT) + W(t), \tag{7.55}$$

where $g(t)$ is the waveform of each UWB pulse after passing the communication channel, which could consist of the replicas along multiple paths, and $W(t)$ is the noise. The received signal is then compressed. In [332], the signal compression is carried out via the correlator. The M outputs of the correlator array, $y_1, \ldots, y_M$, are used to recover the K information bits. Note that the outputs $\mathbf{y} = (y_1, \ldots, y_M)$ can be written as a linear transform of a virtual discrete time information signal $\mathbf{x} = (x_0, \ldots, x_{N-1})$; i.e.,

$$\mathbf{y} = \mathbf{H}\mathbf{x} + \mathbf{w}, \tag{7.56}$$

where N is the total number of time intervals for the transmitted signal and each time interval is approximately the width of one pulse. It is assumed that there exists an uncertainty of the time of arrival, denoted by Γ. Then, we have

$$N = \Gamma + (K - 1)N_{baud}, \tag{7.57}$$

where N_{band} is the number of time intervals between two consecutive UWB pulses (more detailed analysis can be found in [337]). Obviously, $\mathbf{x}$ can only take values in $\{+1, -1, 0\}^N$.

7.4.2 Demodulation

Equation (7.56) looks to be a standard CS problem since the observation $\mathbf{y}$ is a linear transformation of the symbol sequence $\mathbf{x}$, if we ignore the noise $\mathbf{w}$. However, the difference is that the elements in $\mathbf{x}$ can only take discrete values (+1, –1 or 0), while in standard CS the non-zero elements in $\mathbf{x}$ are continuous. The signal reconstruction approaches introduced in the previous sections can still be applied to reconstruct $\mathbf{x}$ by considering $\mathbf{x}$ as a continuous vector. However, the direct application of standard CS ignores the a priori information that $\mathbf{x}$ must be discrete; hence, there could be a performance loss.

In [337], the reconstruction of $\mathbf{x}$, or equivalently the demodulation, is carried out by incorporating the binary value of elements in $\mathbf{x}$. The following two approaches are proposed for the demodulation.

- Maximum likelihood (ML) demodulation: In this approach, it is assumed that the noise $\mathbf{w}$ is an additive Gaussian one with covariance matrix Σ, which can be obtained from the correlator setup and the covariance of noise $W(t)$ (the details can be found in [337]). Then, the conditional probability of $\mathbf{y}$ given $\mathbf{x}$ (or the likelihood) is given by

$$P(\mathbf{y}|\mathbf{x}) = \exp\left\{-(\mathbf{y} - \mathbf{H}\mathbf{x})^T \Sigma^{-1} (\mathbf{y} - \mathbf{H}\mathbf{x})\right\}. \tag{7.58}$$

  Thus, according to the ML criterion, $\mathbf{x}$ can be chosen to maximize the likelihood in (7.58); i.e.,

$$\mathbf{x}_{ML} = \arg \min_{\mathbf{x}\in\{+1,-1,0\}^N} (\mathbf{y} - \mathbf{H}\mathbf{x})^T \Sigma^{-1} (\mathbf{y} - \mathbf{H}\mathbf{x}). \tag{7.59}$$

  Note that the ML demodulation can minimize the BER. However, when N and K are large, it is difficult to find the optimal $\mathbf{x}$ that maximizes the likelihood. Hence, we need to devise suboptimal but computationally feasible approaches for the demodulation.
- Suboptimal demodulation: To make the demodulation computationally efficient, a suboptimal demodulation algorithm is proposed in [337]. The key idea is to relax the constraint $\mathbf{x} \in \{+1, -1, 0\}^N$ to $\mathbf{x} \in \mathcal{R}^N$ while incorporating the *prior* information about the non-zero elements locations into account. To that end, a penalty is defined as

$$\xi_n = \begin{cases} 1, & n \in \Omega \\ \text{a large number}, & \text{otherwise} \end{cases}, \tag{7.60}$$

  where Ω is the set of possible non-zero elements. In [337], Ω is defined as

$$\Omega = \{n | n = l + kN_{band}, l \in [a + l_1, a + l_2], k \in \mathcal{N}\}, \tag{7.61}$$

  where a, l_1 and l_2 are parameters such that $[a + l_1, a + l_2]$ represents the ranges of the initial time of arrival.

Let us assume that the parameters a, l_1 and l_2 are determined; i.e., the range of the arrival time is fixed. Then, the optimization problem in (7.59) is reformulated as

$$\hat{\mathbf{x}} = \arg \min_{\|\Gamma \mathbf{x}\|_1 = K, \mathbf{x} \in \mathcal{R}^N} (\mathbf{y} - \mathbf{H}\mathbf{x})^T \Sigma^{-1} (\mathbf{y} - \mathbf{H}\mathbf{x}), \tag{7.62}$$

where $\Gamma = \text{diag}\{\xi_n\}$. Note that here we know the exact value of K. If not, we can replace the constraint $\|\Gamma \mathbf{x}\|_1 = K$ with $\|\Gamma \mathbf{x}\|_1 \leq K_{upp}$, where K_{upp} is an upper bound for K. An intuitive explanation for the constraint $\|\Gamma \mathbf{x}\|_1 = K$ is that no elements out of the possible set Ω are assigned non-zero values.

It is easy to verify that (7.62) can be rewritten as [337]

$$\hat{\mathbf{x}} = \hat{\mathbf{z}}_n - \hat{\mathbf{z}}_{n+N}, \qquad n = 0, 1, 2, \ldots, N, \tag{7.63}$$

where

$$\hat{\mathbf{z}} = \min_{\mathbf{z} \geq 0, [\xi^T, \xi^T]\mathbf{z} = K} \mathbf{f}^T \mathbf{z} + \frac{1}{2} \mathbf{z}^T \mathbf{Q} \mathbf{z}, \tag{7.64}$$

and

$$\mathbf{Q} = \begin{pmatrix} \mathbf{H}^T \Sigma^{-1} \mathbf{H} & -\mathbf{H}^T \Sigma^{-1} \mathbf{H} \\ -\mathbf{H}^T \Sigma^{-1} \mathbf{H} & \mathbf{H}^T \Sigma^{-1} \mathbf{H} \end{pmatrix}, \tag{7.65}$$

and

$$\mathbf{f} = \left(\mathbf{y}^T \boldsymbol{\Sigma}^{-1} \mathbf{H}, \mathbf{y}^T \boldsymbol{\Sigma}^{-1} \mathbf{H} \right). \tag{7.66}$$

Obviously, the above optimization is essentially a Quadratic Programming (QP) problem, which can be efficiently solved with a time complexity $O(K^3)$. However, the optimization is based on known parameters a, l_1 and l_2. Below, the following algorithm is applied to refine the estimation of the arrive time interval and thus estimate the parameters a, l_1 and l_2 [337].

- Step 1: Initialize $a = 0$, $l_1 = 0$ and $l_2 = \Gamma$. Carry out the QP optimization in (7.63). The result is denoted by $\mathbf{x}^1$.
- Step 2: We estimate the beginning time using

$$\hat{t} = \arg \max_{n' \in 0,1,2,\ldots,\Gamma} \sum_n \left| \hat{x}^1(n - n') \right| \xi(0, 0, 0)(n), \tag{7.67}$$

  which means to find an optimal time offset that maximizes the correlation between $\hat{x}^1$ and the template $\xi(0, 0, 0)$.
- Step 3: We set $a = \hat{t}$, $l_1 = 0$ and $l_2 = 0$, and carry out the QP optimization in (7.63) again. The result $\mathbf{x}^2$ is used to generate the demodulation result; i.e.,

$$\hat{b}_k = \hat{x}^2(\hat{t} + kN_{baud}). \tag{7.68}$$

Note that the above reconstruction algorithm for demodulation is significantly different from traditional CS algorithms such as BP or OMP. The reason for this is that the uncertainty of the non-zero elements in the context of demodulation lies in the unknown arrival time. Once the timing of the first impulse is obtained, the locations of all non-zero elements are known. This makes the task of demodulation here easier than the classical

CS. However, if the data arrivals are random and the transmitter ceases transmitting when there is no data (i.e., $b_k = 0$ means no data; otherwise, b_k equals +1 or –1), there will be much more uncertainty in the locations of non-zero elements, thus making the task much more challenging.

7.5 Conclusions

In this chapter, we have discussed the application of CS in UWB systems. The key feature of a UWB system is the narrow pulse width in the time domain, which makes the direct sampling in the time domain very difficult. Hence, we can compress the UWB signal such that it is dispersed in the time domain and fewer samples can reconstruct the original UWB signal. The signal compression can be carried out either at the transmitter or at the receiver, as follows:

- At the transmitter, we can use a filter to compress the UWB signal.
- At the receiver, we can use either a microstrip circuit or an array of correlators to complete the signal compression.

Once the UWB signal is compressed, we can reconstruct the original UWB signal from the samples. Traditional reconstruction algorithms in CS, such as BP or OMP, can be applied directly. However, it is better to take the features of UWB signals into account. Hence, we have discussed the following two approaches as being suitable for UWB systems:

- Block CS, in which we fully utilize the feature of blocked locations of non-zero elements in UWB signals.
- Bayesian CS, in which the Gaussian noise is taken into account.

We have also discussed a special case of UWB communications with uncertain arrival times. In this context, the signal reconstruction can be formulated as a QP problem and can then be solved efficiently.

8 Positioning

8.1 Introduction to positioning

In this chapter, we discuss the application of CS in the task of positioning. It is easy to imagine that there are many applications of geographical positioning. For example, in a battlefield, it is very important to know the location of a soldier or a tank. In cellular networks, the location of the mobile user can be used for Emergency-911 services. For goods and item tracking, it is of key importance to track their locations. The precision of positions also ranges from subcentimeters (e.g., robotic surgery) to tens of meters (e.g., bus information).

Several classifications of positioning technologies are given below [338]:

- Classified by signaling scheme: In positioning, the target needs to send out signals to base stations or receive signals from base stations in order to determine the target's position. Essentially, the signal needs to be wireless. Radio-frequency (RF), infrared or optical signals can be used.
- Classified by RF waveforms: Various RF signals can be used for positioning, such as UWB, CDMA and OFDM.
- Classified by positioning-related metrics: The metrics include time of arrival (TOA), time difference of arrival (TDOA), angle of arrival (AOA) and received signal strength (RSS).
- Classified by positioning algorithm: When triangulation-based algorithms are used, the positioning is obtained from the intersections of lines, based on metrics such as AOA. In trilateration-based algorithms, the position is obtained from the intersections of circles, based on metrics such as TDOA or TOA. In fingerprinting-based (also called pattern-matching) algorithms, a training period will be spent to establish a mapping between the location and the received signal fingerprinting (or pattern).

CS can be used in various types of positioning systems. The application of CS is classified into the following two categories:

- Direct application: In this category, the positioning problem is formulated as a standard CS one. The key challenge is how to find a linear transformation between the position information and observation, as well as how to find the sparsity.

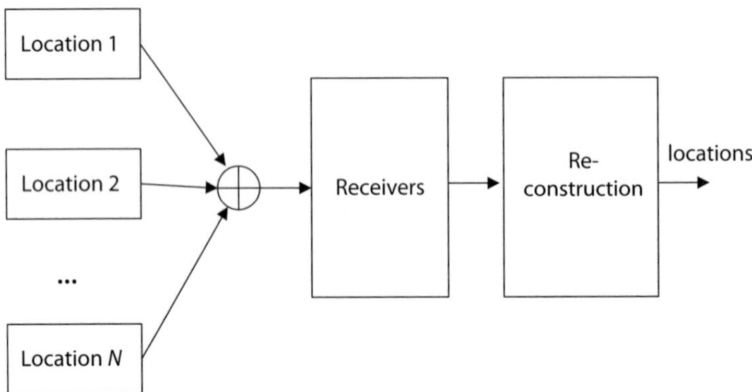

Figure 8.1 An illustration of the direct application of CS in positioning.

- Indirect application: The positioning is not formulated as a CS problem. Here, the CS is used to improve the precision of signal acquisition, thus increasing the accuracy of positioning. However, the CS for signal acquisition can be jointly carried out with the positioning algorithm.

In this chapter, we will use three examples, namely, the positioning in WLAN, cognitive radio and UWB systems, to illustrate the application of CS. The first two belong to the direct application, while the latter is an indirect one. These applications also involve various types of positioning systems introduced above; e.g., the positioning in WLAN uses RF signal with RSS base fingerprinting.

8.2 Direct application of compressive sensing

In this section, we introduce the direct application of CS for the purpose of positioning of multiple targets. We first introduce the general principle of directly applying CS for positioning. Then, we discuss how to apply this general principle in the contexts of wireless local area networks (WLAN) and cognitive radio networks. Note that the number of targets is unknown. However, we can assume that the number of targets is small, compared with the possible positions that the targets could be located. This assumption implies the sparsity of signal and thus facilitates the application of CS.

8.2.1 General principle

The general principle of applying CS for positioning is illustrated in Figure 8.1. First, we assume that there are L receivers receiving the signals for positioning. Then, we need to partition the region in which the targets are located. For example, we can divide the region into N subregions. If a target lies in a subregion, we can assume that the location of the target is the center of the subregion. The larger N is, the more precise the location is; meanwhile, a large N also increases the requirement of the number of receivers. Hence, a tradeoff should be considered between the positioning precision and the cost.

The following two assumptions are needed for the CS-based positioning:

- Fingerprinting: We assume that, if a target is located at subregion n, the receivers can receive an M-vector, coined *fingerprinting vector*, denoted by $\mathbf{f}_n$. We assume that the set of fingerprinting vectors are all known to the system.
- Signal superposition: We assume that the fingerprinting vectors from multiple targets are additively superimposed at the receivers. Hence, the received signal is given by

$$\begin{aligned} \mathbf{r} &= \sum_{n=1}^{N} a_n \mathbf{f}_n \\ &= \mathbf{Fa}, \end{aligned} \tag{8.1}$$

where a_n is the amplitude of the target located in subregion n (if $a_n = 0$, there is no target in the subregion), $\mathbf{F} = (\mathbf{f}_1, \ldots, \mathbf{f}_n)$ and $\mathbf{a} = (a_1, \ldots, a_N)^T$.

Based on the above two assumptions, we obtain the standard form of CS in (8.1): a known L-vector $\mathbf{r}$, a known observation matrix $\mathbf{F}$ and an unknown N-vector $\mathbf{a}$. The assumption of few targets implies the sparsity of $\mathbf{a}$, which facilitates the application of CS. Once we obtain the vector $\mathbf{a}$, the non-zero elements in $\mathbf{a}$ indicate the locations of the targets and the amplitudes bring a bonus to the positioning.

The same argument can also be applied in the case that the target wants to determine its own location. At this time, the L receivers becomes L transmitters. The target receives the signals from the L transmitters and then carries out the positioning task using (8.1), where $\mathbf{a}$ contains the information of its own position.

8.2.2 Positioning in WLAN

The generic principle discussed above can be applied in the positioning in WLAN, which has been studied in [339]. We assume that there are L access points (APs) in the WLAN. A user carrying a mobile device that can measure the radio signal strength (RSS). It is also assumed that there are N reference points (RPs). The task is to find how a mobile device estimates its own location using the readings of RSS from the APs.

The positioning in WLAN has two stages, namely offline and online stages, which are illustrated in Figure 8.2:

- Offline stage: In this stage, the radio map, which will be introduced soon, is estimated and maintained.
- Online stage: Based on the radio map, the mobile device first has a coarse estimation of its own position. Then, CS is applied to refine the positioning.

The details of both stages will be introduced soon.

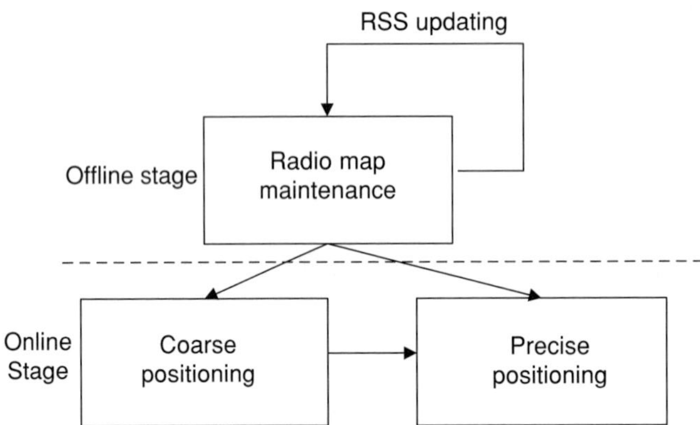

Figure 8.2 Procedure of positioning in WLAN.

Radio map

For such a scenario, the radio map is defined as

$$\boldsymbol{\Psi} = \begin{pmatrix} \psi_{11} & \psi_{12} & \cdots & \psi_{1N} \\ \psi_{21} & \psi_{22} & \cdots & \psi_{2N} \\ \vdots & \vdots & \ddots & \vdots \\ \psi_{L1} & \psi_{L2} & \cdots & \psi_{LN} \end{pmatrix}, \tag{8.2}$$

where ψ_{ij} is the average RSS reading of AP i for RP j. We denote by ψ_n the nth column of the matrix $\boldsymbol{\Psi}$, namely, the RSS readings at RP n. The matrix Ψ is known to the positioning system. Moreover, it is assumed that the variance of RSS of AP i at RP j, denoted by Δ_{ij} is also known. The online maintenance of the information is discussed in detail in [339].

After obtaining the radio map, a clustering procedure will be carried out in order to reduce the number of RSS fingerprinting vectors, since the number of APs is usually very large. The set of clusters is denoted by $\mathcal{H}$ and each cluster is represented by an exemplar RP. The cluster represented by RP j is denoted by $\mathcal{C}_j$.

Coarse positioning

When the mobile device wants to find its own position, it measures the RSS readings from the APs, which are denoted by $\mathbf{r}$. It first carries out a coarse positioning by comparing the distance between the RSS reading $\mathbf{r}$ and the exemplar readings. The set of possible RPs is given by

$$\mathcal{C} = \bigcup_{j \in \mathcal{S}} \mathcal{C}_j, \tag{8.3}$$

where

$$\mathcal{S} = \{j | d(r, j) \geq \alpha, j \in \mathcal{H}\}, \tag{8.4}$$

where $d(r, j)$ is the distance between $\mathbf{r}$ and ψ_j and α is a predetermined threshold. Intuitively, the set $\mathcal{S}$ collects the RP exemplars that are close to the RSS reading of the mobile device; moreover, $\mathcal{C}$ consists of the RPs in the corresponding clusters. In [339], the threshold α is set to be a high fixed percentage of the exemplars, which is given by

$$\alpha = 0.95 \max_{l \in \mathcal{H}} \left\{ d(r, l) - \min_{j \in \mathcal{H}} d(r, j) \right\}. \tag{8.5}$$

Fine positioning

Based on the coarse positioning, we use CS to obtain a finer estimation of the position. In the coarse positioning, we have narrowed the scope of the possible RPs to the set $\mathcal{C}$. Then, the RSS reading can be written as

$$\mathbf{r} = \tilde{\Psi}\mathbf{x} + \mathbf{n}, \tag{8.6}$$

where $\mathbf{n}$ is some measurement noise, $\tilde{\boldsymbol{\Psi}}$ is the submatrix of $\boldsymbol{\Psi}$ corresponding to the RPs in the the set $\mathcal{C}$, and $\mathbf{x}$ is the unknown vector indicating the position of the mobile device. In the ideal case (i.e., the RSS reading should be exactly the same as one of the readings at the N RPs), $\mathbf{x}$ should have a single 1 at the dimension corresponding to the correct RP, and all other elements should be zero. In [339], a mechanism of AP selection is also used to reduce the dimension of $\mathbf{r}$ and $\tilde{\boldsymbol{\Psi}}$ since the number of AP could be very large and thus incurs significant computational cost. The following two criteria are applied to select the APs:

- Strongest APs: The set of APs that have the largest RSS readings should be selected. The main purpose is to increase the SNR.
- Fisher's criterion: For each AP in $\mathcal{C}$, the following metric is computed:

$$\xi_i = \frac{\sum_{j \in \mathcal{C}} \left(\psi_{i,j} - \frac{1}{N} \sum_{j \in \mathcal{C}} \psi_{i,j} \right)^2}{\sum_{j \in \mathcal{C}} \Delta_{ij}}, \tag{8.7}$$

  where Δ_{ij} is the corresponding variance. The higher the metric is, the greater is the discrimination capability of the AP for the location of the mobile device. Hence, APs with higher metrics should be incorporated.

After selecting the APs, (8.6) is solved using various methods in CS. In the ideal case, only one element in $\mathbf{x}$ should be non-zero. However, because of the limited number of RPs and the difference between the actual position of the mobile device and the location where the RSS readings are obtained for the RP, as well as measurement noise, $\mathbf{x}$ has multiple non-zero elements. Then, the final position estimation is obtained by averaging the positions of the RPs corresponding to the non-zero elements in $\mathbf{x}$.

Experiment results

In [339], the proposed positioning scheme is tested in a 12 m $\times$ 36 m area in a building at the University of Toronto. There are a total of 17 APs. The area is partitioned into 118 RPs approximately formed from a 1-m spacing grid. Twenty seconds are spent on measuring the RSS readings and forming the radio map. In the offline training stage, 10 clusters are formed for the 118 RPs. In the experiment, a PDA is used as the mobile device. The SNR is 10 dB for the RSS readings. In the stage of coarse positioning, the average positioning error is less than 2 m. The error decreases if more clusters are formed in the offline stage. Meanwhile, the error also decreases if more APs are used in the positioning; however, the performance almost saturates when more than 5 APs are used. In the stage of fine positioning, the positioning error can be reduced to less than 1 m if sufficient APs are used. The positioning experiment is also carried out in a public area. With 12 APs, the positioning error can be less than 2 m.

A very similar positioning algorithm is proposed in [340]. The corresponding experiment is carried out in the building 21 of INRIA, Paris. IEEE802.11 APs are used for the positioning. The area is divided into 0.76 m × 0.76 m cells. In each cell, averagely 5 APs are used for the RSS readings. In total, 32 RPs are tested in the experiment. The Bayesian CS [340] is used in the reconstruction. Experimental results show that the positioning error is around 1.4 m, which is similar to that in [340].

8.2.3 Positioning in cognitive radio

As has been discussed in previous chapters, a key task in cognitive radio system is spectrum sensing; i.e., how to determine whether primary users exist. In a traditional spectrum sensing algorithm, only the existence of primary users is considered, while the positions of the primary users are not obtained. However, the information of primary-user positions can further improve the performance of cognitive radio. For example, the position information can be broadcast throughout the secondary users. Then, secondary users can combine the geographical information (e.g., the secondary user has a GPS and can determine its own position) with the spectrum-sensing result, thus obtaining a higher precision of the detection of primary users; or the spectrum sensing can be carried out less frequently, which can significantly reduce the overhead spent on spectrum sensing, since the secondary users can use the locations to determine whether they will cause significant interference to the primary users.

Static primary users

We first consider the case of primary users with static positions by following the model and analysis in [341]. As illustrated in Figure 8.3, we assume that there are an unknown number of primary users within an area, whose locations are also unknown. There are totally N_{ch} channels. Each primary user occupies one distinct channel. N_c RF sensors are distributed in this area, whose measurements will be sent to a fusion center. The task is to detect the existence of primary users in different channels, their transmitting power and their locations.

For the purpose of joint spectrum sensing and positioning, the following path-loss model of a certain channel is used:

$$L(f, d) = P_0 + 2\log_{10}(f) + 10\alpha\log_{10}(d)(dB), \tag{8.8}$$

where $L(f, d)$ is the path loss, P_0 is a constant dependent on antenna gain, f is the center frequency of the corresponding channel, α is the path-loss exponent, and d is the distance between the transmitter and receiver.

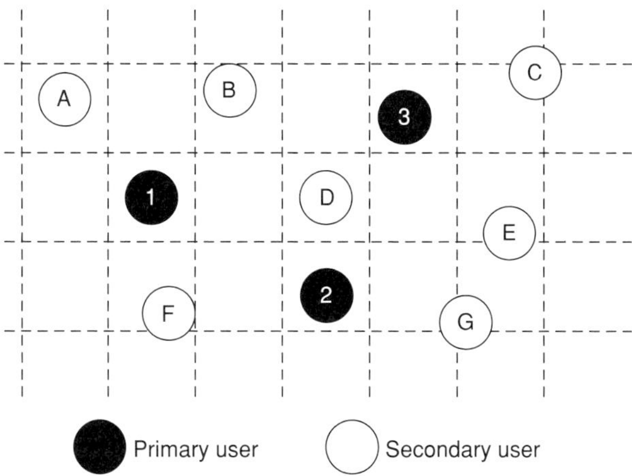

Figure 8.3 An illustration of static primary users.

As explained in the generic principle, the area needs to be discretized into a grid with M columns and N rows, thus resulting in MN cells. The distance intervals of the horizontal and vertical axes are Δx and Δy, respectively.

Now, we denoted by d_{ki} the power received by secondary user k at channel 1. From the time-domain receive signal, the vector of correlations $\mathbf{r}_k$ can be obtained, which satisfies

$$\mathbf{r}_k = \mathbf{F}\mathbf{d}_k, \tag{8.9}$$

where $\mathbf{d}_k = \left(d_{k1}, \ldots, d_{kN_{ch}}\right)^T$ is the vector of received power and $\mathbf{F}$ is the matrix representing the inverse discrete Fourier transform (IDFT). We further link $\mathbf{r}$ obtained from the measurements to the location and transmit power information of the primary users. It is easy to verify that

$$\mathbf{d}_k = \mathbf{L}_k\mathbf{p}, \tag{8.10}$$

where (here f_m is the center frequency of channel m and $d(k, n)$ is the distance between secondary user k and cell n)

$$(\mathbf{L}_k)_{mn} = 10^{L(f_m, d(k,n))/10}, \tag{8.11}$$

and

$$\mathbf{p} = (P_1, \ldots, P_{MN}), \tag{8.12}$$

where P_r is the transmit power of primary user at cell r and $P_r = 0$ if there is no primary user at cell r. We notice that the information of location and transmit power of primary users has been incorporated into the unknown vector $\mathbf{p}$: if P_r is non-zero, there is a primary user at cell r; P_r also provides the value of transmit power.

By combining (8.9) and (8.10) and stacking all equations for different secondary users, we have

$$\mathbf{r} = \mathbf{\Psi}\mathbf{p}, \tag{8.13}$$

where $\mathbf{r} = \left(\mathbf{r}_1^T, ..., \mathbf{r}_{N_c}^T\right)^T$ and

$$\boldsymbol{\Psi} = \begin{pmatrix} \mathbf{F} & 0 & 0 & 0 \\ 0 & \mathbf{F} & 0 & 0 \\ \vdots & \vdots & \ddots & \vdots \\ 0 & 0 & 0 & \mathbf{F} \end{pmatrix} \begin{pmatrix} \mathbf{L}_1 \\ \mathbf{L}_2 \\ \vdots \\ \mathbf{L}_{N_c} \end{pmatrix}. \tag{8.14}$$

Note that, in (8.14), $\mathbf{r}$ is known from the time-domain measurements, $\boldsymbol{\Psi}$ is obtained from the path-loss model, and $\mathbf{p}$ is the unknown vector consisting of all information we need. When the number of primary users is small, compared with MN, $\mathbf{p}$ is sparse, thus facilitating the application of CS. Hence, once we solve (8.14) by employing CS, the task of joint spectrum sensing and positioning is accomplished.

In [341], the measurement noise is considered in the reconstruction of $\mathbf{p}$. By assuming that the noise is Gaussian, Bayesian CS is applied. The details of Bayesian CS can be found in previous chapters. Then, simulations that demonstrate the performance of the positioning algorithm are carried out for 20 channels and primary users within a 10×10 grid. When the noise power is 0.0288 and the transmit power is uniformly random between 10 and 20, no missed detections and no false alarms are found in the simulation for five secondary users. More simulation results can be found in [341].

Mobile primary users

In the previous discussion, the primary users are assumed to be fixed. In practice, the primary users could be mobile; e.g., a primary user could be a mobile cellular phone that has a license for the spectrum band. In the case of mobile primary users, the algorithm of joint spectrum sensing and positioning can still be applied if we simply consider the positions of primary users in each time snapshot. However, such a straightforward application ignores the fact that the speeds of primary users are limited (unless the primary user is a jet fighter). Hence, in [342], the joint spectrum sensing and positioning are studied using Kalman filtering, similarly to the tracking of moving targets in radar.

When the primary users are mobile, the dynamics of the primary users can be described for the following difference equation, if we consider the discrete time domain:

$$\mathbf{x}(t+1) = \mathbf{A}\mathbf{x}(t) + \mathbf{n}, \tag{8.15}$$

where $\mathbf{x} = (\mathbf{p}, \mathbf{v})$ is the system state (where $\mathbf{v}$ is the vector of change of power, related to the motions of primary users, $\mathbf{n}$ is the Gaussian noise with expectation zero and covariance matrix $\mathbf{N}$, and $\mathbf{p}$ is the vector of transmit power in (8.12)) and

$$\mathbf{A} = \begin{pmatrix} \mathbf{I} & \mathbf{I} \\ \mathbf{0} & \mathbf{I} \end{pmatrix}. \tag{8.16}$$

Suppose that the observation is $\mathbf{r}$ (the correlation in the time domain). Then, the observation process is given by

$$\mathbf{r}(t) = \boldsymbol{\Psi}\mathbf{x}(t) + \mathbf{w}, \tag{8.17}$$

where $\mathbf{w}$ is the observation noise, which is Gaussian with expectation zero and covariance matrix $\mathbf{W}$.

The following implicit assumptions are used in the system dynamics. In practice, more complicated models could be used to describe the dynamics of motion.

- We assume that the expectation of $\mathbf{v}$ does not change; i.e., the expectation of the change of power does not change. This is reasonable since the best prediction of the motion speed is still the same one as that in the previous time-slot.
- The power and power change are both affected by Gaussian noise.

By applying the Kalman filtering, the expectation of the system state is given by

$$\mathbf{x}_{t+1|t} = \mathbf{A}\mathbf{x}_{t|t}, \tag{8.18}$$

where $\mathbf{x}_{t+1|t}$ is the expected system state at time-slot $t+1$ given the observations until time-slot t, while $\mathbf{x}_{t|t}$ is the expectation at time-slot t. The expectation at time t given the observations until time t, is given by

$$\mathbf{x}_{t|t} = \mathbf{x}_{t|t-1} + \mathbf{K}_t \left(\mathbf{r}(t) - \mathbf{\Psi}\mathbf{x}_{t|t-1}\right), \tag{8.19}$$

which is due to the Kalman filtering procedure, where the computation of $\mathbf{K}_t$ will be left until later.

The covariance matrix of the system state (or equivalently the covariance matrix of system state estimation error) at time t, given the observations until time $t-1$, is given by

$$\mathbf{\Sigma}_{t+1|t} = \mathbf{A}\Sigma_{t|t}\mathbf{A}^T + \mathbf{N}. \tag{8.20}$$

The covariance matrix at time t, given the observations until time t, is given by

$$\mathbf{\Sigma}_{t|t} = \Sigma_{t+1|t} - \mathbf{K}_t \Psi \Sigma_{t|t-1}. \tag{8.21}$$

In Kalman filtering, we have

$$\mathbf{K}_t = \Sigma_{t|t-1}\mathbf{\Psi}^T \left(\mathbf{\Psi}\Sigma_{t|t-1}\mathbf{\Psi}^T + \mathbf{W}\right)^{-1}. \tag{8.22}$$

From the above discussion, we see that the CS is actually embedded in the procedure of Kalman filtering; i.e., given a dynamical process in which the system state vector is sparse, how can we estimate the system state given a small number of observations? This is an interesting but still open problem. In [342], a heuristic approach is proposed to solve this problem, which is illustrated in Figure 8.4 and explained as follows. Note that it is based on the algorithm of dynamic compressive spectrum sensing, which will be introduced later.

1. In each time-slot, the secondary users compute the difference of the norms of observations; i.e.,

$$\Delta r(t) = \mathbf{r}(t) - \mathbf{r}(t-1). \tag{8.23}$$

   If $|\Delta(t)|$ is larger than a threshold, then the system believes that the spectrum occupation has been changed.

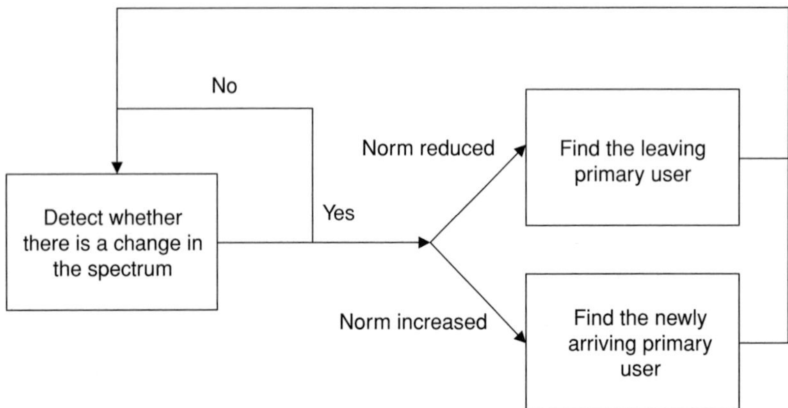

Figure 8.4 An illustration of detecting the primary users in a dynamic manner.

2. Upon the change of spectrum occupancy, we use dynamic CS, which will be introduced in the next subsection, to solve the following problem:

$$\min_{n \in \{1,\dots,MNN_c\}} \min_{p_n} \|\boldsymbol{\Psi}_n p_n - \Delta r\|_2, \tag{8.24}$$

which means that we want to find the most probable location and frequency of the disappearing primary user in order to explain the change in the observation. The following two possible situations will be considered separately:

- If $|\mathbf{r}(t)| < |\mathbf{r}(t-1)|$, we claim that a primary user leaves. Then, we can find out the changing entry p_n by searching in the set of active primary users; i.e.,

$$n^* = \arg\min_{n \in \Omega} \|\boldsymbol{\Psi}_n p_n - \Delta r\|_2, \tag{8.25}$$

  where Ω is the set of active primary users detected before.
- If $|\mathbf{r}(t)| > |\mathbf{r}(t-1)|$, we claim that a new primary user arrives. Then, we can use (8.24) to find the transmit power and location of the new primary user. Note that the difference between (8.24) and (8.25) is the range of the search: in (8.24) the search is over all possible locations since the new primary user could emerge at any location, while the search is limited to the existing primary users in (8.25).

3. Remove the leaving primary user from the active set or adding the arriving primary user to the active set.

Numerical simulations are carried out in [342] to test the performance proposed above. In the simulation, the number of channels is assumed to be 64. Twenty secondary users are randomly deployed in a 1000 m × 1000 m square that consists of 100 uniformly distributed reference points. The motion of the primary user satisfies a symmetric lattice random walk. Simulation shows that, when there are three primary users, the missed detection probability is 0.018, while the detection rate is 0.985.

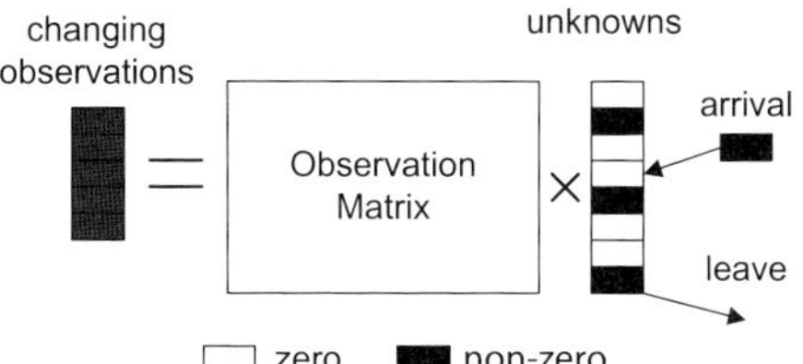

Figure 8.5 An illustration of dynamic CS.

8.2.4 Dynamic compressive sensing

In the previous subsection, we have seen that the non-zero elements in the unknown vector could be changing in different time-slots. This is a key feature for the positioning when the existence of targets could be time varying. Hence, we introduce the dynamic CS, which was proposed in [324]. The system model is illustrated in Figure 8.5.

Problem formulation

In [324], it is assumed that the relationship between the observation and the unknowns is linear; i.e.,

$$\mathbf{R}(t) = \mathbf{\Psi X}(t), \tag{8.26}$$

where $\mathbf{R}$ is a $P \times M$ matrix consisting of the observations, $\mathbf{X}$ is the unknown $N \times M$ matrix, Ψ is the $P \times M$ observation matrix, t is the time index and we assume that there is no noise at the observation. We consider the discrete-time case. In each time slot, one certain row of $\mathbf{X}$ could be changed. This is reasonable if the time-slot is sufficiently small such that it is almost impossible for multiple changes to happen simultaneously. We define

$$\begin{aligned} \Delta\mathbf{R}(t) &= \mathbf{R}(t) - \mathbf{R}(t-1) \\ &= \psi_{n^*}\delta\mathbf{X}_{n^*}(t), \end{aligned} \tag{8.27}$$

where $\delta\mathbf{X}_{n^*}$ is the change in the n^*th row of $\mathbf{X}$, n^* is the row of $\mathbf{x}$ in which the change occurs, and ψ_{n^*} is the n^*th column of $\mathbf{\Psi}$. Moreover, we assume that not all elements in $\Delta\mathbf{R}(t)$ can be observed. The set of indices that can be observed is denoted by Ω; i.e., if $(m, n) \in \Omega$, $\Delta\mathbf{R}_{mn}(t)$ can be observed. The task is to find n^* and $\delta x_{n^*}(t)$ from the observation $\Delta\mathbf{R}_\Omega(t)$, where the subscript Ω specifies the elements in $\mathbf{R}$ that can be observed.

Solution

The solution to the dynamic CS is illustrated in Figure 8.6. The recovery of $\Delta\boldsymbol{r}(t)$ and n^* can be solved by the following least-squares problem:

$$\min_{n\in\{1,\ldots,N\}} \min_{\mathbf{y}} \| \{\psi_n \mathbf{y}\}_\Omega - \Delta\mathbf{R}_\Omega\|_2, \tag{8.28}$$

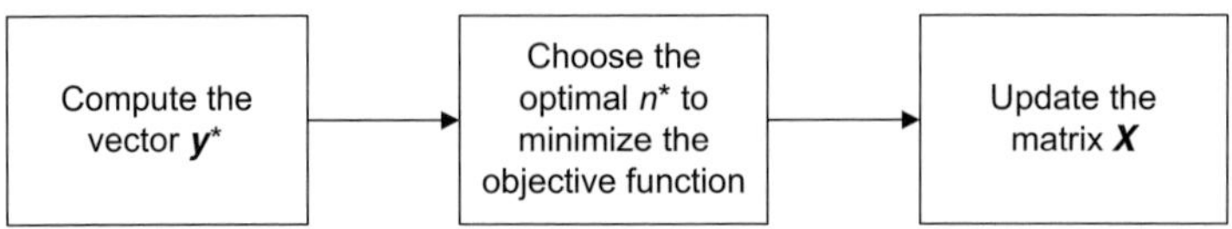

Figure 8.6 An illustration of the solution to the dynamic CS.

where $\mathbf{y}$ is a row vector and $\{\psi_n\mathbf{y}\}_\Omega$ means the elements of matrix $\psi_n\mathbf{y}$ in the set Ω. Equivalently, the objective function in (8.28) can be written as

$$\| \{\psi_n\mathbf{y}\}_\Omega - \Delta\mathbf{R}_\Omega\|_2 = \sum_{m=1}^{M} \|y_m\psi_{I_m,n} - \Delta\mathbf{R}_{I_m,m}\|_2, \tag{8.29}$$

where I_m is the set $\{p : (p, m) \in \Omega\}$.

We first fix an n and try to find the optimal $\mathbf{y}_n$ that minimizes the objective function in (8.28). It is easy to verify that the minimizer of (8.28) is given by

$$(y_n^*)_m = \frac{\psi_{I_m}^T \Delta\mathbf{R}_{I_m,m}}{\psi_{I_m}^T\psi_{I_m}}. \tag{8.30}$$

We can compute $(y_n^*)_m$ one by one using (8.30). However, a more efficient algorithm, which can utilize the matrix operation in Matlab, is proposed in [324]. It is easy to verify that the matrix $\mathbf{Y}^*$ consisting of the elements $(y_n^*)_m$ is given by

$$\mathbf{Y}^* = \frac{\boldsymbol{\Psi}\Delta\bar{\mathbf{R}}}{\hat{\boldsymbol{\Psi}}\mathbf{P}}, \tag{8.31}$$

where

$$\left(\Delta\bar{\mathbf{R}}\right)_{p,m} = \begin{cases} \Delta\mathbf{R}, & (p, m) \in \Omega \\ 0, & \text{otherwise} \end{cases}, \tag{8.32}$$

and

$$\{\mathbf{P}\}_{pm} = \begin{cases} 1, & (p, m)\, in\, \Omega \\ 0, & \text{otherwise} \end{cases}, \tag{8.33}$$

and

$$\hat{\boldsymbol{\Psi}}_{p,n} = A_{p.n}^2. \tag{8.34}$$

From $\mathbf{Y}^*$, we can obtain all optimal values $\mathbf{y}_n^*$ for $n = 1, \ldots, N$. To obtain n^*, we simply compute

$$\| \left\{\psi_n\mathbf{y}_n^*\right\}_\Omega - \Delta\mathbf{R}_\Omega\|_2, \tag{8.35}$$

for $n = 1, \ldots, N^*$. We choose the n^* that minimizes (8.35) as the optimal one. Then, we can add $\mathbf{y}_n^*$ to the nth row of $\mathbf{X}$ to obtain its new estimation.

The following theorem provides a sufficient condition for the perfect recovery of $\mathbf{X}$.

Theorem 11. *Suppose that* $\mathbf{X}(t-1)$ *and* $\mathbf{R}_\Omega(t-1)$ *are exact, and the entries of* Ψ *are in general positions.* $\Omega_m \neq \phi$, *for* $m = 1, \ldots, M$. *There exists one or more* m *such that* $\|\Omega_m\| \geq 2$. *Then,* $\mathbf{X}$ *obtained from the above procedure is exact.*

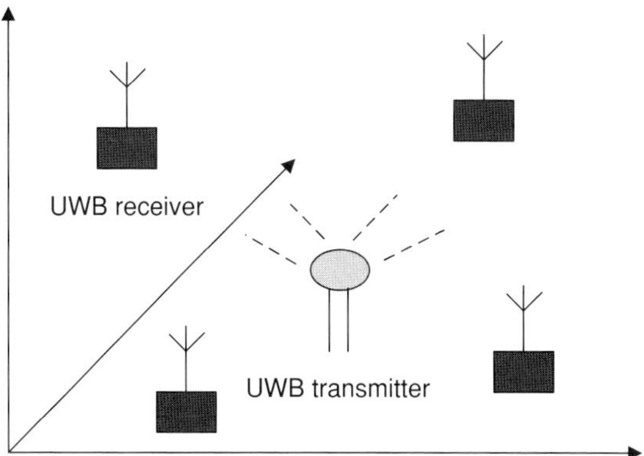

Figure 8.7 An illustration of a UWB-based positioning system.

8.3 Indirect application of compressive sensing

In the previous section, we established a linear mapping from the information indicating the positions to the observations and then carried out the positioning by using CS to solve the unknowns in the mapping. In this section, we consider more traditional positioning techniques in which CS is used to enhance the signal acquisition for positioning. Although it is applied in an indirect manner, the CS is carried out jointly with the positioning procedure, thus improving the efficiency of the positioning algorithm. As a case study, we will introduce UWB-based precise positioning in which CS can make a substantial contribution.

8.3.1 UWB positioning system

A UWB positioning system is illustrated in Figure 8.7. At the target, there is a UWB transmitter. Multiple UWB receivers are located at the base stations, which receive the UWB signals sent out by the target and then determines the corresponding location. Because of the very narrow pulse width of UWB signals, which can provide very precise timing information, the accuracy of the UWB-based positioning can be very high, e.g., in the order of centimeters.

When the UWB signals are acquired and the arrival times are determined, we can use the TDOA algorithm to recover the position of the target. Below, we give a brief introduction on the TDOA algorithm. Suppose that there are I base stations. Let (x_i, y_i, z_i), $i = 1, 2, \ldots, I$, be the coordinates of the ith base station and (x_t, y_t, z_t) be the location of the target, respectively. The distance from the ith base station to the target is denoted by D_i. Then, between the first and the ith base stations, the difference of the UWB pulse arrival time τ_{1i} can be derived from the received UWB signals. Then, we have

$$\tau_{1i} = cD_{1i}, \quad i = 2, 3, \ldots, I, \tag{8.36}$$

where c is the light speed and

$$D_{1i} = \sqrt{(x_1 - x_t)^2 + (y_1 - y_t)^2 + (z_1 - z_t)^2} - \sqrt{(x_i - x_t)^2 + (y_i - y_t)^2 + (z_i - z_t)^2}. \tag{8.37}$$

Taking the derivative on both sides of the equation, we obtain

$$dD_{1i} = \frac{(x_i - x_t)dx_t + (y_i - y_t)dy_t + (z_i - z_t)dz_t}{\sqrt{(x_i - x_t)^2 + (y_i - y_t)^2 + (z_i - z_t)^2}} - \frac{(x_1 - x_t)dx_t + (y_1 - y_t)dy_t + (z_1 - z_t)dz_t}{\sqrt{(x_1 - x_t)^2 + (y_1 - y_t)^2 + (z_1 - z_t)^2}}. \tag{8.38}$$

Stacking the expression for dD_{1i}, we have the matrix form, which is given by

$$\begin{pmatrix} dD_{12} \\ dD_{13} \\ \vdots \\ dD_{1I} \end{pmatrix} = \begin{pmatrix} \alpha_{11} & \alpha_{12} & \alpha_{13} \\ \alpha_{21} & \alpha_{22} & \alpha_{23} \\ \vdots & \vdots & \vdots \\ \alpha_{I1} & \alpha_{I2} & \alpha_{I3} \end{pmatrix} \begin{pmatrix} dx_t \\ dy_t \\ dz_t \end{pmatrix}, \tag{8.39}$$

where we have

$$\begin{cases} \alpha_{i1} = \frac{x_1 - x_t}{D_1} - \frac{x_i - x_t}{D_i} \\ \alpha_{i2} = \frac{y_1 - y_t}{D_1} - \frac{y_i - y_t}{D_i} \\ \alpha_{i3} = \frac{z_1 - z_t}{D_1} - \frac{z_i - z_t}{D_i} \end{cases}. \tag{8.40}$$

The TDOA algorithm is initiated by a conjecture of the position of the target. By iteratively updating the solution to (8.39), TDOA will converge to the true position.

Principally, we can apply the UWB acquisition technique for an individual receiver in the task of positioning, which has been discussed in previous chapters, and then use an independent TDOA procedure. However, such a simple scheme has the following disadvantages:

- There are multiple UWB receivers for the positioning, which share many common information. Hence, it is desirable to exploit the redundancies in the received signals at the base stations.
- The serial operation of acquisition+TDOA approach needs to wait for the completion of the UWB signal reconstruction, which is not efficient.

To address the above two disadvantages, the approaches coined space-time CS and joint CS and TDOA are proposed in [330] and will be discussed in the next two subsections.

8.3.2 Space-time compressive sensing

Individual Bayesian CS

We assume that the signal in the kth frame (a frame means one pulse repetition) after compression at base station i is given by

$$\mathbf{y}_k^i = \mathbf{\Psi}^i \mathbf{x}_k^i + \mathbf{n}_k^i, \tag{8.41}$$

where $\mathbf{y}_k^i$ is the output of the compression, $\mathbf{\Psi}^i$ is the observation matrix at base station i, which may be achieved by using filters, $\mathbf{x}_k^i$ is the originally received signal in the time domain and $\mathbf{n}_k^i$ is the Gaussian noise. We assume that the unknown vector $\mathbf{x}_k^i$ is random and satisfies

$$P(\mathbf{x}_k^i|\alpha_k^i) \sim \mathcal{N}\left(\mathbf{x}_k^i|\mathbf{0}, \left(\text{diag}\left(\alpha_k^i\right)\right)^{-2}\right), \tag{8.42}$$

namely, the elements in $\mathbf{x}_k^i$ are mutually independent and the nth element is Gaussian distributed with expectation zero and variance α_{nn}^{-2}. Intuitively, when α_{nn} is large, the variance is small and the nth element in $\mathbf{x}_k^i$ is more likely to be a zero one; in contrast, if α_{nn} is small, the corresponding element is more likely to be a non-zero one. For a single base station and a single frame, we can use the Bayesian CS to compute the optimal estimation for $\mathbf{x}_k^i$.

Redundancies in space and time

Although the Bayesian CS is an off-the-shelf application for reconstructing $\mathbf{x}_k^i$, completely independent signal reconstruction at different base stations may not be optimal. The reason for this is that it overlooks the redundancies of the received signals in the time and the space, which are explained as follows:

- Spatial redundancy: The received signals at the base stations are from the same source. Hence, the arrival times of the received signals should be similar within the same time frame.
- Temporal redundancy: Since the target cannot move very fast, the signal arrival times in two consecutive frames should also be similar.

Therefore, if $\mathbf{x}_k^i(t)$ is non-zero (or zero), it is highly possible that $\mathbf{x}_k^j(t)$, $j \neq i$, and $\mathbf{x}_{k+1}^i(t)$ are also non-zero (or zero). Hence, if we can utilize this redundancy and carry out the signal reconstruction jointly in both space and time, the performance of the positioning is expected to be improved.

To exploit the spatial and temporal redundancies in the received UWB signals, we can use prior parameters to convey information among the reconstructions at different base stations. Intuitively speaking, we can carry out a rough reconstruction at a base station, using the signal it receives, and then use it as the prior information for the reconstructions at other base stations. Those base stations will also feed back their estimations, as the prior information for this base station. The information is exchanged among the base stations in an iterative manner, which hopefully converges to a satisfying solution. Note that such a scheme is motivated by the famous Turbo coding scheme [343] (we can also

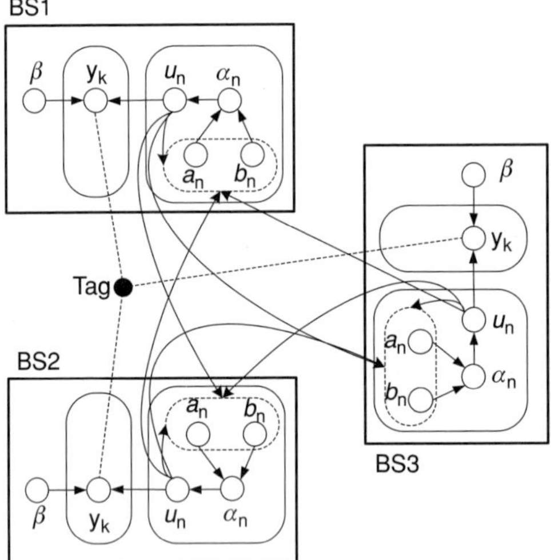

Figure 8.8 An illustration of the space message exchange in the UWB signal reconstruction.

consider the reconstruction as an analog decoding), in which information is exchanged between two component decoders as prior parameters.

Spatial collaboration

We fix one time frame and consider the collaboration of the base stations (i.e., the spatial collaboration), which is illustrated in Figure 8.8. To that end, α_k^i is assumed to be random and a gamma priori is imposed on the α_k^i with hyperparameters a_k^i and b_k^i. Note that the reason why we focus on α_k^i is that it indicates whether an element is zero or not, as we have seen above. We assume that different α_k^is are mutually independent, thus resulting in

$$
\begin{aligned}
& P(\alpha_k^i | a_k^i, b_k^i) \\
&= \prod_{j=1}^{N} \Gamma(\alpha_{j,k}^i | a_{j,k}^i, b_{j,k}^i) \\
&= \prod_{j=1}^{N} \frac{(b_{j,k}^i)^{a_{j,k}^i} (\alpha_{j,k}^i)^{(a_{j,k}^i - 1)} \exp(-b_{j,k}^i \alpha_{j,k}^i)}{\Gamma(a_{j,k}^i)},
\end{aligned}
\tag{8.43}
$$

where $a_{j,k}^i$ and $b_{j,k}^i$ are the parameters controlling the distribution of $\alpha_{j,k}^i$ for the jth signal element in the ith BCS procedure at the kth frame.

Now assume that we have the jth reconstructed signal element, e.g., $u_{d,j}^c$ from the reconstruction procedure at the cth BCS in the dth time frame. In order to facilitate the information exchange among the reconstruction procedures at different base stations, we assume that, for different i and c, we have $\alpha_j^i = \alpha_j^c$, $\forall j$. Then, by abbreviating $\alpha_{j,k}^i$

as α_j for convenience, we compute

$$
\begin{aligned}
& P(u_d^c | a_k^i, b_k^i) \\
&= \int P(u_{d,j}^c | \alpha_j) P(\alpha_j | a_{k,j}^i, b_{k,j}^i) d\alpha_j \\
&= \frac{(b_{k,j}^i)^{a_{k,j}^i} \Gamma(a_{k,j}^i + \frac{1}{2})}{(2\pi)^{\frac{1}{2}} \Gamma(a_{k,j}^i)} \left(b_{k,j}^i + \frac{(u_{k,j}^i)^2}{2} \right)^{-(a_{k,j}^i + \frac{1}{2})},
\end{aligned}
\tag{8.44}
$$

where $\Gamma(\cdot)$ is the gamma function, defined as $\Gamma(x) = \int_0^\infty t^{x-1} e^{-t} dt$.

Given the message $u_{d,j}^c$, we can update the parameters $a_{k,j}^i$ and $b_{k,j}^i$ to incorporate the information from the ith base station as prior parameters. We have

$$
\hat{a}_{j,k}^i = \arg\max_{a_{j,k}^i} \prod_{c,d=1}^{n} P(u_{d,j}^c | a_{j,k}^i, b_{j,k}^i), \tag{8.45}
$$

$$
\hat{b}_{j,k}^i = \arg\max_{b_{j,k}^i} \prod_{c,d=1}^{n} P(u_{d,j}^c | a_{j,k}^i, b_{j,k}^i). \tag{8.46}
$$

It is challenging to estimate the parameters by maximizing the likelihood function. Alternatively, we update the parameters in the following manner:

$$
P(\alpha_j | u_{d,j}^c, a_{j,k}^i, b_{j,k}^i) \tag{8.47}
$$

$$
\begin{aligned}
&= \frac{P(u_{d,j}^b | \alpha_j) P(\alpha_j | a_{j,k}^i, b_{j,k}^i)}{\int P(s_j^b | \alpha_j) P(\alpha_j | a_j^b, b_j^b) d\alpha_j} \\
&= \frac{\left(\tilde{b}_{j,k}^i\right)^{\tilde{a}} (\alpha_j)^{(\tilde{a}_{j,k}^i - 1)} \exp\left(-\tilde{b}_{j,k}^i \alpha_j\right)}{\Gamma(\tilde{a}_{j,k}^i)}.
\end{aligned}
\tag{8.48}
$$

By comparing (8.47) with (8.43), we can update the parameters in the following way:

$$
\begin{cases}
\tilde{a}_{j,k}^i = a_{j,k}^i + \frac{1}{2}, \\
\tilde{b}_{j,k}^i = b_{j,k}^i + \frac{(u_{d,j}^c)^2}{2}.
\end{cases}
\tag{8.49}
$$

Based on the signal element $u_{d,j}^c$, the parameter $a_{j,k}^i$ and $b_{j,k}^i$ are updated to $\tilde{a}_{j,k}^i$ and $\tilde{b}_{j,k}^i$. Obviously, the posterior still has a gamma distribution, which provides substantial convenience for parameter updating. When multiple reconstructed signal elements are transferred, e.g., $u_{d_1,j}^{c_1}$, $u_{d_2,j}^{c_2}$, and $u_{d_n,j}^{c_n}$, the parameters are updated to

$$
\begin{cases}
\tilde{a}_{j,k}^i = a_{j,k}^i + \frac{n}{2}, \\
\tilde{b}_{j,k}^i = b_{j,k}^i + \frac{\sum_{i=1}^n (u_{d_i,j}^{c_i})^2}{2}.
\end{cases}
\tag{8.50}
$$

Therefore, given the messages from other reconstruction procedures, parameters are updated as if the prior distribution has been changed. Note that, initially, the parameters $a_{j,k}^i$ and $b_{j,k}^i$ are set to zero, which implies that there will be no prior information, thus resulting in the traditional BCS algorithm without any collaborations.

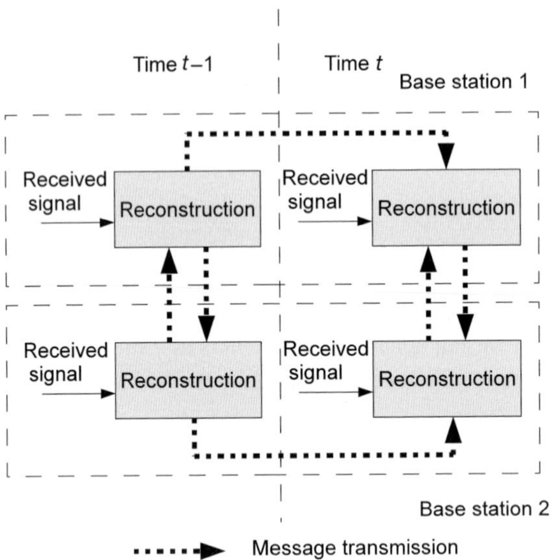

Figure 8.9 An illustration of the space-time message exchange in the UWB signal reconstruction.

Space-time collaboration

In the above discussion, we considered only the spatial redundancies and the message exchanges among different reconstruction procedures at different base stations. As we have found, there also exist temporal redundancies in the received UWB signals in different time frames. Hence, the above mechanism of message exchange can also be applied for different reconstruction procedures for different time-slots. An illustration for two base stations is given in Figure 8.9. Besides the message exchange among the reconstruction procedures at different base stations in the same time-slot, the reconstruction procedure in time-slot $t-1$ also sends messages to that of the same base station in time-slot t. In this way, we can effectively utilize both the spatial and temporal redundancies.

8.3.3 Joint compressive sensing and TDOA

As we have explained, it is inefficient to carry out the CS-based reconstruction and the TDOA algorithm in a serial manner. We propose to interleave both procedures in an interative manner. This parallel CS-TDOA algorithm is able to calculate the position of the tag at each iteration while the UWB signal reconstruction procedures are still ongoing. It is based on the fact that the acquisition of the pulse arrival time does not require a full signal recovery. Even though the pulse arrival time is not accurate when the signal is far from being well reconstructed, the approximate value will help the TDOA converge quickly. The algorithm is summarized in Procedure 3 and is illustrated in Figure 8.10.

The performance of the above CS-assisted positioning algorithm has been tested using simulations [330]. Experimental IEEE 802.15.4a UWB propagation standards have been

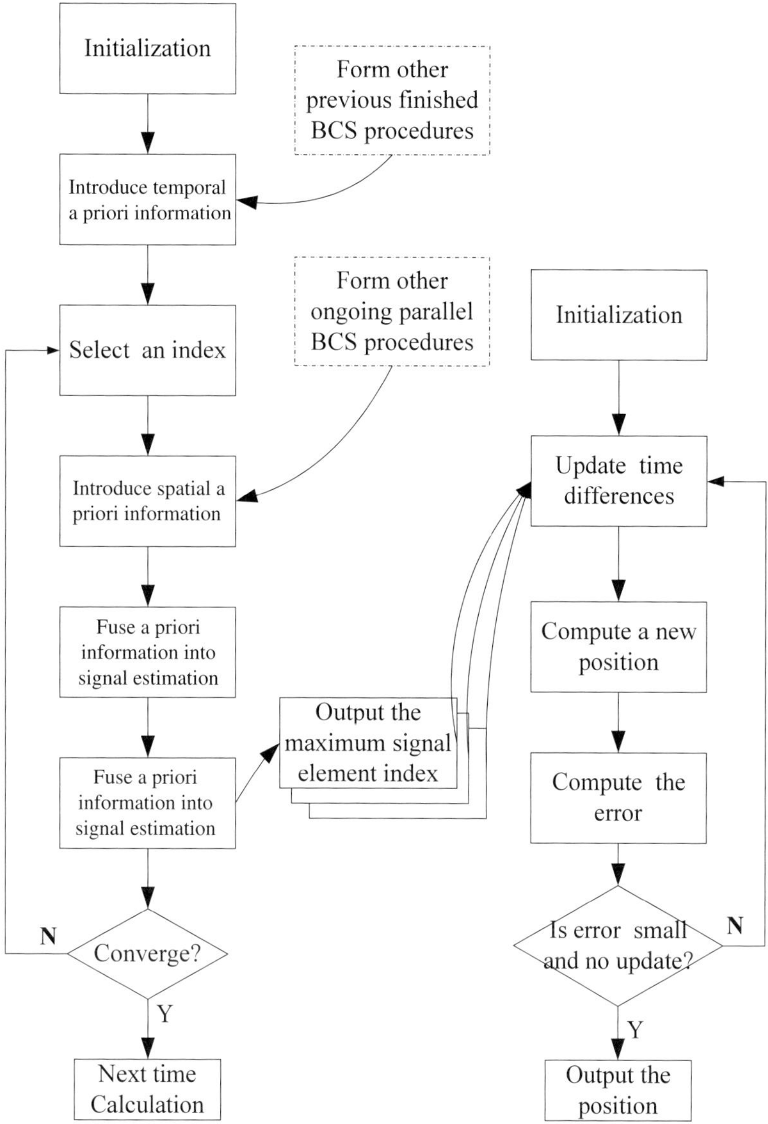

Figure 8.10 The algorithm for joint CS and TDOA.

used. The simulation is carried out for a 5 m × 5 m × 4 m room with four base stations. The SNR is around 15 dB. The expected norm of positioning error is 0.8 mm when the proposed algorithm is used, while it is equal to 5.52 mm when a traditional sequential sampling algorithm is applied. Moreover, the simulation shows that the positioning error does not change with the position of the target when the proposed algorithm is used; when the traditional positioning algorithm is used, the error is much larger when the target is close to a base station than when it is closer to the center of the room.

Algorithm 3 STBCS-TDOA Algorithm

1: Initialization: the hyperparameter α is set to $\alpha = \{\infty\}$; the non-zero signal index set $\Omega = \emptyset$; the parameters $a, b : a, b = \{0, 0\}$; the position of the tag is set to an initial position.
2: Introducing temporal a priori information: update a, b using Eq.(8.45), (8.46), (8.49) and (8.49) from the previous reconstructed non-zero signal elements.
3: **repeat**
4: Introducing spatial a priori information: receive the ongoing reconstructed signal elements from other simultaneous BCS procedures to update the parameter a, b.
5: Calculate the components g_j and h_j and form the candidate set Λ (the details of computation can be found in [330]).
6: Select an index from the index set Λ. Assume it is the jth index. Then add it into index set Ω: $\Omega = \Omega \cup j$;
7: Update α_j if the parameters are updated so that $a_j \neq 0, b_j \neq 0$. Otherwise update α_j.
8: **if** $\Omega \cup \Lambda \neq \Omega$ **then**
9: Delete the index: $\Omega = \Omega \backslash \{j\}$, where $j \in \Omega$ but $j \notin \Lambda$.
10: **end if**
11: Compute the signal elements whose indices are in Ω. Output the index of the maximum signal based on the current reconstructed signal vector as the pulse arrival time for the TDOA algorithm for computing the position of the tag.
12: Send out the ongoing reconstructed signal elements to other BCS procedures as spatial a priori information.
13: **until** converged
14: Send out the reconstructed non-zero signal elements for the next frame utilization as temporal a priori information.
15: In parallel, the TDOA algorithm receives values of pulse arrival times from different BCS signal reconstruction procedures on base stations. With respect to one base station, the differences of pulse arrival and corresponding distance differences are calculated based on (8.36).
16: The tag position is calculated using (8.39) with an initial start position.
17: Go to step 16 to check any new updates. If there are new updates, utilizing the current calculated tag position as the initial position value recalculate the tag position. Otherwise, iteratively compute the position of the tag until TDOA converge.

8.4 Conclusions

In this chapter, we have discussed the application of CS in geographical positioning. Both the direct and indirect applications of CS were explained. For the direct application, the key problem is how to convert the positioning problem into the problem of solving a linear equation with sparse unknown vector. This is typically achieved by discretizing the space and considering the observation as the superposition of features corresponding to different locations. As we have seen, CS can be directly applied in the positioning of WiFi and the spectrum sensing of cognitive radio. CS can also be used to assist the signal acquisition in positioning, e.g., UWB-based positioning. Although the application of CS is indirect in these scenarios, interleaving the reconstruction in CS and the positioning algorithm can improve the speed of the positioning.

9 Multiple access

9.1 Introduction

In this chapter, we discuss multiple access using CS in the context of wireless communications. Essentially, multiple access means that multiple users send their data to a single receiver that needs to distinguish and reconstruct the data from these users, as illustrated in Figure 9.1. It exists in many wireless communication systems, such as

- Cellular systems: In cellular systems, the mobile users within a cell are served by a base station. The users may need to transmit their data during the same time period; e.g., multiple cellular phone users make phone calls simultaneously.
- Ad hoc networks: In such systems, there is no centralized base station. However, one node may need to receive data from multiple neighbors simultaneously; e.g., a relay node in a sensor network receives the reports from multiple sensors and forwards them to a data sink.

The key challenge of multiple access is how to distinguish the information from different transmitters. Hence, there are basically two types of multiple-access schemes:

- Orthogonal multiple access: This type of multiple access includes time division multiple access (TDMA), orthogonal frequency division multiple access (OFDMA) and carrier sense multiple access (CSMA). In TDMA, each transmitter is assigned a time-slot and it can only transmit over the assigned time-slot. Hence, the signals from different transmitters are distinguished in the time. OFDMA has a similar scheme to TDMA. The only difference is that OFDMA allocates different frequency channels to different transmitters and thus separates the transmitters in the frequency domain. CSMA also separates the transmitters in the time domain. However, there is no dedicated time-slot for each transmitter. Each transmitter needs to sense the environment first. If there is no transmission, it will go ahead to transmit; otherwise, it backs off for a random period of time and then senses the environment again, until it sends out the packet.
- Non-orthogonal multiple access: A typical non-orthogonal multiple-access scheme is the code division multiple access (CDMA) [344], which is the fundamental signaling technique in 3G systems. In CDMA, the signals of all transmitters are mixed in both

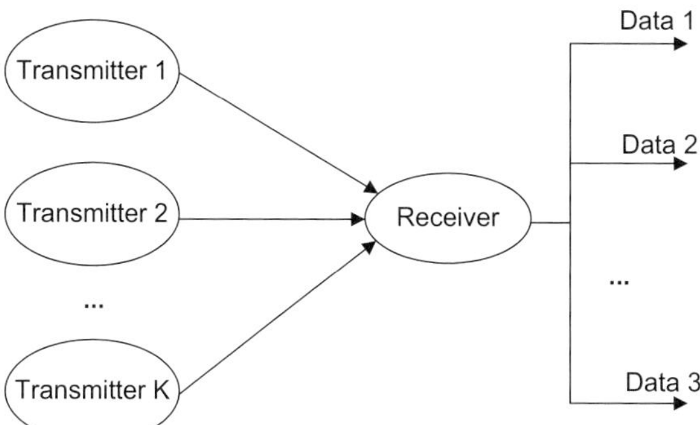

Figure 9.1 An illustration of multiple access.

time and frequency domains. Each transmitter is assigned a code based on which the receiver can distinguish the signals from different transmitters.

As we have seen, it is very easy to distinguish the signals for different transmitters in the orthogonal multiple access. The main challenge of these multiple-access schemes is how to keep the data channels (time-slot, or frequency channel) orthogonal (e.g., how to keep the transmitters synchronized in the time domain). Hence, we study the non-orthogonal multiple access scheme since the CS technique may play an important role in the signal separation/reconstruction.

When we mention CS, the first question is where the sparsity is from. In the context of multiple access, the sparsity may come from:

- Data sparsity: At each time, only a small fraction of the transmitters have data to transmit; the set of active transmitters is unknown in advance since the packet arrivals at different transmitters could be random.
- Observation sparsity: When the packets at the transmitters are mutually correlated (e.g., they are the observations on the temperatures in a region), the received signal may also be written as a linear transformation of a sparse vector in a certain domain.

We will consider both types of sparsity in this chapter. Note that the CS-based multiple access has a tight relationship to the technique of multiple-user detection in CDMA systems. Hence, we will provide a brief introduction on multiuser detection and then discuss the CS-based multiple access in both cellular and sensor networks.

9.2 Introduction to multiuser detection

Multiuser detection is a powerful technology for CDMA receivers, which can effectively separate signals from different users. It can also be applied in any vector multiple access channels; i.e., both the transmitted and received signals are vectors. For example, in multiple-input–multiple-output (MIMO) systems, in which both the transmitters and receivers are equipped with multiple antennas, the technique of multiuser detection can also be applied if different messages are sent out from different transmit antennas.

In this section, we will first provide a signal model for CDMA systems. Then, we will introduce various multiuser detection techniques. Finally, we will make a comparison between multiuser detection and CS. A comprehensive introduction to multiuser detection can be found in [345].

9.2.1 System model for CDMA

For simplicity, we consider synchronous CDMA system; i.e., all users have the same timing and keep their information symbols perfectly synchronized. It is not difficult to extend the algorithms in multiuser detection to the asynchronous case. We also assume that all users use binary phase shift keying (BPSK); i.e., the information symbol can only take value from +1 or –1. Moreover, we assume that the wireless channel is frequency flat, although it is more reasonable to consider a frequency-selective channel for CDMA systems; the analysis can be extended to frequency-selective channels (or equivalently, channels with multiple paths).

We assume that there are totally K users. The information symbol of user k is denoted by b_k. Each user is assigned an N-dimensional spreading code, denoted by $\mathbf{s}_k$ for user k. Denoting by g_k the channel gain of user k (recall that we consider a frequency flat channel; hence one channel gain can represent the wireless channel), we can obtain the expression for the received signal:

$$\mathbf{r} = \sum_{k=1}^{K} g_k b_k \mathbf{s}_k + \mathbf{n}, \tag{9.1}$$

where $\mathbf{n}$ is additive white Gaussian noise. The task of multiuser detection is to reconstruct the information symbols $\{b_k\}_{k=1,\ldots,K}$ from the received signal $\mathbf{r}$.

The received signal in (9.1) can be rewritten as the following vector form:

$$\mathbf{r} = \mathbf{SAb} + \mathbf{n}, \tag{9.2}$$

where $\mathbf{A} = \text{diag}(g_1, ..., g_K)$ and $\mathbf{S} = (\mathbf{s}_1, ..., \mathbf{s}_K)$. The goal of multiuser detection is to estimate $\mathbf{b}$, given $\mathbf{r}$, $\mathbf{A}$ and $\mathbf{S}$ (of course, $\mathbf{n}$ is unknown).

9.2.2 Comparison between multiuser detection and compressive sensing

If we compare the expression in (9.2) with the standard CS problem $\mathbf{y} = \mathbf{Hx}$, we can find that they are very similar to each other. In both cases, the known signal is the linear transformation of the unknown signal ($\mathbf{b}$ in multiuser detection and $\mathbf{x}$ in CS), perhaps contaminated with noise. Hence, in [346] Donoho et al. pointed out the similarity between multiuser detection and CS: "There are strong connections to stagewise/stepwise regression in statistical model building, successive interference cancelation multiuser detectors in digital communications and iterative decoders in error-control coding."

However, there are also significant differences between multiuser detection and CS:

- The elements in $\mathbf{b}$ are binary, while $\mathbf{x}$ is a continuous real or complex vector.
- The non-zero elements in $\mathbf{x}$ are sparse, while there is no sparsity assumption in multiuser detection.
- In multiuser detection, the dimension of the observation N is usually comparable to the dimension of unknown vector K (or equivalently the number of users). However, in CS, the observation has a much lower dimension than the original signal.

Hence, although looking similar to each other, multiuser detection and CS have different challenges. The difficulty of multiuser detection is how to handle the curve of dimensions incurred by the binary unknowns, while the CS is mainly challenged by the low-dimensional observation. Meanwhile, they can also share many ideas with each other, as we will see in the next subsection.

9.2.3 Various algorithms of multiuser detection

In this subsection, we provide a brief introduction to various algorithms for multiuser detection, and meanwhile explain its similarity to algorithms in CS.

9.2.4 Optimal multiuser detector

The optimal multiuser detector maximizes the likelihood of the received signal $\mathbf{x}$. There are two types of optimal multiuser detectors:

- Joint Optimal Multiuser Detector (JO-MUD): In this case, the unknown signal $\mathbf{b}$ is chosen to maximize the joint conditional probability of $\mathbf{y}$ given $\mathbf{b}$, on assuming that the noise $\mathbf{n}$ is white Gaussian with variance σ_n^2; i.e.,

$$\hat{\mathbf{b}} = \arg\max_{\mathbf{b}} \exp\left[-\frac{1}{2\sigma_n^2}\|\mathbf{r} - \mathbf{SAb}\|^2\right]. \tag{9.3}$$

- Individual Optimal Multiuser Detector (IO-MUD): Different from the JO-MUD, the IO-MUD considers the detection of each individual bit and maximizes the marginal distribution; i.e.,

$$\hat{b}_k = \arg\max_{b_k} P(\mathbf{r}|b_k), \tag{9.4}$$

where

$$P(\mathbf{r}|b_k = b) \propto \sum_{\mathbf{b}, b_k = b} \exp\left[-\frac{1}{2\sigma_n^2}\|\mathbf{r} - \mathbf{SAb}\|^2\right]. \tag{9.5}$$

In both (9.3) and (9.4), the optimization problems are discrete, thus requiring exponentially complex computations and being infeasible when K and N are large. Note that there are no counterparts in the context of CS since the unknown vector $\mathbf{x}$ in CS is real.

However, it is possible to extend the real $\mathbf{x}$ to discrete ones in CS when the underlying application requires discrete variables.

Linear multiuser detector

In this family of multiuser detectors, the unknown vector $\mathbf{b}$ is first considered as a real vector and then is transformed by a linear vector. The final decision is obtained from the output of the linear transform. Take user k, for example, we first carry out the linear transformation using a vector $\mathbf{v}_k$ and the output z_k is given by

$$z_k = \mathbf{v}_k^T \mathbf{r}, \tag{9.6}$$

and the decision is given by

$$\hat{b}_k = \text{sgn}(z_k), \tag{9.7}$$

where $\text{sgn}(z_k)$ is the sign of number z_k.

We have the following choices for the vector $\mathbf{v}_k$

- Matched filter (MF): In MF, we set $\mathbf{v}_k = \mathbf{s}_k$. Hence, the output of the linear multiuser detector is the projection of the received signal on the spreading code of user k.
- Decorrelator: We set $\mathbf{v}_k$ as the kth column vector of $\mathbf{S}^T\mathbf{S}$. It is easy to verify that, when there is no noise and $K \le N$, the decorrelator can perfectly reconstruct the original bits.
- Minimum Mean Square Error (MMSE) Detector: We set

$$\mathbf{v} = \left(\sigma_n^2 \mathbf{I} + \sum_{n=1}^{K} g_n \mathbf{s}_n \mathbf{s}_n^T \right)^{-1} \mathbf{s}_k. \tag{9.8}$$

It has been shown that the MMSE detector can maximize the signal-to-interference-and-noise (SINR) ratio.

There is no counterpart in CS corresponding to the above algorithms. However, they are similar to some steps in MP or OMP algorithms in CS. For example, the MF detector can be considered as the step of finding the dictionary element $\mathbf{h}_k$ and the corresponding unknown x_k. The MMSE detector is similar to the one of finding the unknown x_k in OMP.

Interference cancelation

For detecting the bit of user k, we can rewrite the received signal in (9.1) as

$$\mathbf{r} = g_k b_k \mathbf{s}_k + \sum_{n=1, n\neq k}^{K} g_n b_n \mathbf{s}_n + \mathbf{n}, \tag{9.9}$$

where the term $\sum_{n=1,n\neq k}^{K} g_n b_n \mathbf{s}_n$ is the multiple access interference (MAI) of other users on user k. If $\{b_n\}_{n\neq k}$ are known, we can easily reconstruct and subtract the MAI from the received signal and thus convert the problem into a single-user receiver one. Since we do not know $\{b_n\}_{n\neq k}$ in advance, we can use an intermediate result as the estimation and thus

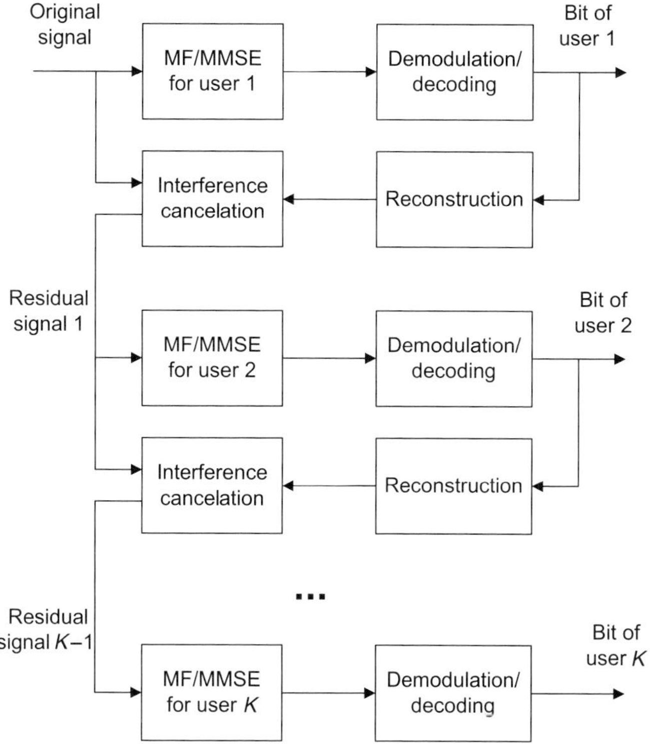

Figure 9.2 An illustration of successive interference cancelation.

improve the performance in iterations. We call the philosophy interference cancelation. As we will see, the procedure of interference cancelation can be very similar to MP or OMP.

There are two types of interference cancelations:

- Parallel interference cancelation (PIC) [347]: In this case, the bits of all users are estimated simultaneously. Then, they are fed back to cancel the interference. The new estimation will be used for further iterations.
  - Step 0: Initialize $\hat{b}_k^0 = 0$, $k = 1, \ldots, K$.
  - Step l ($l > 0$): Cancel the interference and estimate the bits using

$$\hat{b}_k^l = \text{sgn}\left(\mathbf{s}_1^T\left(\mathbf{r} - \sum_{n \neq k} g_n b_n^{l-1}\mathbf{s}_n\right)\right). \tag{9.10}$$

- Successive Interference Cancelation (SIC) [348]: As illustrated in Figure 9.2, the SIC scheme detects the bits of users in a successive way. Suppose that the order of the detection is from user 1 to user K. The original signal will be used to detect b_1 using either MF or MMSE. Then, the output $\hat{b}_1$ is fed back to reconstruct the signal of user 1; i.e., $g_1\hat{b}_1\mathbf{s}_1$, which is used to cancel the MAI brought by user 1. The residual signal $\mathbf{r} - g_1\hat{b}_1\mathbf{s}_1$ is then used to detect the bit of user 2. This procedure is repeated until the bit of user K is detected. Note that the order of successive detection is important. Usually the detection is sorted with a decreasing order of channel gains; i.e., the larger the channel gain is, the earlier the detection is. The rationale is that it is easier to obtain a correct detection outcome for users with larger channel gain since the corresponding SINR is higher. It has been shown that the SIC+MMSE scheme can achieve the channel capacity from the viewpoint of Shannon theory.

There is no counterpart of PIC in CS. The main reason is that PIC requires obtaining a somewhat reliable estimation for all bits in the first iteration. However, this is prohibitive in CS since $K \leq N$ in CS.

However, SIC is very similar to MP or OMP in CS. The similarities are:

- Both SIC and MP/OMP detect or reconstruct the signal in a successive way; i.e., one element is reconstructed in each step.
- Both SIC and MP/OMP take the action of interference cancelation; i.e., the current estimation is fed back to cancel the interference to still unknown elements.
- In SIC and MP/OMP, the successive detection follows the order of descending power. In SIC, the order is obtained from the channel gains, which can be obtained from channel estimation; in MP/OMP, the order is obtained from estimating the "power" of elements; i.e., projecting the residual signal to the atom vectors.
- When SIC uses MF to estimate the bit, it is similar to MP; when MMSE is used in SIC, it is similar to OMP.

Meanwhile, SIC and MP/OMP also have differences since SIC does not have the advantage of signal sparsity while MP/OMP has to handle the unknown locations of non-zero elements. The detailed differences are

- In OMP, the non-zero elements that have been detected will be re-estimated in each step when new non-zero elements are found; however, in SIC, once a bit is estimated, it will not be further modified.
- A stopping criterion is needed for MP/OMP since the number of non-zero elements may not be known. This is not necessary in SIC since the number of users is known in advance.
- The detection order in SIC can be explicitly obtained since the channel gains are usually known in advance. However, the order in MP/OMP has to be adaptive to the detection procedure.

There have not been any algorithms in multiuser detection that are similar to BP in CS. The main reason is that the signal sparsity is explicitly formulated in BP; however, there is no assumption of signal sparsity in multiuser detection. Note that the sparsity assumption may be true in many scenarios of multiuser detection; e.g., in wireless sensor networks, if the data traffic is random and only a small portion of users are simultaneously transmitting. In these scenarios, BP may find its position in the multiuser detection.

9.3 Multiple access in cellular systems

In this section, we study the CS-based multiple access in cellular systems; i.e., there is a base station that carries out centralized processing and each transmitter sends data to the base station in a single-hop manner. We will discuss the uplink (transmitters to base station) and downlink (base station to transmitters) separately.

9.3.1 Uplink

We first discuss the uplink in cellular systems. We follow the framework in [349], which considers a random on–off access channel. In this framework, the transmitters are allowed to transmit simultaneously, similarly to the CDMA systems introduced in the previous section. However, in traditional multiple-access systems, the number of active users is known (e.g., the number of cellular phones in call is known according to the management procedure). In sharp contrast, the set of active transmitters in the scenario considered in [349] is random and is not known in advance. The base station has to determine the set of active users in a manner similar to determining the non-zero elements in CS. Moreover, in [349], it is assumed that each user transmits at most one bit, and thus the channel coding techniques are irrelevant.

Problem formulation

We assume that there are K transmitters sharing a single wireless channel. Each transmitter is assigned an N-dimensional codeword $\mathbf{s}_k$ (for transmitter k). Note that this codeword is equivalent to the spreading codes in the CDMA systems discussed in the previous section. Only a fraction of the transmitters are transmitting their data and the ratio, coined activity ratio, is denoted by λ. We assume that $M = \lambda K$ is an integer and is known in advance. The set of active users is denoted by I_{true}.

The received signal at the base station is then given by

$$\begin{aligned} \mathbf{r} &= \sum_{k=1}^{K} \mathbf{s}_k x_k + \mathbf{w} \\ &= \mathbf{S}\mathbf{x} + \mathbf{w}, \end{aligned} \tag{9.11}$$

where x_k is the complex symbol transmitted by transmitter k and $\mathbf{w}$ is the noise. When a transmitter is not active, $x_k = 0$. Similarly to the CDMA system, $\mathbf{S}$ contains the codewords $\mathbf{s}_1, \ldots, \mathbf{s}_K$, while $\mathbf{x}$ is the vector containing the symbols of the transmitters. Note that an implicit assumption here is the perfect time synchronization among the transmitters. For simplicity of analysis, we assume that the noise $\mathbf{w}$ is Gaussian with expectation zero and covariance matrix $\frac{1}{N}\mathbf{I}$ such that the total noise power of $\mathbf{w}$ is one. Then, the SNR is given by $\|\mathbf{x}\|^2$. We also define the minimum-to-average ratio (MAR), which is given by

$$MAR = \frac{\min_{j \in I_{true}} |x_j|^2}{\|\mathbf{x}\|^2 / \lambda K}, \tag{9.12}$$

where the numerator is the minimum power of non-zero elements while the denominator is the average power in $\mathbf{s}$. Note that the metric MAR is used to measure the near–far effect among the transmitters.

Performance analysis of traditional approaches

Now we discuss the performance of reconstructing the information symbols $x_1, \ldots, x_K$ for various receiver policies. Before evaluating the performance of different algorithms, we need to define asymptotic reliable detection [349]:

Definition: Consider deterministic sequences $N = N(K)$ and $\mathbf{x} = \mathbf{x}(K)$ with respect to K. The probability of error of a given detection algorithm is given by

$$p_{err} = Pr\left(\hat{I} \neq I_{true}\right), \tag{9.13}$$

where $\hat{I}$ is the estimated set of active transmitters, while I_{true} is the true set. The probability is over the randomness of the code matrix $\mathbf{S}$. Then, we say that the algorithm achieves an asymptotic reliable detection if $p_{err}(K) \to 0$ if $K \to \infty$.

We notice that the success of the detection is measured by the set of active transmitters. Even if the demodulations for some active transmitters are wrong (actually the probability is non-zero if the SNR is bounded), we still claim that it is a perfect detection.

- Optimal detection without noise: When there is no noise, the optimal detection algorithm is to exhaustively search for the non-zero elements in $\mathbf{x}$, or equivalently the active transmitters. This requires $N \geq M$. Note that here we have no assumption on the modulation scheme of x_k and simply consider it as an arbitrary complex scalar. If the modulation of x_k is known, e.g., BPSK, the condition $N \geq M$ may no longer be necessary since we can exploit the prior knowledge about $\mathbf{x}$.
- Maximum likelihood (ML) detection with noise: We consider $\mathbf{x}$ as an unknown but deterministic vector. Hence, the ML criterion can be used to estimate $\mathbf{x}$. Intuitively, the ML detector looks for the subspace that is spanned by the codes of active transmitters and has the larger power. Again, we do not assume any knowledge about the modulation of $\mathbf{x}$.

  It has been shown by Wainwright [350] that there exists a constant C such that, if

$$N \geq C \max\left\{\frac{1}{SNR_{\min}} \log\left(K(1-\lambda)\right), \lambda K \log\left(\frac{1}{\lambda}\right)\right\}, \tag{9.14}$$

  the ML detector can detect the correct set of active users asymptotically, where $SNR_{\min}$ is called the minimum component SNR, which is defined as

$$SNR_{\min} = \min_{j \in I_{true}} |x_j|^2, \tag{9.15}$$

  namely, the minimum power of non-zero elements.

  Note that (9.14) is a sufficient condition for the perfect detection of non-zero elements in $\mathbf{x}$. A necessary condition for the perfect detection is found in [349], which is given by

$$N \geq \frac{1-\delta}{SNR_{\min}} \log\left(K(1-\lambda)\right) + \lambda K, \tag{9.16}$$

  for any $\delta > 0$.

- Single-user detection: This is very similar to the MF detection in multiuser detection in CDMA. In such a single-user detection, a threshold μ is set such that the estimated set of active transmitters is given by

$$\hat{I} = \left\{ k \left| \frac{\left|\mathbf{s}_k^T \mathbf{y}\right|^2}{\|\mathbf{s}_k\|^2 \|\mathbf{y}\|^2} > \mu \right. \right\}, \tag{9.17}$$

i.e., we claim that transmitter k is active if the correlation coefficient between the code of transmitter k, namely, $\mathbf{s}_k$, and the received signal $\mathbf{y}$ is larger than the threshold μ. It has been shown in [349] that, when the following inequality holds, we have

$$N > \frac{(1+\delta)L(\lambda, K)(1+SNR)}{SNR_{\min}}, \tag{9.18}$$

where $L(\lambda, K) = \left(\sqrt{\log(K(1-\lambda))} + \sqrt{\log(K\lambda)}\right)^2$, then there exists a sequence of thresholds $\mu(n)$ such that the single-user detection can detect the true set of active users asymptotically.

A disadvantage of the single-user detection is the saturated performance in the high-SNR regime; i.e., when $SNR_{\min} \to \infty$, p_{err} does not converge to 0. This performance saturation is due to the MAI that exists only when SNR is sufficiently high. Another feature of the single-user detection scheme is the constant offset $L(\lambda, K)$.

- Lasso and OMP: The Lasso estimator [351] can also be used for reconstructing the sparse signal by solving the following optimization problem:

$$\hat{\mathbf{x}} = \arg\min_{\mathbf{x}} \left(\|\mathbf{y} - \mathbf{A}\mathbf{x}\|_2^2 + \mu\|\mathbf{x}\|_1\right), \tag{9.19}$$

where μ is a weighting factor. It has been shown in [349] that, if K, N and λK all tend to infinity, and $SNR_{\min} \to \infty$ (the minimum SNR also tends to infinity), the following inequality is necessary and sufficient for asymptotic reconstruction of the signal:

$$m > \lambda K \log(K(1-\lambda)) + \lambda K + 1. \tag{9.20}$$

When the greedy OMP algorithm is applied, a sufficient condition for the asymptotically reliable reconstruction is given by

$$N > 2\lambda K \log(K) + C\lambda K, \tag{9.21}$$

where $C > 0$ is a constant, when $\mathbf{S}$ has Gaussian entries.

Compared with the single-user detection, we find that the constant offset $L(\lambda, K)$ is removed by the Lasso and greedy OMP. The performance is also limited in the high-SNR regime due to the redundant MAI.

Sequential orthogonal matching pursuit

As shown in the previous discussion, there is a significant gap between the ML detector and the existing traditional CS approaches in the context of multiple access. In [351], a sequential orthogonal matching pursuit algorithm is proposed to significantly bridge this gap. Moreover, it does not saturate in the high-SNR regime This algorithm is similar to the OMP algorithm in CS. However, it is carried out in a sequential manner.

The detailed algorithm of the sequential OMP algorithm is given by

1. Initialize the iteration index as $l = 1$ and make the set of active transmitters $\hat{I}(0)$ as an empty set.
2. Compute $\mathbf{P}(j)\mathbf{s}_j$, where $\mathbf{P}(j)$ is an operator that projects $\mathbf{s}_j$ onto the orthogonal complement of the subspace spanned by $\{\mathbf{s}_l, l \in \hat{I}(j-1)\}$.
3. Compute the correlation, which is given by

$$\rho(j) = \frac{\left|\mathbf{s}_j^T \mathbf{P}(j)\mathbf{y}\right|^2}{\|\mathbf{P}(j)\mathbf{s}_j\|^2 \|\mathbf{P}(j)\mathbf{y}\|^2}. \tag{9.22}$$

4. If $\rho(j) > \mu$, where μ is a threshold, add the index j to the set of active transmitters $\hat{I}(j-1)$.
5. Increase $j = j + 1$. If $j \leq K$, go to Step 2.
6. Output the set of active transmitters $\hat{I}(K)$.

The sequential OMP algorithm is similar to a single-user detector (or MF detector in CDMA), since it uses the correlation to detect the existence of nonzero elements. The difference is that, in the sequential OMP algorithm, it is $\mathbf{P}(j)\mathbf{s}_j$, instead of $\mathbf{s}_j$, that is used to compute the correlation and detect the non-zero elements.

Unfortunately, numerical simulations show that the performance of the sequential OMP algorithm is generally much worse than the standard OMP. However, the value of the sequential OMP algorithm is that it can achieve a better scaling in the high-SNR regime. Here, the high-SNR regime is represented by the definition of minimum signal-to-interference-and-noise ratio (SINR), which is given by

$$\gamma = \min_{l=1,\ldots,K} \frac{p_l}{\hat{\sigma}^2(l)}, \tag{9.23}$$

where

$$\hat{\sigma}^2(l) = 1 + \lambda \sum_{j=l+1}^{K} p_j. \tag{9.24}$$

An intuitive explanation for $\hat{\sigma}^2(l)$ is the expected power of interference plus noise when the signals of transmitters 1 to l have been perfectly detected and canceled, where the power of transmitter j is scaled by λ because the probability that it is active is λ. Then, γ is the corresponding SINR.

The asymptotic performance is demonstrated in the following theorem [351]:

Theorem 12. *Let $\lambda = \lambda(K)$ and $N = N(K)$ and the power profile $\{p_j\}_j = 1^K = \{|x_j|^2\}_{j=1}^K$ be deterministic such that $N - \lambda K$, λK and $(1-\lambda)K \to \infty$ and $\gamma \to 0$. It* *is also assumed that the sequence of power profile satisfies*

$$\lim_{n\to\infty} \max_{i=1,\ldots,K-1} \log(n)\hat{\sigma}^{-4}(i) \sum_{j>i}^{n} p_j^2 = 0. \tag{9.25}$$

We also assume that

$$N \geq \frac{(1+\delta)L(K,\lambda)}{\gamma} + \lambda K, \qquad \forall K, \tag{9.26}$$

for some $\delta > 0$ *(note that the function* $L(K,\lambda)$ *is defined in (9.18)). Then, there exists a sequence of thresholds* $\mu = \mu(n)$*, such that the sequential OMP algorithm is asymptotical reliable; i.e.,*

$$p_{err} = Pr\left(\hat{I}_{SOMP} \neq I_{true}\right) \to 0. \tag{9.27}$$

Based on the conclusion in Theorem 12, we can discuss the following properties of sequential OMP:

- Near–far resistance: Suppose that the power ordering is known. Then, the receiver can detect the signals in the descending order of power (recall that it is more desirable to decode the users with the same order in the SIC scheme of multiuser detection in CDMA). Then, it can be shown that the sufficient condition (9.26) can be written as

$$N \geq (1+\delta)\lambda K L(K,\lambda) + \lambda K, \tag{9.28}$$

  which is independent of the MAR. Hence, the sequential OMP algorithm is resistent to the near–far effect, given the known order of powers.
- Power control – equal receive power: We assume that the transmitters are able to control their transmit powers such that the receive powers are all equal; i.e., $p_1 = p_2 = \ldots = p_K = p$. Then, the sufficient condition for the reliable detection becomes

$$N > \frac{(1+\delta)(1+SNR)L(\lambda,K)}{SNR}\lambda K + \lambda K. \tag{9.29}$$

  It is easy to verify that this is exactly the same as the condition of the single-user case when $MAR = 1$. Hence, in the equal receive power case, the sequential OMP algorithm does not bring any performance gain.
- Power control – optimal power profile: Again, we assume that the transmitters can precisely adjust their transmit powers. However, the receive powers are not equal to each other; they are designed to satisfy the following power profile:

$$p_l = \gamma\,(1+\lambda\gamma)^{K-l}, \tag{9.30}$$

  where the SINR γ satisfies

$$\begin{aligned}\gamma &= \frac{1}{\lambda}\left[(1+SNR)^{1/n} - 1\right] \\ &\approx \frac{1}{\lambda K}\log\left(1+SNR\right).\end{aligned} \tag{9.31}$$

  Obviously the optimal power profile satisfies an exponential power shaping. We can prove that this exponential power profile can maximize the SINR in (9.31) given a constant SNR. Transmitters that are detected in the early stages have more receive power, and thus are more reliable in the detection. This provides an intuitive explanation as to why it performs better than the case of equal receive power. Note that (9.30) satisfies the following recursive equation:

$$p_l = \gamma \left(1 + \lambda \sum_{j=l+1}^{K} p_j \right). \tag{9.32}$$

Then, we can obtain the sufficient condition for the reliable detection, which is given by

$$m \geq \frac{(1+\delta)L(K,\lambda)}{\log(1+SNR)} \lambda K + \lambda K. \tag{9.33}$$

We can also see that the exponential power file in (9.30) results in a performance that does not saturate in the high-SNR regime, since as $SNR \to \infty$ the right-hand side of (9.33) tends to zero.

- Robust power control: The asymptotic analysis in Theorem 12 requires infinite codeword length, which guarantees the reliable detection of the transmitters. However, in practice, the codeword length is always bounded; hence, it is possible that the active transmitters are not perfectly detected, thus introducing some residual interference.

9.3.2 Downlink

In the downlink of cellular systems, the information is sent from the base station to different users. Suppose that there are totally K users served by a base station. However, the base station can serve M users simultaneously. A special case is $M = 1$; i.e., time division multiplexing (TDM). When code division multiplexing (CDM) or orthogonal frequency division multiplexing (OFDM) are used, M could be larger than one. When K is very large and the number of corresponding channels (e.g., codes in CDM or subcarriers in OFDM) is limited, M could be much less than K. Then, the base station needs to schedule the traffic of the users.

An efficient scheduling algorithm is the opportunistic scheduling [352], in which users with good channel qualities are scheduled. This scheduling algorithm can improve the throughput of data traffic. A challenge of the opportunistic scheduling is that the base station needs to know the channel qualities of different users. However, the downlink channel qualities must be reported by the users; e.g., the base station can broadcast pilot signals from which the users can estimate the channel qualities. If all users report their channel qualities in each scheduling period, the large number of K may cause a high overhead for the control channel or uplink system.

In [353], a CS-based scheme is proposed for the channel quality report mechanism in the downlink scheduling. In this scheme, only users having good channel qualities will report to the base station; e.g., there could exist a threshold and only when the channel quality is better than the threshold can the user send its report. We denote by K_0 the number of users that send channel-quality reports. When K is sufficiently large and the channel quality is an i.i.d. random variable for different users, K_0 is almost a constant with some random fluctuations. The base station can also have a good estimate for K_0 due to the channel-quality distribution and the threshold. We can assume that $K_0 \ll K$ if K is very large and M is much smaller. However, the base station is unable to know which set of users have good channel qualities and send their reports. This is very similar to the unknown non-zero elements in CS.

Because of the similarity to CS, a reporting period can be set before the downlink scheduling. Similarly to CDMA, each user is assigned a code $\mathbf{s}_k$ (say, for user k). In the reporting period, if user k has a channel quality better than the threshold, it transmits $\mathbf{s}_k x_k$, in which x_k is an information symbol containing the channel quality. Then, the received signal at the base station is given by

$$\mathbf{r} = \sum_{k \in I_{act}} \mathbf{s}_k x_k + \mathbf{w}, \qquad (9.34)$$

where I_{act} is the set of users that plan to report the channel qualities.

9.4 Multiple access in sensor networks

Sensor networks are usually used to monitor certain physical environments such as temperature fields or spectrum activities. The corresponding data traffic is characterized either by the corresponding physical environment or by the multihop transmissions. Various sparsities may emerge in sensor networks, thus justifying the application of CS.

9.4.1 Single hop

We first consider a single-hop sensor network, in which each sensor can send reports to a data fusion center directly. An illustration of such a sensor network monitoring the physical environment is given in Figure 9.3. We follow the work in [354], which considers a grid topology of the sensor network; i.e., the sensors form an $I \times J$ array. The measurement of sensor (i, j) is denoted by x_{ij}. Hence, the reports from the IJ sensors form a matrix $\mathbf{X}$.

We denote by $\mathbf{Y}$ the two-dimensional discrete Fourier transform (DFT) of $\mathbf{X}$, namely,

$$\mathbf{Y} = \mathbf{W}_I \mathbf{X} \mathbf{W}_J, \qquad (9.35)$$

where $\mathbf{W}_I$ is the matrix containing the DFT coefficients, namely, $\mathbf{W}_I(m, k) = e^{-j\frac{2\pi mk}{I}}$.

It is difficult to handle the two-dimensional matrix in the framework of CS. Hence, we convert the matrix into a vector by stacking the columns of the matrix into a vector; this operation is denoted by *vec*. We denote by $\mathbf{y} = vec(\mathbf{Y})$ and $\mathbf{x} = vec(\mathbf{X})$. Then, we have

$$\mathbf{y} = (\mathbf{W}_J \otimes \mathbf{W}_I)\mathbf{x}, \qquad (9.36)$$

where $\otimes$ is the Kronecker product.

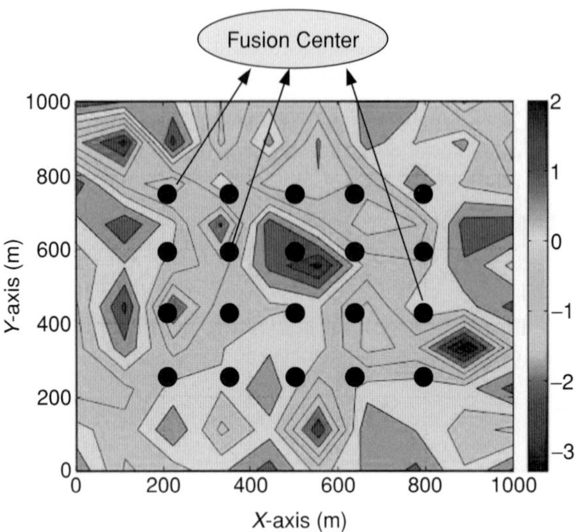

Figure 9.3 An illustration of a sensor network monitoring the environment.

Suppose that the fusion center has received the reports from M sensors. Then, the received signal at the fusion center is given by

$$\mathbf{r} = \mathbf{R}\mathbf{y} + \mathbf{w}, \tag{9.37}$$

where $\mathbf{w}$ is noise and $\mathbf{R}$ is an $M \times IJ$ matrix, which represents the selection of the M sensors and consists of M rows of an $IJ \times IJ$ identity matrix. By taking (9.36) into account, we have

$$\mathbf{r} = \mathbf{R}\Psi\mathbf{x} + \mathbf{w}, \tag{9.38}$$

where $\Psi = \mathbf{W}_J \otimes \mathbf{W}_I$.

Then, the following l_1-norm optimization is used for reconstructing the original signal $\mathbf{x}$, namely,

$$\min_{\mathbf{x}} \|\mathbf{x}\|_1, \qquad s.t. \ \|\mathbf{R}\Psi\mathbf{x} - \mathbf{r}\|_2 \leq \epsilon, \tag{9.39}$$

where $\epsilon > 0$ is a predetermined number.

In [354], the fusion center considers a very simple approach for collecting the reports from the sensors; i.e., letting each sensor transmit its report once it generates a measurement and a packet. The fusion center simply discards the packets that are unsuccessfully decoded due to collisions or noise. It accumulates the successfully decoded packets and waits for sufficiently many reports assuring a reliable reconstruction of $\mathbf{x}$.

It is assumed that each sensor sends packets in a Poisson manner with average arrival rate λ_1. It is proposed in [354] that the arrival of the "useful" packets, after discarding the undecoded or duplicated packets, can be approximated by a Poisson process with arrival rate given by

$$\lambda' = \frac{IJ\left(1 - e^{\lambda_1 T}\right) e^{-2I\lambda_1 T_p}(1 - P_E)}{T}, \tag{9.40}$$

where P_E is the error probability of channel decoding due to noise, T_p is the time needed to transmit one packet and T is the size of the observation window. This conjecture is justified by the following reasons:

- The aggregated packet arrival rate is $IJ\lambda_1$. Then, the probability of no collision at the fusion center is given by

$$P(\text{no collision}) = e^{-2IJ\lambda_1 T_p}. \tag{9.41}$$

  The probability of no decoding error due to noise is given by $1 - P_E$.
- During the observation window T, if a sensor generates more than one packet, only one will be accepted by the fusion center. Hence, the "effective" number of packets generated by one sensor is given by $1 - P(\text{no packet}) = 1 - e^{\lambda_1 T}$.
- Summarizing the above two effects, the average arrival rate of "useful" (no collision, no decoding error, no duplication) packets is approximated by

$$\lambda' = IJ\lambda_1 e^{-2IJ\lambda_1 T_p}(1 - P_e)\frac{1 - e^{\lambda_1 T}}{\lambda_1 T}. \tag{9.42}$$

9.4.2 Multiple hops

When there are multiple hops between the sensors and the data sink, the multiple-access problem also concerns the routing issue; i.e., how to choose the routes from the sensors to the data sink. In this chapter, we use [355], which discusses the CS-based routing and multiple access in sensor network, as a case study.

Simple routing scheme

We consider an n-sensor network, whose diameter is denoted by d (in terms of hops). We denote by x_n the measurement at sensor n. The average distance of nodes to the data sink is of the order of $O(d)$. In traditional sensor networks, each sensor chooses the shortest path to the sink and the transmission of each sensor is independent of other sensors. In this case, the average cost of each reading in the network is of the order of $O(nd)$. As we will see, if we conduct routing and compression, the cost will be significantly reduced.

CS-based routing

We assume that a spanning tree has been formed. Each sensor sends its data to the data sink along the spanning tree. When a sensor receives a packet from a neighbor in the lower level of the spanning tree, it adds its own measurement scaled by +1 or –1 to the packet.

An illustration of the joint routing and CS is shown in Figure 9.4, where two paths are shown. In the path 1-2-3-4-sink, the four sensors add their measurements sequentially to the packet. Only one packet is transmitted throughout the path. The same procedure

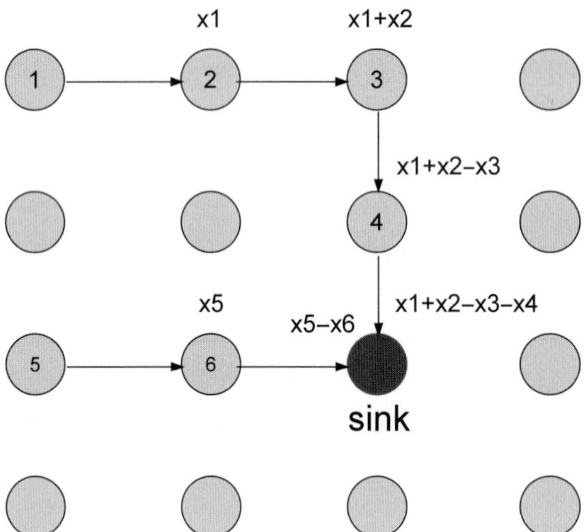

Figure 9.4 An illustration of joint routing and compression in a sensor network.

is carried out along the path 5-6-sink. Then, the sink receives $x_1 + x_2 - x_3 - x_4$ and $x_5 - x_6$, instead of receiving the individual measurements. Hence, totally two packets are transmitted through the network in the example, instead of six.

It is easy to verify that the received packets at the sink can be written as

$$\mathbf{y} = \mathbf{Mx}, \tag{9.43}$$

where $\mathbf{x}$ is the n-vector of the measurements, $\mathbf{y}$ is the vector of received packets at the data sink, whose dimension is equal to the number of children of the sink in the spanning tree, and $\mathbf{M}$ is the measurement matrix. Obviously, the non-zero elements in the mth row of $\mathbf{M}$ corresponds to the sensors on the subtree under the mth sensor directly connected to the link. In the above example, we have

$$\mathbf{M} = \begin{pmatrix} +1 & +1 & -1 & -1 & 0 & 0 \\ 0 & 0 & 0 & 0 & +1 & -1 \end{pmatrix}, \tag{9.44}$$

where we ignore the other sensors.

Suppose that the measurement $\mathbf{x}$ is sparse under certain linear transformations; i.e., there exists a linear transformation $\mathbf{F}$ such that

$$\mathbf{x} = \mathbf{Fz}, \tag{9.45}$$

where $\mathbf{z}$ is sparse. Then, substituting (9.45) into (9.43), we obtain

$$\mathbf{y} = \mathbf{MFz}, \tag{9.46}$$

which is the standard form of CS and $\mathbf{H} = \mathbf{MF}$ is the measurement matrix. Once we obtain $\mathbf{z}$ from $\mathbf{y}$, we can recover all the measurements $\mathbf{x}$ at the sensors. It is easy to verify that the total communication cost is decreased from $O(nd)$ to $O(kd \log n)$, where k is the sparsity of $\mathbf{z}$.

From the example, we can observe that the routing scheme determines the measurement matrix, since each row of $\mathbf{M}$ corresponds to one data path. A bad routing scheme may make the CS not work; an extreme example is that all sensors use the same data path such that $\mathbf{y}$ becomes a scalar. Hence, the routing algorithm can considerably affect the performance. In [355], a heuristic routing algorithm is proposed based on the coherence of $\mathbf{H}$, which is defined as

$$c = \max_{i \neq j} \langle \mathbf{h}_i, \mathbf{h}_j \rangle, \tag{9.47}$$

where $\mathbf{h}_i$ is the ith column of $\mathbf{H}$. It is also easy to verify

$$\langle \mathbf{h}_i, \mathbf{h}_j \rangle = \mathbf{f}_i^T \mathbf{M}^T \mathbf{M} \mathbf{f}_j, \tag{9.48}$$

where $\mathbf{f}_i$ is the ith column of $\mathbf{F}$. Informally speaking, a lower coherence c means a better capability of recovering the sparse signal from the observations. Hence, the purpose of routing is to make the coherence of $\mathbf{H}$ as small as possible. In [355], it is proposed to make $\mathbf{M}^T\mathbf{M} \approx \mathbf{I}$ such that the coherence is mainly determined by $\mathbf{F}$, provided that $\mathbf{F}$ has a low coherence.

In [355], a greedy algorithm is proposed to minimize the coherence. The routing algorithm begins from a randomly selected node. The next node in the path is selected as the one minimizing the intermediate coherence. The following two heuristic constraints are placed to reduce the energy consumptions:

- The next hop node must be closer to the sink than the current one.
- "Crossing over" the sink is forbidden.

The greedy selection is repeated until all the paths have been found. The algorithm is numerically evaluated in 16×16 and 32×32 sensor networks, while the data is generated using both 2D discrete cosine transform (DCT) and multiresolution 2D Haar basis. The numerical simulation results show that the proposed greedy routing and compression algorithm can significantly reduce the cost, when compared with traditional routing and transmission schemes.

9.5 Conclusions

In this chapter, we have discussed the application of CS in multiple access in communication systems. The similarity between the multiuser detection in CDMA systems and CS has been analyzed. Many multiuser detection algorithms can find their counterparts in CS. Then, the applications of CS are studied in both the uplink and downlink of cellular systems. It is also applied in sensor networks, in which both scenarios of single hop and multiple hops have been explained. In particular, the joint routing and compression are demonstrated to be very useful in sensor networks.

10 Cognitive radio networks

10.1 Introduction

Ever since the 1920s, every wireless system has been required to have an exclusive license from the government in order not to interfere with other users of the radio spectrum. Today, with the emergence of new technologies that enable new wireless services, virtually all usable radio frequencies are already licensed to commercial operators and government entities. According to former U.S. Federal Communications Commission (FCC) chair William Kennard, we are facing a "spectrum drought" [356]. On the other hand, not every channel in every band is in use all the time; even for premium frequencies below 3 GHz in dense, revenue-rich urban areas, most bands are quiet most of the time. The FCC in the United States and the Ofcom in the United Kingdom, as well as regulatory bodies in other countries, have found that most of the precious, licensed radio-frequency spectrum resources are inefficiently utilized [37, 357].

In order to increase the efficiency of spectrum utilization, diverse types of technologies have been deployed. Cognitive radio is one of those that leads to the greatest technological gain in wireless capacity. Through the detection and utilization of the spectra that are assigned to the licensed users but standing idle at certain times, cognitive radio acts as a key enabler for spectrum sharing. Spectrum sensing, aiming at detecting spectrum holes (i.e., channels not used by any primary users), is the precondition for the implementation of cognitive radio. The Cognitive Radio (CR) nodes must constantly sense the spectrum in order to detect the presence of the Primary Radio (PR) nodes and use the spectrum holes without causing harmful interference to the PRs. Hence, sensing the spectrum in a reliable manner is of vital importance and constitutes a major challenge in CR networks. However, detection is compromised when a user experiences shadowing or fading effects or fails in an unknown way. To get a better understanding of the problem, consider the following example: a typical Digital TV receiver operating in a 6 MHz band must be able to decode a signal level of at least −83 dBm without significant errors [358]. The typical thermal noise in such bands is −106 dBm. Hence, a CR that is 30 dBm more sensitive has to detect a signal level of −113 dBm, which is below the noise floor [359]. In such cases, one CR user cannot distinguish between an unused band and a deep fade. In order to combat such effects, recent studies suggest collaboration among multiple CR nodes for improving spectrum-sensing performance.

Collaborative spectrum sensing (CSS) techniques have been introduced to improve the performance of spectrum sensing. By allowing different secondary users to collaborate and share their information, PR detection probability can be greatly increased. There are many results that address cooperative spectrum sensing schemes and challenges. The performance of hard-decision combining schemes and soft-decision combining schemes is investigated in [360, 361]. In these schemes, all secondary users send sensing reports to a common decision center. Cooperative sensing can also be done in a distributed way, where the secondary users collect reports from their neighbors and make the decision individually [362–364]. Optimized cooperative sensing is studied in [365, 366]. When the channel that forwards sensing observations experiences fading, the sensing performance degrades significantly. This issue is investigated in [367, 380]. Furthermore, energy efficiency in collaborative spectrum sensing is addressed in [368].

The existing literature mostly focuses on the CSS performance examination when the centralized fusion center receives and combines *all* CR reports. In an n-channel cognitive radio network with m CR nodes, the fusion center has to deal with $n \times m$ reports and combine them wisely to form a channel-sensing result. However, it is known that wireless channels are subject to fading and shadowing. When secondary users experience multipath fading or happen to be shadowed, the reports transmitted by CR users are subject to transmission loss. As a result, in practice, no entire report data set is available at the fusion center. Besides, because each cognitive radio can only sense a small proportion of the spectrum with limited hardware, each CR user gathers only very limited information about the entire spectrum.

We sought to release CRs from sending, and the central control unit from gathering, an excessively large number of reports, also targeted at the situations where there are only a few CR nodes in a large network and thus unable to gather enough sensing information for the traditional CSS. We proposed to equip each cognitive radio node with a frequency-selective filter, which linearly combines multiple channel information. The linear combinations are sent as reports to the fusion center, where the occupied channels are decoded from the reports by CS algorithms. As a consequence, the amount of channel sensing at CRs and the number of reports sent from the CRs to the fusion center are both significantly reduced.

Following previous work [260, 369] on CS, we proposed two approaches to collaborative spectral sensing. The first approach is based on solving a matrix completion problem [88, 149, 150, 370, 371], which seeks to efficiently reconstruct a matrix (typically low rank) from a relatively small number of revealed entries. In this approach, the entries of the underlying matrix are linear combinations of channel powers. Each CR node takes its local spectrum measurements, but instead of directly recording channel powers, it uses its frequency-selective filters to take p *linear combinations* of channel powers and reports them to the fusion center. The total $p \times m$ linear combinations taken by m CRs form a $p \times m$ matrix at the fusion center. Considering transmission loss, we allow the matrix to be incomplete. We show that this matrix is low rank and has the properties enabling its reconstruction from only a small number of its entries, and therefore, information about the complete spectrum usage can be recovered from a small number of reports from the CR nodes. This approach significantly reduces the amount of sensing and communication workload. The second approach is based on joint sparsity recovery [128, 372–375], which is motivated by the observation that the spectrum-usage information the CR nodes collect has a common sparsity pattern: each of the few occupied channels is typically observed by multiple CRs. We develop a novel algorithm for joint sparsity signal recovery, which is more effective than existing algorithms in the CS literature since it can accommodate a large dynamic range of channel gains. In both approaches, every CR senses all channels (by taking random linear projections of the powers of all channels), and the CRs do not communicate. While they work independently, their measurements are analyzed jointly by the detection algorithms running at the fusion center. Therefore, our approaches are very different from the existing collaborative spectrum sensing schemes in which different CRs are assigned to different channels. Our approaches move from collaborative sensing to collaborative computation and shift coordination from the sensing phase to the postsensing phase.

Based on the above approaches, we have two extensions. The first one is to consider the dynamic case in which the primary user comes and goes. Instead of recalculating the whole algorithm, an updating algorithm is proposed so as to reduce the computation and signaling. The second extension is for joint spectrum sensing and localization, where the problem is formulated and the Kalman filter is utilized for tracking.

The rest of this chapter is organized as follows: The literature review of the current state-of-the-art is given in Section 10.2. The CS-based collaborative spectrum sensing is explained in Section 10.3. Dynamic spectrum sensing is then elaborated in Section 10.4. With combination of localization, the problem is studied in Section 10.5. Finally, Section 10.6 summarizes the whole chapter.

10.2 Literature review

There has been an increasing research interest in cooperation and learning for dynamic spectrum access during the last few years. A few results are briefly explained here. In [360], the CRs collaborate by sharing their sensing decisions through a centralized fusion center that combines the CRs' sensing bits. A similar approach is used in [376] using different decision-combining methods. In [377], spatial diversity techniques are proposed for improving the performance by combating the error probability due to fading on the reporting channel between the CRs and the FC. Other performance aspects are studied in [359, 367, 378].

It has been shown that, through collaborative spectrum sensing (CSS) among CRs, the effects of the hidden terminal problem can be reduced and the probability of detecting the primary radio (PR) can be improved [359, 360, 367, 376–379]. In [360], the CRs collaborate by sharing their sensing decisions through an FC, which combines the CRs' sensing bits using the OR-rule for data fusion. A similar approach is used in [376] using different decision-combining methods. In [379], an evolutionary game model for CSS is proposed in order to inspect the strategies of the CRs and their contribution to the sensing process. In [377], spatial diversity techniques are proposed for improving the performance of CSS by combating the error probability due to fading on the reporting channel between the CRs and the FC. Other performance aspects of CSS are studied in [359, 367, 378].

CSS can be classified into two categories. The first category involves multiple users exchanging information [360, 368], and the second uses relay transmission [363, 364]. Some recent studies on collaborative spectrum sensing include cooperative scheme design guided by game theory [362] and random matrix theory [378], cluster-based cooperative CSS [380], and distributed rule-regulated CSS [381]; studies concentrating on CSS performance improvement include [377] introducing spatial diversity techniques to combat the error probability due to fading on the reporting channel between the CR nodes and the central fusion center. There are also studies concerning other interesting aspects of CSS performance under different constraints [367, 378, 382, 383]. Very recently, there are emerging applications of the CS concept for CSS [384].

There are several recent applications of CS in wireless communications. In [385], a two-step compressed spectrum sensing scheme for efficient wideband sensing is proposed. In [386], a Bayesian CS framework reduces the sampling requirement and computational complexity by bypassing signal reconstruction. In [387], to collect spatial diversity against wireless fading, multiple CRs collaborate and establish consensus among local spectral estimates through running a decentralized consensus optimization algorithm. In [388], compressive sampling is performed at local CRs to scan wide spectra, and measurements from multiple CR detectors are fused to collect spatial diversity gain, which improves the detection quality, in particular, under fading channels. In [389], joint dynamic resource allocation and waveform adaptation is proposed under the sparsity constraints for cognitive radio networks.

Dynamic CS work [311, 390–392] develops efficient algorithms for updating the solutions of l_1 minimization for streaming measurements, where the underlying signal changes and new measurements of a fixed signal are sequentially added to the system. The proposed algorithms are based on homotopy continuation, which breaks down the solution update into a small sequence of linear steps. The idea of [393] is to estimate the initial signal support by CS, run a reduced-order Kalman filter over the estimated support, and apply CS to the Kalman innovations or filtering error upon any change that occurs to the support.

Exploiting the CS technique, the spectrum sensing and PR localization problem are well studied in [387, 394, 395]. In [394], the spectrum sensing problem and PR location detection problem are combined together and formulated as a sparse vector. The information of the PRs is reconstructed using the CS technology. In [395] a decentralized way is proposed to solve the spectrum sensing and localization problem. However, existing work mostly investigated the CR network with static PRs. In general, many factors in practice (such as the mobility of PRs and the dynamical spectrum usage) may significantly compromise the detection performance in spectrum sensing. Thus, a more efficient sensing mechanism is needed to handle the dynamics in a practical cognitive radio network.

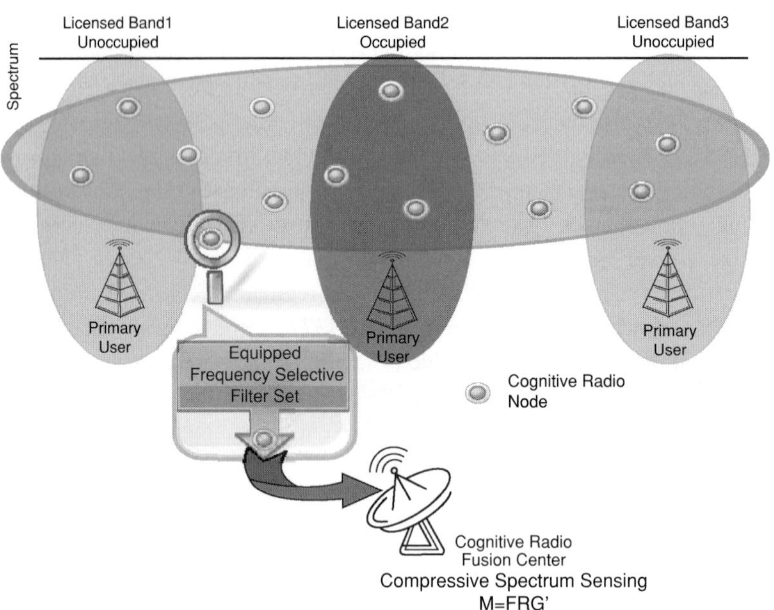

Figure 10.1 System model.

10.3 Compressive sensing-based collaborative spectrum sensing

In this section, we study how to employ CS to the collaborative spectrum sensing in CR networks. First, we give the system model in Section 10.3.1. Two algorithms are proposed in Section 10.3.2 and Section 10.3.3, respectively. The properties of the proposed algorithms are discussed in Section 10.3.4. Simulations are illustrated in Section 10.3.5.

10.3.1 System model

We consider a cognitive radio network with m CR nodes that locally monitor a subset of n channels. A channel is either occupied by a PR or unoccupied, corresponding to the states 1 and 0, respectively. We assume that the number s of occupied channels is much smaller than n. The goal is to recover the occupied channels from the CR nodes' observations. Since each CR node can only sense limited spectrum at a time, it is impossible for limited m CRs to observe n channels simultaneously.

To overcome this problem, we propose the scheme depicted in Figure 10.1. Instead of scanning all channels and sending each channel's status to the fusion center, using its frequency-selective filters, a CR takes a small number of measurements that are linear combinations of multiple channels. The filter coefficients can be designed and implemented easily. In order to mix the different channel-sensing information, the filter coefficients are designed to be random numbers. Then, these filter outputs are sent to the fusion center. Suppose that there are p frequency-selective filters in each CR node sending out p reports regarding the n channels. For the non-ideal cases, where we have relatively fewer measurements $pm < n$; i.e., the number of reports sent from all CRs is less than the total number of channels. The sensing process at each CR can be represented by a $p \times n$ filter coefficient matrix $\mathbf{F}$. Let an $n \times n$ *diagonal* matrix $\mathbf{R}$ represent the states of all the channel sources using 0 and 1 as diagonal entries, indicating the unoccupied or occupied states, respectively. There are s non-zero entries in diag($\mathbf{R}$). In addition, channel gains between the CRs and channels are described in an $m \times n$ channel gain matrix $\mathbf{G}$ given by [1]

$$\mathbf{G}_{i,j} = P_i(d_{i,j})^{-\alpha/2}|h_{i,j}|, \tag{10.1}$$

where P_i is the ith primary user's transmitted power, $d_{i,j}$ is the distance between the primary transmitter using the jth channel and the ith CR node, α is the propagation loss factor, and $h_{i,j}$ is the channel fading gain. For AWGN channel, $h_{i,j} = 1$, for all $\{i, j\}$; for the Rayleigh channel, $|h_{i,j}|$ follows the independent Rayleigh distribution; and for shadowing fading, $|h_{i,j}|$ follows the log-normal distribution [395]. Without loss of generality, we assume that all PRs use unit transmit power (otherwise, we can compensate by altering the corresponding channel gains). The measurement reports sent to the fusion center can be written as a $p \times m$ matrix

$$\mathbf{M}_{p\times m} = \mathbf{F}_{p\times n}\mathbf{R}_{n\times n}(\mathbf{G}_{m\times n})^T. \tag{10.2}$$

Note that due to loss or errors, some of the entries of $\mathbf{M}$ are possibly missing. The binary numbers on the diagonal of $\mathbf{R}$ are the n-channel states that we shall estimate from the available entries of $\mathbf{M}$.

10.3.2 CSS matrix completion algorithm

It is typically difficult for the fusion center to acquire all entries of $\mathbf{M}$ due to transmission failure, which means that our observation is a subset $\mathbf{E} \subseteq [p] \times [m]$ of $\mathbf{M}$. However, it is possible to recover the missing entries in $\mathbf{M}$ since it holds the following two important properties [88] required for matrix completion:

1. **Low Rank**: rank(M) equals s, which is the number of prime users in the network and is usually very small.
2. **Incoherent Property**: Generate $\mathbf{F}$ randomly (subject to hardware limitation). From (10.1) and the fact that $\mathbf{R}$ has only s non-zeros on the diagonal, $\mathbf{M}$'s SVD factors $\mathbf{U}$, Σ, and $\mathbf{V}$ satisfy the *incoherence condition* [370].
   - There exists a constant $\mu_0 > 0$ such that for all $i \in [p]$, $j \in [m]$, we have $\sum_{k=1}^{s} \mathbf{U}_{i,k}^2 \leq \mu_0 s$, $\sum_{k=1}^{s} \mathbf{V}_{i,k}^2 \leq \mu_0 s$.
   - There exists μ_1 such that $|\sum_{k=1}^{s} \mathbf{U}_{i,k}\Sigma_k \mathbf{V}_{j,k}| \leq \mu_1 s^{1/2}$.

$\mathbf{M}$ is in general incomplete because of transmission failure. Moreover, each CR might only be able to collect a random (up to p) number of reports due to the hardware limitation. Therefore, the fusion center receives a subset $\mathbf{E} \subseteq [p] \times [m]$ of $\mathbf{M}$'s entries.

We assume that the received entries are uniformly distributed with high probability.[1] Hence, we work with a model in which each entry shows up in $\mathbf{E}$ identically and independently with probability $\epsilon/\sqrt{p \times m}$. Given $\mathbf{E}_{p\times m}$, the partial observation of $\mathbf{M}$ is defined as a $p \times m$ matrix given by

$$\mathbf{M}_{ij}^{E} = \begin{cases} \mathbf{M}_{ij}, & \text{if } (i,j) \in \mathbf{E}; \\ 0, & \text{otherwise.} \end{cases} \tag{10.3}$$

We first recover the unobserved elements of $\mathbf{M}$ from $\mathbf{M}^E$. Then, we reconstruct $(\mathbf{RG}^T)$ from the given $\mathbf{F}$ and $\mathbf{M}$ using the fact that all but s rows of $(\mathbf{RG}^T)$ are zero. These non-zero rows correspond to the occupied channels. Since p and m are much smaller than n, our approach requires a much smaller amount of sensing and transmission, compared to traditional spectrum sensing in which each channel is monitored separately.

In previous research on matrix completion [149, 150, 370, 371], it was proved that under some suitable conditions, a low-rank matrix can be recovered from a random, yet small subset of its entries by nuclear-norm minimization

$$\min_{\mathbf{M}\in\mathbb{R}^{p\times n}} \tau\|\mathbf{M}\|_* + \frac{1}{2}\sum_{(i,j)\in\mathbf{E}} \left|\mathbf{M}_{i,j} - \mathbf{M}_{i,j}^E\right|^2, \tag{10.4}$$

where $\|\mathbf{M}\|_*$ denotes the nuclear-norm of matrix $\mathbf{M}$ and τ is a parameter discussed in Section 10.3.2 below. For notational simplicity, we introduce the linear operator $\mathcal{P}$ that selects the components $\mathbf{E}$ out of a $p \times n$ matrix and forms them into a vector such that $\|\mathcal{P}\mathbf{M} - \mathcal{P}\mathbf{M}^E\|_2^2 = \sum_{(i,j)\in\mathbf{E}} |\mathbf{M}_{i,j} - \mathbf{M}_{i,j}^E|^2$. The adjoint of $\mathcal{P}$ is denoted by $\mathcal{P}^*$.

Recent algorithms developed for (10.4) include, but are not limited to, the singular-value thresholding (SVT) algorithm [149] and the fixed-point continuation iterative algorithm (FPCA) [150] for fast completion of large-scale matrices (e.g., more than 1000×1000), a special trimming step introduced by Keshavan and Oh in [370].

For our problem, we adopt FPCA, which appears to run very well for our small-dimensional tests. In the following subsections, we describe this algorithm and the steps we take for nuclear-norm minimization. Also, we study how to use the approximate singular-value decomposition (SVD)-based iterative algorithm introduced in [150] for fast execution. We further discuss the stopping criteria for iterations to acquire optimal recovery. Finally, we show how to obtain $\mathbf{R}$ from the estimation $\tilde{\mathbf{M}}$ of $\mathbf{M}$.

Nuclear-norm min. via fixed-point iterative algorithm

FPCA is based on the following fixed-point iteration

$$\begin{cases} \mathbf{Y}^k = \mathbf{M}^k - \delta_k\mathcal{P}^*(\mathcal{P}\mathbf{M}^k - \mathcal{P}\mathbf{M}^E), \\ \mathbf{M}^{k+1} = S_{\tau\delta_k}(\mathbf{Y}^k), \end{cases} \tag{10.5}$$

where δ_k is step size and $S_\alpha(\cdot)$ is the matrix shrinkage operator defined as follows:

[1] Depending on the different channel gain, the CRs will select different coding/modulation/power control schemes so that the received signal-to-noise ratio can be maintained about a certain threshold. Because of this reason, we can assume that the loss of information is uniformly distributed.

Definition 9. ***Matrix shrinkage operator*** $S_\alpha(\cdot)$*: Assume* $\mathbf{M} \in \mathbb{R}^{p\times m}$ *and its SVD is given by* $\mathbf{M} = \mathbf{U}diag(\sigma)\mathbf{V}^T$*, where* $\mathbf{U} \in \mathbb{R}^{p\times r}$*,* $\sigma \in \mathbb{R}_+^r$*, and* $\mathbf{V} \in \mathbb{R}^{m\times r}$*. Given* $\alpha > 0$*,* $S_\alpha(\cdot)$ *is defined as*

$$S_\tau(\mathbf{M}) := \mathbf{U}\mathrm{diag}\left(s_\alpha(\sigma)\right)\mathbf{V}^T, \tag{10.6}$$

with the vector $s_\alpha(\sigma)$ *defined as*

$$s_\alpha(x) := \max\{x - \alpha, 0\}, \ \textit{component-wise.} \tag{10.7}$$

Simply speaking, $S_\tau(\mathbf{M})$ reduces every singular values (which is non-negative) of $\mathbf{M}$ by τ; if one is smaller than α, it is reduced to zero. In addition, $S_\alpha(\mathbf{M})$ is the solution of

$$\min_{\mathbf{X}\in\mathbb{R}^{m\times n}} \alpha\|\mathbf{X}\|_* + \frac{1}{2}\|\mathbf{X} - \mathbf{M}\|_F^2, \tag{10.8}$$

where $\|\cdot\|_F$ is the Frobenius norm.

To understand (10.5), observe that the first step of (10.5) is a gradient-descent applied to the second term in (10.4) and thus reduces its value. Because the previous gradient-descent generally increases the nuclear-norm, the second step of (10.5) involves solving (10.8) to reduce the nuclear-norm of $\mathbf{Y}^k$. Iterations based on (10.5) converge when the step sizes δ_k are properly chosen (e.g., less than 2, or select by line search) so that the first step of (10.5) is not expansive (the other step is always non-expansive).

Approximate SVD-based fixed-point iterative algorithm

As stated in [150], the second step of (10.5) requires computing the SVD decomposition of $\mathbf{Y}^k$, which is the main computational cost of (10.5). However, if one can predetermine the rank of the matrix $\mathbf{M}$, or have the knowledge of the approximate range of its rank, a full SVD can be simplified to computing only a rank-r approximation to $\mathbf{Y}^k$. Combined with the above fixed-point iteration, the resulting algorithm is called a fixed-point continuation algorithm with approximate SVD (FPCA). Specifically, the approximate SVD is computed by a fast Monte Carlo algorithm developed by Drineas et al., [396]. For a given matrix $\mathbf{A} \in \mathbb{R}^{m\times n}$ and parameters k_s, this algorithm returns approximations to the largest k_s singular values corresponding to the left singular vectors of the matrix $\mathbf{A}$ in a linear time.

Stopping criterion for iterations

We tune the parameters in FPCA for a better overall performance. Continuation is adopted by FPCA, which solves a sequence of instances of (10.4), easy to difficult, corresponding to a sequence of large to small values of τ. The final τ is the given one but solving the easier instances of (10.4) gives intermediate solutions that warm start the more difficult ones so that the entire solution time is reduced. Solving each instance of (10.4) requires proper stopping. Because our ultimate goal is to recover 0/1 values on the diagonal of $\mathbf{R}$, accurate solutions of (10.4) are not required. Therefore, we use the criterion

$$\frac{\|\mathbf{M}^{k+1} - \mathbf{M}^k\|_F}{\max\{1, \|\mathbf{M}^k\|_F\}} < mtol, \tag{10.9}$$

where *mtol* is a small positive scalar. Experiments show that $1e^{-6}$ is good enough for obtaining the optimal **R**.

Channel availability estimation based on the complete measurement matrix

Since **F** has more columns than rows, directly solving $\mathbf{X} := \mathbf{R}\mathbf{G}^T$ in (10.1) from a given **M** is underdetermined. However, each row X_i of **X** corresponds to the occupancy status of channel i. Ignoring noise in **M** for now, X_i contains a positive entry if and only if channel i is used. Hence, most rows of **X** are completely zero, so every column $X_{\cdot,j}$ of **X** is sparse and all $X_{\cdot,j}$s are jointly sparse. Such sparsity allows us to reconstruct **X** from (10.1) and to identify the occupied channels, which are the non-zero rows of **X**.

Since the channel fading decays fast, the entries of **X** have a large dynamic range, which none of the existing algorithms can deal with sufficiently well. Hence, we develop a novel joint-sparsity algorithm briefly described as follows. The algorithm is much faster than matrix completion and typically needs 1–5 iterations. At each iteration, every column $X_{\cdot,j}$ of **X** is independently reconstructed using the model $\min\{\sum_i w_i|X_{i,j}| : FX_{\cdot,j} = M_{\cdot,j}\}$, where $M_{\cdot,j}$ is the jth column of **M**. For noisy **M**, we instead use the constraint $\|FX_{\cdot,j} - M_{\cdot,j}\| \leq \sigma$. The same set of weights w_i is shared by all j at each iteration. w_i is set to 1 uniformly at iteration 1. After channel i is detected in an iteration, w_i is set to 0. Through w_i, joint sparsity information is passed to all j. Channel detection is performed on the reconstructed $X_{\cdot,j}$s at each iteration. It is possible that some reconstructed $X_{\cdot,j}$ is wrong, so we let larger and sparser $X_{\cdot,j}$s have more say. If there is a relatively large $X_{i,j}$ in a sparse $X_{\cdot,j}$, then i is detected. We have found this algorithm to be very reliable. The detection accuracy is determined by the accuracy of **M** provided.

10.3.3 CSS joint sparsity recovery algorithm

Next, we describe a new, highly effective algorithm for recovering **X** (in (10.10))

$$\mathbf{X}_{n\times m} = \mathbf{R}_{n\times n} \times (\mathbf{G}_{m\times n})^T, \tag{10.10}$$

and then recovering **R** by thresholding **X**. The algorithm allows but does not require the same **F** for all CRs; i.e., each CR can use a different sensing matrix **F**. The design of **F** is discussed in Section 10.3.4 below.

In **X**, each column (denoted by $X_{\cdot,j}$) corresponds to the channel occupancy status received by CR j, and each row $X_{i,\cdot}$ corresponds to the occupancy status of channel i. *Ignoring noise* for now, a row has a positive value (i.e., $|X_{i,\cdot}| > 0$) if and only if channel i is used. Since there are only a small number of used channels, **X** is sparse in terms of the number of non-zero rows. In each non-zero row $X_{i,\cdot}$, there is typically more than one non-zero entry; in other words, if $X_{i,j} \neq 0$, other entries in the same row are likely to be non-zero. Therefore, **X** is *jointly sparse*. In the case that the true **X** contains noise, it is approximately, rather than exactly, jointly sparse.

Joint sparsity is utilized in our algorithm to recover **X**. While there are existing algorithms for recovering jointly sparse signals in the literature (e.g., in [372–374]), our algorithm is very different and more effective for our underlying problem. None of the existing algorithms work well to recover **X** because the entries of **X** have a very large dynamic range. This is because, in any channel fading model, channel gains decay rapidly with distance between CRs and PRs. Most of the existing algorithms are based on minimizing $\sum_i \|X_{i,\cdot}\|_p$ for $p \geq 1$ and $p = \infty$. If $p = 1$, it is the same as minimizing the 1-norm of each column independently, so joint sparsity is not used for recovery. If $p > 1$ or $p = \infty$, joint sparsity is considered, but it penalizes a large dynamic range since the large values in a non-zero row of **X** contribute superlinearly, more than the small values in that row, to the minimizing objective. In short, p close to 1 loses joint sparsity and p bigger than 1 penalizes large dynamic ranges. Our new algorithm not only utilizes joint sparsity but also takes advantage of the large dynamic range of **X**.

The large dynamic range has its pros and cons in CS recovery. It makes it easy to recover the locations of large entries, which can be achieved even without recovering the locations of smaller ones. On the other hand, it makes it difficult to recover both the locations and values of the smaller entries. This difficulty has been studied in our previous work [397], where we proposed a fast and accurate algorithm for recovering 1D signals x by solving several (about $5 - 10$) subproblems in the form of

$$\text{Truncated } \ell_1 \text{ minimization:} \quad \min\{\sum_{i \in T} |x_i| : Ax = b\}, \tag{10.11}$$

where the index set T is formed iteratively as $\{1, \ldots, n\}$ excluding the identified locations of large entries of x. With techniques such as early detections and warm starts, it achieves both the state-of-the-art speed and least requirement on the number of measurements. We integrate the idea of this algorithm with joint sparsity into the new algorithm below.

The framework of the proposed algorithm is shown in Algorithm 4. At each iteration, every channel is first subject to independent recovery. Unlike minimizing $\sum_i \|X_{i,\cdot}\|_p$,

Algorithm 4 Joint detection algorithm

$T \leftarrow \{1, \ldots, n\}$
repeat
  Independence recovery:
  $\mathbf{X} \leftarrow 0$
  $X_{\cdot,j} \leftarrow \min\{\sum_{i \in T} X_{i,j} : A_j X_{\cdot,j} = b_j, X_{\cdot,j} \geq 0\}$ for every CR j with enough measurements (in the presence of measurement noise, $A_j X_{\cdot,j} = b_j$ is replaced by $\|A_j X_{\cdot,j} - b_j\| \leq \sigma$)
  Channel detection:
  select trusted $X_{\cdot,j}$ and detect used channels from the selections
  Update of T**:**
  Update T according to detected channels and **X**
until the tail of **X** is small enough
Report **X**, and **R** by thresholding **X**

which ties all CRs together, independent recovery allows large entries of $\mathbf{X}$ to be quickly recovered. Joint sparsity information is passed among the CRs through a shared index set T, which is updated iteratively to exclude the used channels that are already discovered. Below, we describe each step of the above algorithm in more detail.

In the **independence recovery** step, for every qualified CR, a constrained problem in the form of (10.11) with constraints $A_j X_{\cdot,j} = b_j$ in the noiseless case, or $\|A_j X_{\cdot,j} - b_j\| \leq \sigma$ in the noisy case, is considered, where σ is an estimated noise level. As problem dimensions are small in our application, solvers are easily chosen: MATLAB's "linprog" for noiseless cases and Mosek [398] for noisy cases. Both of these solvers run in polynomial times. This step dominates the total running time of Algorithm 4, but up to m optimization problems can be solved in parallel. Parallelization is simple for the joint-sparsity approach. At each outer iteration, all LPs are solved independently, and they have small scales relative to today's LP solvers, like Gurobi [399] and its MATLAB interface Gurobi Mex [400], where Gurobi automatically detects and uses all CPU and cores for solving LPs. CRs without enough measurements (e.g., most of their reports are missing due to transmission losses or errors) are not qualified for independent recovery because CS recovery is known to be unstable in such a case. Specifically, we require the number of available measurements from each qualified CR to exceed twice as many as used channels or $n - |T|$.

When measurements are ample, the first iteration will yield exact or nearly exact $X_{\cdot,j}$s. Otherwise, insufficient measurements can cause a completely wrong $X_{\cdot,j}$ that misleads channel detection; neither the locations nor the values of the non-zero entries are correct. The algorithm, therefore, filters trusted $X_{\cdot,j}$s that must be either sparse or compressible. Large entries in such $X_{\cdot,j}$s likely indicate correct locations. A theoretical explanation of this argument based on stability analysis for (10.11) is given in [238].

Used channels are detected among the set of trusted $X_{\cdot,j}$s. To further reduce the risk of false detections, we compute a percentage for every channel in a way that those channels corresponding to larger values in $\mathbf{X}$ and whose values are located in relatively sparser $X_{\cdot,j}$s are given higher percentages. Here, relative sparsity is defined proportionally to the number of measurements; for a fixed number of non-zeros or degree of compressibility, the greater the number of measurements, the higher the relative sparsity. Hence, $X_{\cdot,j}$ corresponding to more reported CR j also tends to have a higher percentage. In short, larger and sparse solutions have more say. The channels receiving higher percentages are detected as used channels.

The index set T is set as $\{1, \ldots, n\}$ excluding the used channels that are already detected. Obviously, T needs to change from one iteration to the next; otherwise, two iterations will result in an identical $\mathbf{X}$ and thus the stagnation of the algorithm. Therefore, if the last iteration posts no change in the set of used channels yet the stopping criterion (see next paragraph) is not met, the channels i corresponding to the larger $\|X_{i,\cdot}\|_2$ are also excluded from T, and such exclusion becomes more aggressive as the iteration number increases. This is not an ad hoc but a rigorous treatment. It is shown in [238] that larger entries in an inexact CS recovery tend to be the true non-zero entries, and furthermore, as long as the new T excludes more true than false non-zero entries by a certain fraction, (10.11) will yield a better solution in terms of a certain norm. In short, used channels leave T, and in the case of no leaves, channels with larger joint values $\|X_{i,\cdot}\|_2$ leave T.

Finally, the iteration is terminated when the tail of $\mathbf{X}$ is small enough. One way to define the tail size of $\mathbf{X}$ is the fraction $\sum_{i \in T} \|X_{i,\cdot}\|_p / \sum_{i \notin T} \|X_{i,\cdot}\|_p$; i.e., the thought-unused divided by the thought-used. Suppose that T precisely contains the unused channels and measurements are noiseless, then every recovered $X_{\cdot,j}$ in channel detection is exact, so the fraction is zero; with noise, the fraction depends on noise magnitude and is small as long as noise is small. If T includes any used channel, the numerator will be large whether or not $X_{\cdot,j}$s are (nearly) exact. In a sense, the tail size measures how well $\mathbf{X}$ and T match the measurements b and expected sparseness. Unless the true number of used channels is known, the tail size appears to be an effective stopping indicator.

10.3.4 Discussion

Complexity

In the worst case, Algorithm 4 reduces the cardinality of T by 1 per iteration, corresponding to recovering at least one additional used channel. Therefore, the number of iterations cannot exceed the number of total channels. However, the first couple of iterations typically recover most of the used channels. At each iteration, the independence recovery step solves up to m optimization problems, which can be independently solved in parallel, so the complexity equals a linear program (or second-order cone program) whose size is no more than n. The worst-case complexity is $O(n^3)$ but it is almost never observed in sparse optimization thanks to solution sparsity. The two other steps are based on basic arithmetic and logical operations, and they run in $O(p \times n)$. In practice, algorithm 4 is implemented and run on a workstation at the fusion center. Computational complexity will not be a bottleneck of the system. As to the matrix completion algorithm, according to [150], FPCA can recover 1000×1000 matrices of rank 50 with a relative error of 10^{-5} in about 3 min by sampling only 20 percent of the elements.

Comparisons between the two approaches

The matrix completion (Section 10.3.2) and joint sparsity recovery (Section 10.3.3) approaches both take linear channel measurements as input and both return the estimates of used channels. On the other hand, the joint sparsity approach takes full advantage of $\mathbf{F}$, so it is expected to work with smaller numbers of measurements. In addition, even though only one matrix completion problem needs to be solved in the matrix completion approach, it still takes much longer than running the entire joint sparsity recovery, and it is not easy to parallelize any of the existing matrix completion algorithms. However, in the small-scale networks, in cases where too much sensing information is lost during transmission or there are too many active PRs in the network, which increase the signal sparsity level, the joint sparsity recovery algorithm with our current settings will experience degradation in performance.

We, however, cannot predict an eventual winner between the two approaches as they are both being studied and improved in the literature. For example, if a much faster matrix completion algorithm is developed that takes advantage of **F**, the disadvantages of the approach may no longer exist.

Frequency-selective filter design and adaptive sensing

The proposed method senses the channels, not by measuring the responses of individual channels one by one, but rather measures a few incoherent linear combinations of all channels' responses through onboard frequency-selective filter set. The filter coefficients that perform as the sensing matrix should have entries independently sampled from a sub-Gaussian distribution, since this is known to be best for CS in terms of the number of measurements (given in order of magnitude) required for exact recovery. In other words, up to a constant, which is independent of problem dimensions, no other type of matrix is yet known to perform consistently better. However, other types of matrices (such as partial Fourier/DCT matrices [67, 69] and other random circulant matrices [401]) have been either theoretically and/or numerically demonstrated to work as effectively in many cases. These latter sensing matrices are often easier to realize physically or electrically. For example, applying a random circulant matrix performs a subsampled convolution with a random vector.

Frequency-selective surfaces (FSSs) can be used to realize frequency filtering. This can be done by designing a planar periodic structure with a unit element size around half-wavelength of the frequency of interests. Both metallic and dielectric materials can be used. To deal with the bandwidth, unit elements in different shapes will be tested.

10.3.5 Simulations

The Probability of Detection (POD) and False Alarm Rate (FAR) are the two most important indices associated with spectrum sensing. We also consider the Miss Detection Rate (MDR) of the proposed system. The higher the POD, the less interference the CRs will bring to the PRs, while from the CRs' perspective, lower FAR will increase their chance of transmission. There is a tradeoff between POD and FAR. While designing the algorithms, we try to balance the CR nodes' capability of transmission and their interferences to the PR nodes. Performance is evaluated in terms of POD, FAR and MDR defined as follows:

$$\begin{aligned}
\text{FAR} &= \text{No. False }/(\text{No. False} + \text{No. Hit});\\
\text{MDR} &= \text{No. Miss}/(\text{No. Miss} + \text{No. Correct});\\
\text{POD} &= \text{No. Hit}/(\text{No. Hit} + \text{No. Miss}),
\end{aligned}$$

where *No. False* is the number of false alarms, *No. Miss* is the number of miss detections, *No. Hit* is the number of successful detections of primary users, and *No. Correct* is the number of correct reports of no appearance of PR. We define sampling rate as:

$$\frac{\text{No. received measurements at the fusion center}}{\text{No. channels} \times \text{No. CRs}},$$

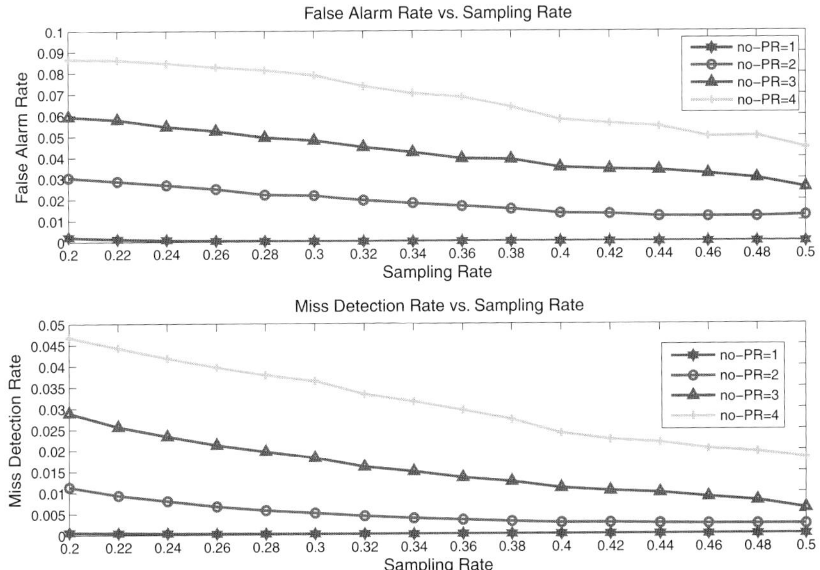

Figure 10.2 False alarm and missing probability vs. sampling rate.

where (No. channel × No. CR) is the amount of total sensing workload in traditional spectrum sensing.

Simulation of matrix completion recovery

According to the FCC and the Defense Advanced Research Projects Agency (DARPA) reports [402, 403] data, we chose to test the proposed matrix completion recovery algorithm for spectrum utilization efficiency over a range from 3% to 12%, which is large enough in practice. Specifically, the number of active primary users is 1 to 4 on a given set of 35 channels with 20 CR nodes.

Figure 10.2 shows the false alarm and miss detection rates at different sampling rates for different numbers of PR nodes. Among all cases, the highest miss detection rate is no more than 5%, and this is from only 20% samples that are supposed to be gathered from the CR nodes regarding all the channels. When the sampling rate is increased to 50% and even when the channel occupancy is relatively high; i.e., 12% of the channels are occupied by the PRs, the miss detection rates can be as low as 2%. From our simulation results, with a moderate channel occupancy at 9%, the false alarm rates are around 3% to 5%.

Figure 10.3 shows the probability of detection at different sampling rates. When the spectrum is lightly occupied by the licensed user at 3% channels being occupied, only 20% samples offer a POD close to 100%, and when there is a slight rise in sampling rate, POD can reach 100%. In the worst case of 12% spectrum occupancy, 20% sampling rate still can offer a POD of higher than 95%, and as the sampling rate reaches 50%, POD can reach 98%.

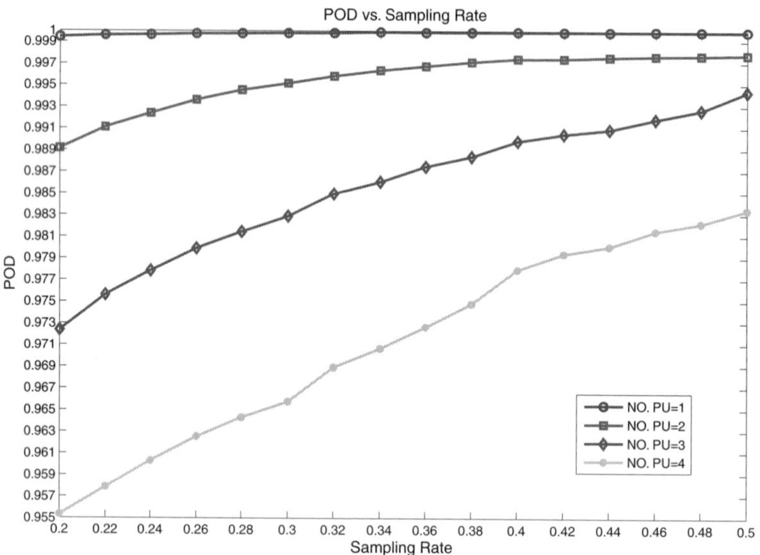

Figure 10.3 POD vs. sampling rate.

Joint sparsity recovery simulation

Joint sparsity recovery is designed for large-scale application, and simulations carried out for larger-dimensional applications with the following settings: We consider a 20-node cognitive radio network within a 500×500 m square area centered at the fusion center. The 20 CR nodes are uniformly randomly located. These cognitive radio nodes collaboratively sense the existence of primary users within a 1000×1000 m square area on 500 channels, which are centered also at the fusion center. We chose to test the proposed algorithm for the number of active PR nodes ranging from 1 to 15 on the given set of 500 channels. Since the fading environments of the cognitive radio networks vary, we evaluate the algorithm performance under three typical channel fading models: AWGN channel, Rayleigh fading channel, and lognormal shadowing channel.

We first evaluate the POD, FAR, and MDR performance of the proposed joint sparsity recovery performance in the noiseless environment. Figure 10.4, Figure 10.5, and Figure 10.6 show the POD, FAR, and MDR performance at different sampling rate, for AWGN channel, Rayleigh fading channel, and log-normal shadowing channel, respectively, when a small number of CR nodes sense the spectrum collaboratively. Figure 10.7, Figure 10.8, and Figure 10.9 show the POD, FAR, and MDR performance at different sampling rates for the aforementioned three types of channel models when there are more CR nodes involved in the collaborative sensing of the spectrum. We observe that the log-normal shadowing channel model shows the best POD, FAR, and MDR performance no matter how many CR nodes are involved in the spectrum sensing. In contrast, the AGWN channel model shows the worst POD, FAR, and MDR performance. With respect to POD, the performance gap between these two models is at most 10%, which happens when the sampling rate is extremely low. For the Rayleigh fading channel model, when the number of samples is 62% of the total number of channels,

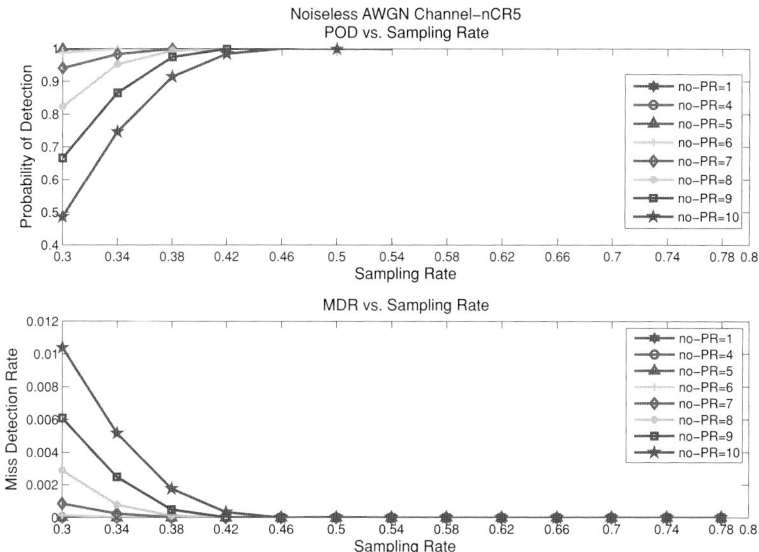

Figure 10.4 Noiseless AWGN channel (No. of CR = 5).

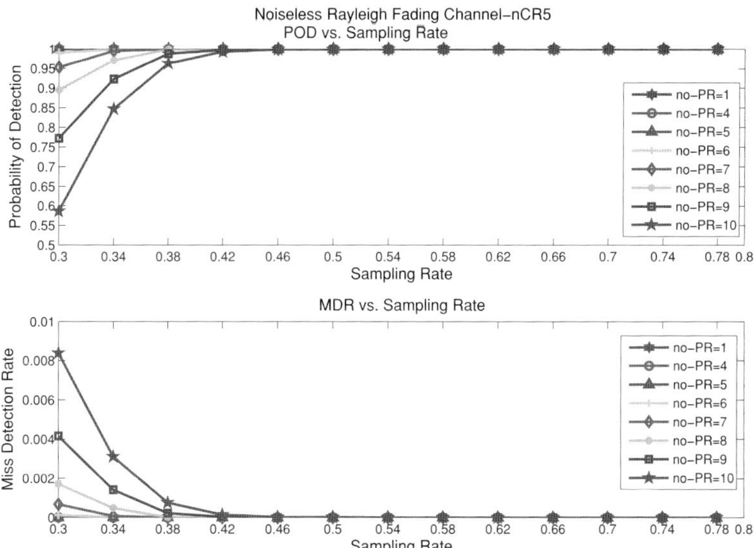

Figure 10.5 Noiseless Rayleigh fading channel (No. of CR = 5).

for all tested cases we achieved 100% POD. If there are fewer active PR nodes in the network, a smaller number of samples is required for exact detection. In essence, the proposed CCS system is robust to severe or poorly modeled fading environments. Cooperation among the CR nodes and the robust recovery algorithm allows us to achieve this robustness without imposing stringent requirements on individual radios.

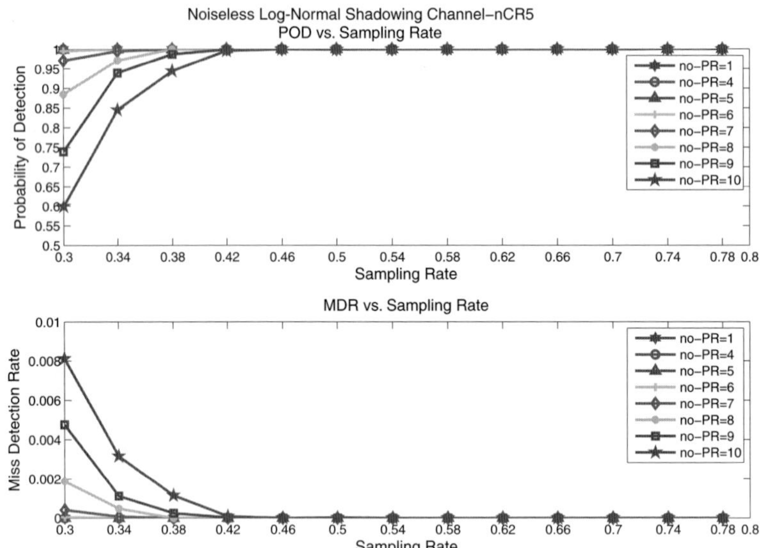

Figure 10.6 Noiseless log-normal shadowing channel (No. of CR = 5).

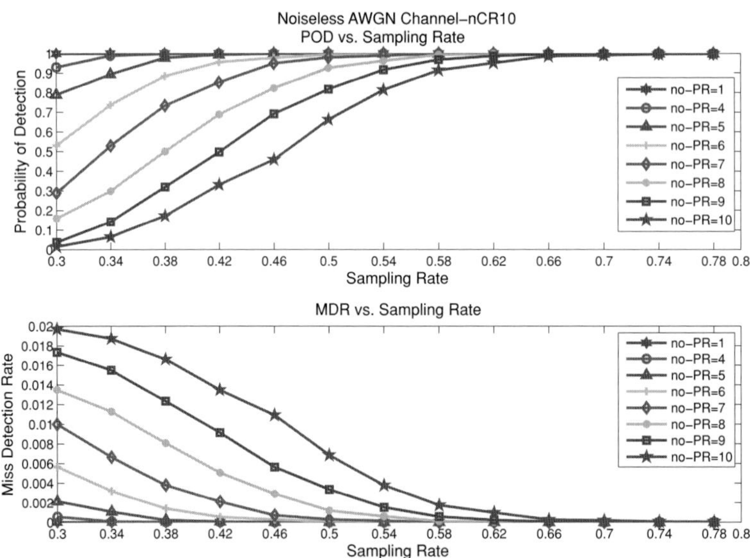

Figure 10.7 Noiseless AWGN channel (No. of CR = 10).

We then evaluated the POD, FAR, and MDR performance of the proposed joint sparsity recovery performance in noisy environments. For all the simulations considering noise, we adopted the Rayleigh fading channel model. Figure 10.10 and Figure 10.11 show the corresponding results. We observed that noise does degrade the performance. However, as shown in Figure 10.10, when the number of active PRs is small enough (e.g., number

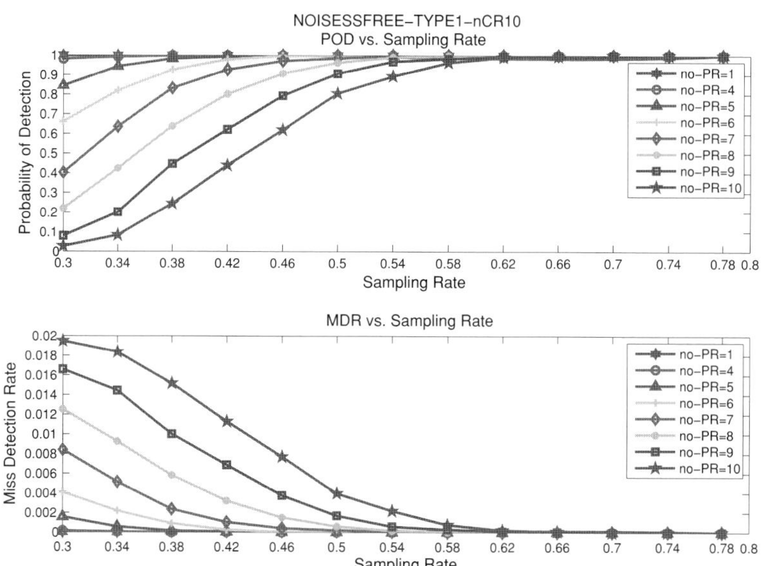

Figure 10.8 Noiseless Rayleigh fading channel (No. of CR = 10).

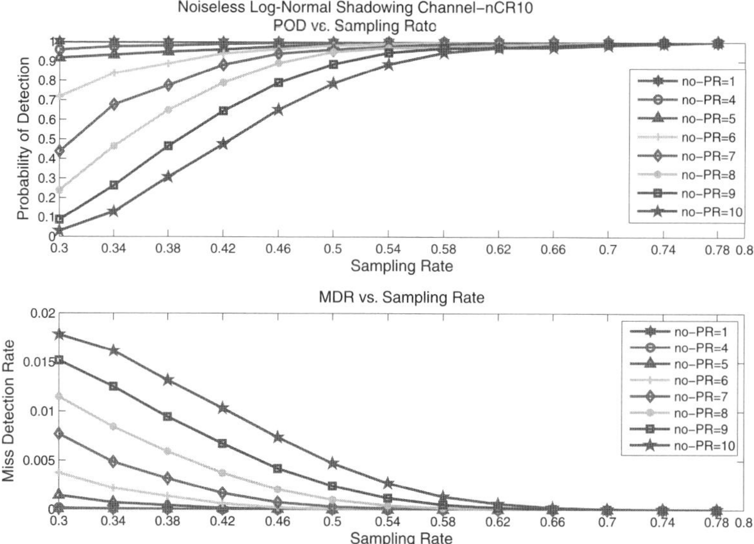

Figure 10.9 Noiseless log-normal shadowing channel (No. of CR = 10).

of PR = 1), even with signal-to-noise ratio as low as 15 dB, we still can achieve 100% POD with a sampling rate of merely 50%. Then, with an increase in the signal-to-noise ratio, lower sampling rate enables more PR nodes to be detected exactly. Figure 10.11 shows the POD, FAR, and MDR performance vs. sampling rate at different noise levels. Each curve for a specific noise level is relatively flat (i.e., performance varies a little

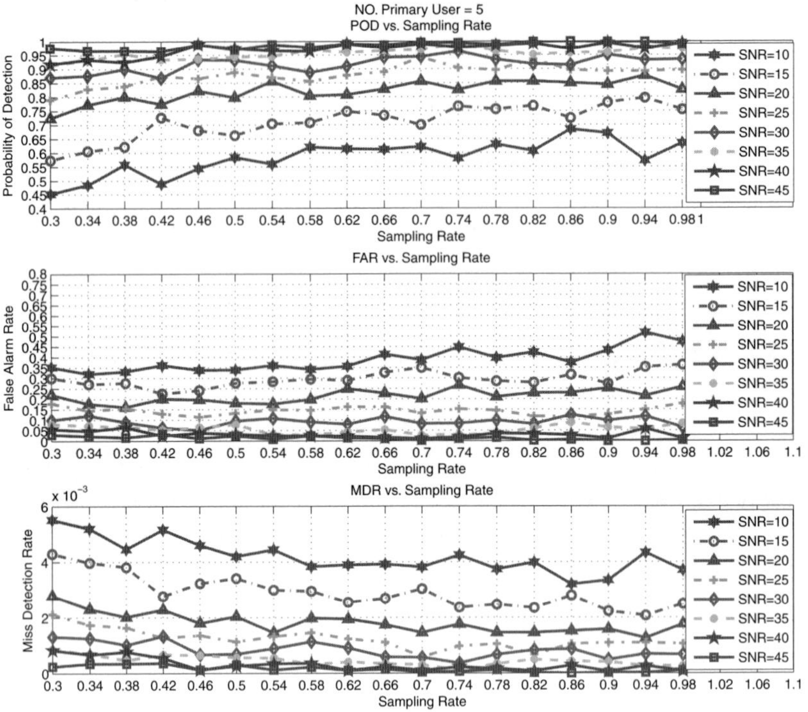

Figure 10.10 POD, FAR, and MDR performance vs. sampling rate at different SNR.

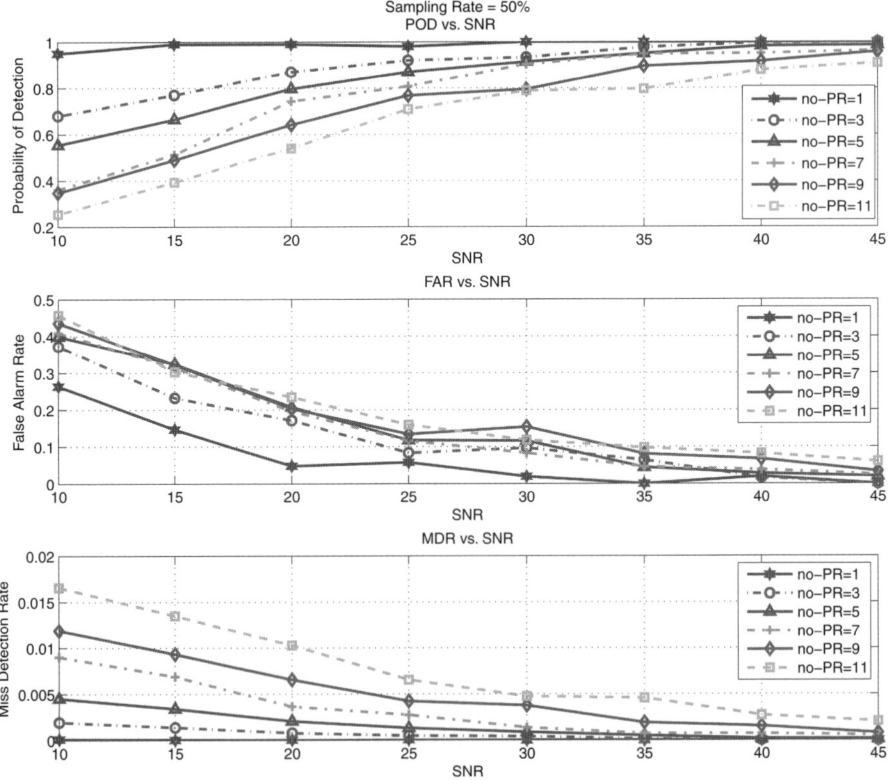

Figure 10.11 POD, FAR, and MDR performance vs. noise level for different number of PR.

as sampling rate changes). This shows that the noise level has greater impact on the spectrum sensing performance rather than the sampling rate. At a low noise level, e.g., SNR = 45 dB, 40% sampling rate enables 100% POD for 4 PR nodes. As SNR reduces to 15 dB, no more than 70% POD will be achieved even when the number of samples equals the number of channels in the network.

Comparison between matrix completion algorithm and joint sparsity recovery algorithm

For comparison, we applied the joint sparsity recovery algorithm on a small-scale network with the same settings as we have used to test the matrix completion recovery. Instead of a 500-channel network, we have a network with only 35 channels. Simulation results show that the joint sparsity recovery algorithm performs better than the matrix completion algorithm in the following aspects:

1. faster computation due to lower computational complexity;
2. higher POD for the spectrum utilization rate between 3% and 12% in the noise-free simulations.

To conclude, the matrix completion algorithm is good for small-scale networks, with relatively high spectrum utilization, while the joint sparsity recovery algorithm has the advantage of low computational complexity, which enables fast computation in large-scale networks.

10.4 Dynamic approach

In the previous section, we proposed *collaborative CS*, in which the cognitive radios (CRs) sense the occupied spectrum channels by measuring linear combinations of channel powers, which is more efficient than sweeping a set of channels sequentially. The measurements are reported to the fusion center, where the occupied channels are recovered by CS algorithms. In this section, we study a method of dynamic CS, which continuously measures channel powers and recovers the occupied channels in a dynamic environment. While standard CS algorithms must recover multiple occupied channels, a dynamic algorithm only needs to recover the recent change, which is either a newly occupied channel or a released one. On the other hand, the dynamic algorithm must recover the change just in time. Therefore, we study a least-squares-based algorithm, which is equivalent to ℓ_0 minimization. We demonstrate its fast speed and robustness to noise and simulation results demonstrate the effectiveness of the proposed scheme. The rest of this section is organized as follows: In Section 10.4.1, the system model is given. Section 10.4.2 studies the dynamic sensing scheme and proposes an algorithm. Simulation results are given in Section 10.4.3.

10.4.1 System model

We consider a cognitive radio network with M CRs monitoring N channels, within which there are S occupied channels. For simplicity, we assume that each occupied channel corresponds to one PR. Each CR is equipped with P frequency-selective filters, which continuously record P different linear combinations of the powers of all the N channels. Their reports are sent to the FC continuously at a certain rate. Here, P and M are much smaller than N. Hence, the system requires a smaller amount of sensing and transmission compared to the traditional approach. The formation of the reports can be described with three matrices $A \in \mathbb{R}^{P\times N}$, $R \in \mathbb{R}^{N\times N}$ and $G \in \mathbb{R}^{M\times N}$. The entries of A are frequency filter coefficients used by the CRs to generate P reports. Each CR can have its own unique A, but we assume a uniform A just for notational simplicity. R is a diagonal matrix with 1 or 0 on its diagonal, where $R_{nn} = 1$ means that channel n is occupied and $R_{nn} = 0$ means unoccupied. G represents the channel gains between the CRs that are given as

$$G_{m,n} = (d_{m,n})^{-\alpha/2}|h_{m,n}|, \tag{10.12}$$

where $d_{m,n}$ is the distance between the mth CR and the primary user using the nth channel, α is the propagation loss factor, and $h_{m,n}$ is the channel fading. The CR reports B are given as

$$B = A(RG^T) = AX \in \mathbb{R}^{P\times M}, \text{ where } X = RG^T. \tag{10.13}$$

The goal of (static) spectrum sensing is to recover X from a subset of entries in B. In X, each column m (i.e., $X_{:,m}$) contains the powers of all channels received by CR m, and each row $X_{n,:}$ corresponds to the powers of channel n. All but S rows of the matrix X are zero. These non-zero rows correspond to the occupied channels. Since there are only a small number of used channels, X is sparse in terms of non-zero rows. In each non-zero row $X_{n,:}$, there are typically more than one non-zero elements; if $X_{n,m} \neq 0$, other entries in the same row are likely non-zero. This joint sparsity property allows X to be recovered from a small number of entries in B by algorithms such as [81, 404]. The algorithm in [397] takes the large dynamic range of channel powers into consideration, so it returns a more stable recovery.

In a dynamic environment, X changes continuously. We consider discrete-times changes, which arise when a certain PR starts or stops using a channel, or the power of a channel changes significantly. Each change corresponds to a change in *one certain row of* X. In the future, we will study cases in which multiple changes can happen simultaneously. The dynamic spectrum sensing problem is handled in a sequence of time periods distinguished by the changes. Let B_{prev} and X_{prev} denote B and X in the previous period. In the current period, we shall recover X given X_{prev} and B_{prev}, as well as the entries of B that have been received since the start of this period. From this X, the current occupied channels and their gains can be computed. Since each period requires solutions from its previous period, the success of dynamic recovery is based on the success in every period, which is in turn based on the received entries in B in every period. First, let us set up the model for the current period.

Let n^* be the index of the changed row of X since the previous period and let $A = [a_1, \cdots, a_N]$. Then, $\Delta B = B - B_{\text{prev}} = AX - AX_{\text{prev}} = a_{n^*}(\Delta X)_{n^*,:}$, where $(\Delta X)_{n^*,:}$ is the n^*th row of ΔX. Let Ω denote the set of the received entries of B (equivalently, those of ΔB). The problem is to recover n^* and $(\Delta X)_{n^*,:}$ from the given A and the received entries $(\Delta B)_{p,m}$, $(p, m) \in \Omega$ following the system model

$$\Delta B = a_{n^*}(\Delta X)_{n^*,:}. \tag{10.14}$$

10.4.2 Dynamic recovery algorithm

The underlying problem is joint-sparsity recovery of only one non-zero row. There is no need to use techniques such as ℓ_1 minimization or greedy pursuit algorithms. Instead, the recovery can be as simple as solving the least-squares problem

$$\min_{n\in\{1,\ldots,N\}} \min_{y_n} \|(a_n y_n)_\Omega - (\Delta B)_\Omega\|_2, \tag{10.15}$$

where y_n is a *row vector* with M entries (thus $(a_n y_n) \in \mathbb{R}^{P\times M}$) and the subscription X_Ω stands for the vector formed by the entries $X_{p,m}$, $(p, m) \in \Omega$. Noting that problem (10.14) seeks a row n of ΔX that best matches the observation $(\Delta B)_\Omega$, we can see that $n = n^*$ and $y_{n^*} = (\Delta X)_{n^*,:}$ attain the minimum zero objective in (10.15).

Parallel implementation

For convenience of linear algebra, define $I_m := \{p : (p, m) \in \Omega\}$, which is the set of B entries that are received from the mth CR. Then, the objective function of (10.15) can be written as

$$\|(a_n y_n)_\Omega - (\Delta B)_\Omega\|_2^2 = \sum_{m=1}^{M} \left\|(y_n)_m a_{I_m,n} - (\Delta B)_{I_m,m}\right\|_2^2, \tag{10.16}$$

where for each m and n given, $a_{I_m,n}$ is the column vector formed by $\{a_{p,n} : (p, n) \in \Omega\}$ and $(\Delta B)_{I_m,m}$ is the column vector formed by $\{(\Delta B)_{pm} : (p, m) \in \Omega\}$. Given k, the minimizer of (10.16) is

$$(y_n^*)_m = \frac{(a_{I_m,n})^T(\Delta B)_{I_m,m}}{(a_{I_n,m})^T a_{I_n,m}}, \quad m = 1, \ldots, M. \tag{10.17}$$

For fast computation, we propose a vectorized algorithm, which returns $Y^* \in \mathbb{R}^{N\times M}$ consisting of all row vectors y_n^*, $n = 1, \ldots, N$. We introduce

$$\overline{\Delta B} \in \mathbb{R}^{P\times M}, \qquad (\overline{\Delta B})_{p,m} = \begin{cases} (\Delta B)_{p,m}, & (p, m) \in \Omega, \\ 0, & \text{otherwise;} \end{cases} \tag{10.18}$$

$$P \in \mathbb{R}^{P\times M}, \qquad P = \begin{cases} 1, & (p, m) \in \Omega, \\ 0, & \text{otherwise;} \end{cases} \tag{10.19}$$

$$\hat{A} \in \mathbb{R}^{P\times N}, \qquad \hat{A}_{p,n} = A_{p,n}^2. \tag{10.20}$$

Then, the numerator $(\mathbf{a}_{I_m,n})^T(\Delta B)_{I_m,m} = (A^T\overline{\Delta B})_{n,m}$ and the denominator $(\mathbf{a}_{I_m,n})^T \mathbf{a}_{I_m,n} = (\hat{A}^T P)_{n,m}$. Therefore, we have

$$Y^* = \frac{A^T(\overline{\Delta B})}{\hat{A}^T P} \quad \text{(component-wise division).} \tag{10.21}$$

Formula (10.21) can be computed by an advance matrix-matrix multiplication algorithm [405] with subcubic complexity.

Note that for (10.21) to be well defined, Ω must have one entry for each CR; i.e., $\{p : (p, m) \in \Omega\} \neq \emptyset$ for $m = 1, \ldots, M$. Under this condition, all entries of $\hat{A}^T P$ are non-zero for all practical choices of CS matrices A.

To solve (10.15) over n, one should compare

$$\|(a_n y_n^*)_\Omega - (\Delta B)_\Omega\|_2, \; n = 1, \ldots, N, \tag{10.22}$$

where y_n^* is the nth row of Y^*. If there is a unique n that gives 0 or a tiny value, then this n is n^*, the row that has changes since the previous period. Therefore, let

$$X_{n,:} \leftarrow (X_{prev})_{n,:} + \begin{cases} y_n^*, & n = n^*; \\ 0, & \text{otherwise,} \end{cases} \quad n = 1, \ldots, N. \tag{10.23}$$

The complexity of (10.22) is $O(|\Omega|N)$ and that of (10.23) is $O(N)$, assuming an in-place update from X_{prev} to X.

Algorithm 5 Dynamic Compressive Spectrum Sensing

Require: A, X_{prev}, $B_{\text{prev}} = AX_{\text{prev}}$, and $B_{p,m}$, $(p, m) \in \Omega$
1: Generate $\overline{\Delta B}$, P, and $\hat{A}$ according to (10.18)–(10.20);
2: Compute Y^* according to (10.21);
3: Set n^* as the minimizer of (10.22);
4: Compute X according to (10.23);
5: Return X.

Required measurements

This subsection studies the required measurements for the correct recovery of X. Let $\Omega_m = \Omega \cap \{(p, m) : p = 1, \ldots, P\}$.

Theorem 13. *Assume that both X_{prev} and $B_{p,m}$, $(p, m) \in \Omega$, are exact, and the entries of A are in general positions.*[2] *If $\Omega_m \neq \emptyset$, $m = 1, \ldots, M$, and there exists at least one m such that $|\Omega_m| \geq 2$, then Algorithm 5 correctly recovers X.*

Proof. First, $n = n^*$ and $y_{n^*} = (\Delta X)_{n^*,:}$ attain the minimum zero objective in (10.15). Secondly, without loss of generality, assume $|\Omega_1| \geq 2$. Since the entries of A are in general positions, the term $\left\|(y_n)_1 a_{I_1,n} - (\Delta B)_{I_1,1}\right\|_2^2$ in (10.16) can reach zero, the minimal possible value, *only* for $n = n^*$. Combining the two points, we conclude that Algorithm 5 returns $n = n^*$. Finally, since every Ω_m is non-empty, every entry of $(y_{n^*})_m$

[2] Any p distinct subsets of $\{A_{ij}\}$ of length $q \geq p$ are linearly independent.

that minimizes the term $\left\|(y_{n^*})_m a_{I_m,n^*} - (\Delta B)_{I_m,m}\right\|_2^2$ in (10.16) is unique and equal to $(\Delta X)_{n^*,m}$. □

In our future work, we will consider a realistic noise scenario and provide the number of measurements for the stable recovery of ΔX. On the other hand, we tested Algorithm 5 in a simulated noisy environment. The results are given in Section 10.5 below.

Dynamic vs. static CS

In a dynamic environment, dynamic CS is not only simpler but also requires fewer measurements than standard (static) CS. Our model of dynamic CS recovers ΔX, which has only one non-zero row. This is a special case of static CS, where one generally recovers multiple non-zero rows. With just one non-zero row, one can afford an exhaustive search for that row, which requires $O(M)$ measurements and is computationally as cheap as solving N simple least-squares problems (which can be solved in parallel, using fast matrix computation, or utilizing group testing). This is equivalent to performing the ℓ_0-minimization over the rows (the ℓ_0 "norm" measures the number of non-zero entries). It is well known that for seeking solutions with a general number of non-zero rows, ℓ_0-minimization is NP-hard. The ℓ_1-minimization, as the best convex approximation of ℓ_0, requires $O(L \log(N/L))$ measurements and runs much slower than N. Therefore, dynamic CS is a much better choice in terms of measurements and computing time required. On the other hand, the success of dynamic CS requires an accurate previous solution. When dynamic CS fails due to either lack of measurements or inaccurate previous solution, one must gather more measurements and fall back on methods for static CS.

10.4.3 Simulations

In this section, we perform numerical simulations to illustrate the performance of the proposed dynamic spectrum sensing algorithm. We study the ability of instant capture of the change in channel occupancy status. As long as no two changes happen simultaneously, our algorithm should be able to detect them one by one. We report the chance of successful change detection over the total number 500 of randomly generated instances in which only one channel status changed.

Simulation settings

Our experiments were based on a cognitive radio network with 100 channels, and 20 CR nodes performing collaborative spectrum sensing. The simulation used two types of channel model: type 1, Rayleigh fading channel; and type 2, log-normal shadowing channel. The number of occupied channels varied from 2 to 20. Each time, we either added or deleted an occupied channel to test if our algorithm could detect the change correctly. The results for significant power change in an occupied channel were similar, so we did not plot them. The SNRs of ΔB were chosen between 10 dB and 30 dB.

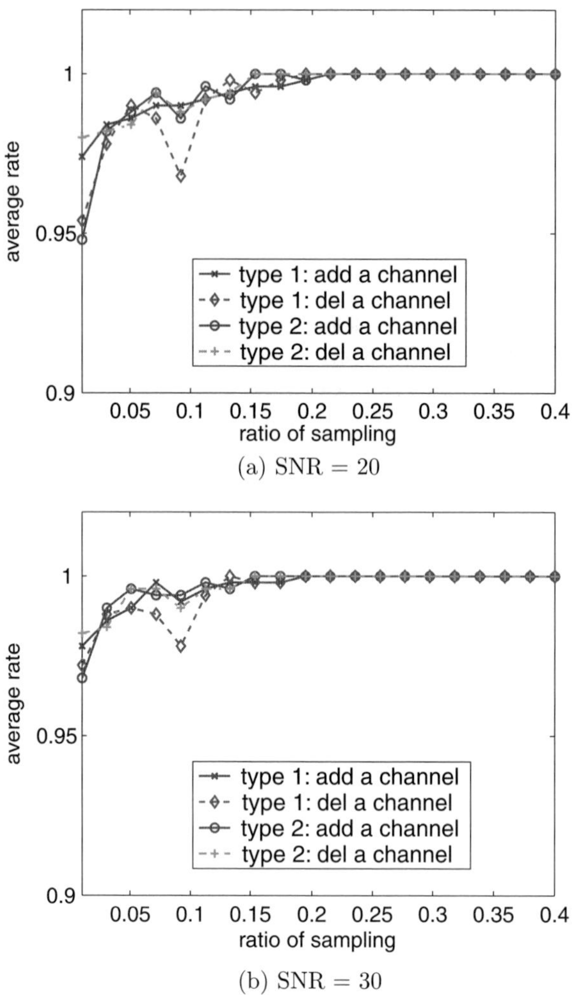

Figure 10.12 Average rate (500 runs each point) of correct recovery for $p = 30$ reports, $n = 100$ total channels, $m = 20$ CRs, $s = 4$ occupied channels before change.

SNRs of ΔB

Our simulations were done with different SNRs of ΔB. In this subsection, we discuss several factors that determine the SNR of ΔB: the SNR of B_{prev} and B, as well as the powers of the changed channel.

The SNR of B_{prev} further depends on the level of thermal noise, the accuracy of the recovered X_{prev} (since some entries of B_{prev} are not directly received but recovered from AX_{prev}), as well as the length of the previous time period, during which there is no significant change in the channel powers. The SNR of B_{prev} usually improves over time as more measurements are received and noise gets averaged over time. Hence, the longer a stationary period, the higher the SNR. Once a change occurs, the new time period starts, and the FC receives entries of B. The SNR of B also depends on the thermal noise level and the number of measurements received. However, the missing entries of B are not used.

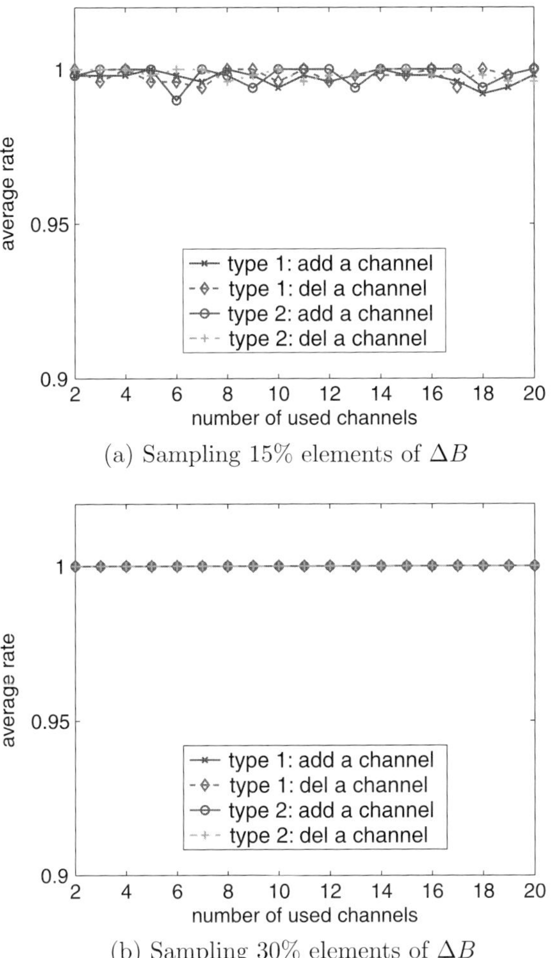

Figure 10.13 Average rate (500 runs each point) of correct recovery for $p = 30$, $n = 100$, $m = 20$ and SNR $= 30$.

For fixed SNRs of B_{prev} and B, the SNR of ΔB will be higher if the change in power (i.e., the signal of ΔB) is stronger, which translates to the unique non-zero row of ΔX containing larger values. Physically, if a nearby or high-power PR starts or stops, the SNR of ΔB will be large. On the contrary, if a distant or low-power PR starts or stops, the SNR of ΔB will be small.

Since the SNR of ΔB depends on several factors, it is tedious to run simulations with all the factors varying. Instead, we directly varied the SNRs of ΔB.

Dynamic CS algorithm performance

In Figure 10.12, we show the chance of successful change detection versus the sampling rate, which is defined as the number of received entries of ΔB over its total number of entries. We varied the sampling rate from 5% to 40%. Note that ΔB is already highly compressed as CS is applied to X, and the reference of our sampling rate is to the size of ΔB, not that of X.

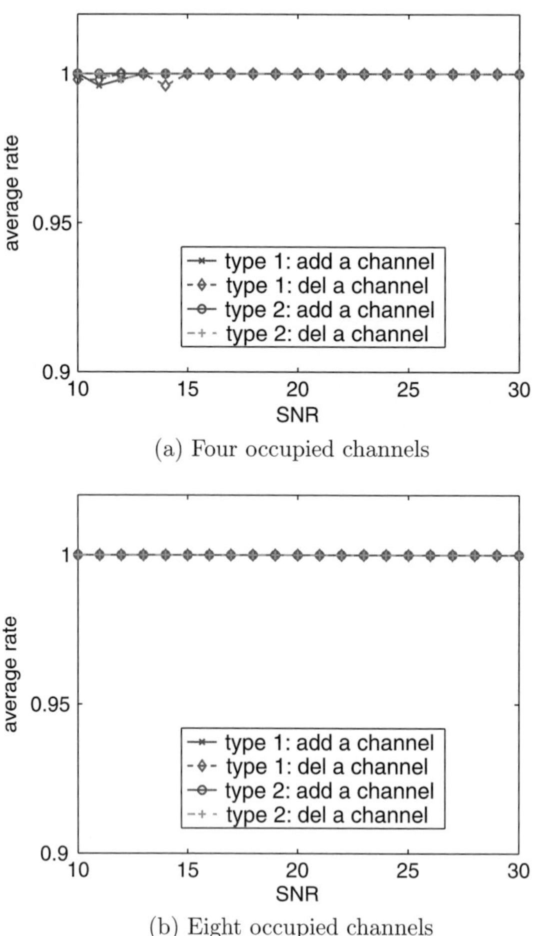

Figure 10.14 Average rate (500 runs each point) of correct recovery for $p = 30$, $n = 100$, $m = 20$. Sampling 30% elements of ΔB.

Figures 10.12(a) and (b) correspond to the performance for SNR = 20 dB and SNR = 30 dB, respectively. When SNR = 20 dB (shown in Figure 10.12(a)), we only need less than 25% of the samples at the fusion center to guarantee a successful detection. Neither channel model nor add/drop type affects the overall detection performance. As SNR increased to 30 dB (shown in Figure 10.12(b)), the sampling rate can be further reduced to 25% for reliable detection.

Figure 10.13 shows the successful detection rate of both adding and deleting one occupied channel as the number of occupied channel changes. In these simulations, we fix the SNR at 30 dB. There is no obvious change in the detection rate as the number of occupied channels increases from 2 to 20 in both cases. Again, the channel models do not affect the detection rate, no matter whether there is an increase or decrease of the number of occupied channels.

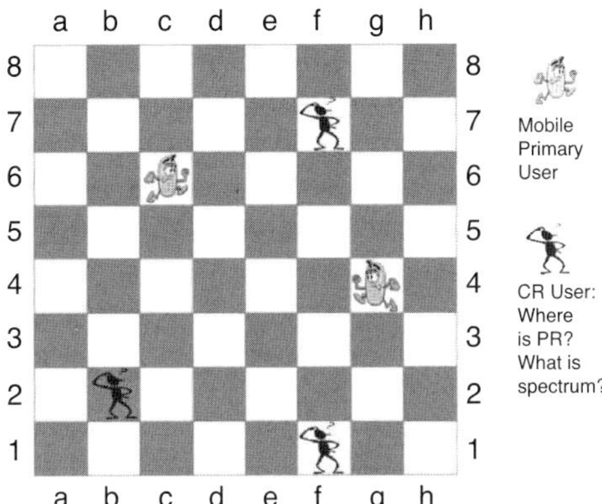

Figure 10.15 System illustration.

Figure 10.13(a) shows a tough situation, in which, only 10% of the reports have been captured by the FC. We have bumpy curves here. When the situation becomes better; i.e., the FC received 20% of the reports from the CRs, the overall detection performance becomes more stable (shown in Figure 10.13(b)).

Figure 10.14 depicts the detection rate at different SNR when we fix the number of occupied channels. Comparing Figures 10.14(a) and (b), we can tell that the increase in the number of occupied channels makes it harder to detect the changes. When the SNR is low and the number of occupied channels is large, the detection rate will be relatively low.

10.5 Joint consideration with localization

The previous two sections studied the cognitive radio networks with static PRs, yet how to deal with the situations for mobile PRs is studied in this section. To effectively and efficiently exploit the two-fold sparsity in spectrum and location, we extend the study in Section 10.3 on compressive spectrum sensing to the spectrum sensing and localization problem. The approach estimates both the power spectrum and locations of the PRs by exploiting the sparsity facts: the relative narrowband nature of the transmitted signals compared with the broad bandwidth of available spectrum and the mobile PRs located sparsely in the operational space. To effectively track mobile PRs, we implement a Kalman filter using the current estimations to update the location information. To handle dynamics in spectrum usage, a dynamic compressive spectrum sensing algorithm is proposed. Joint consideration of the above two techniques is also investigated. Simulation results validate the effectiveness and robustness of the proposed approach. This section is organized as follows: The system model is given in Section 10.5.1. The joint spectrum sensing and localization are studied in Section 10.5.2. Simulations are shown in Section 10.5.3.

10.5.1 System model

Consider a cognitive radio network with S mobile PRs located sparsely in the operational space as shown in Figure 10.15. The PRs can move in arbitrary directions in the space with limited velocity, and they stop/start using their licensed spectrum randomly. N_r CR nodes in the same field work collaboratively to monitor the dynamic spectrum usage for PRs' transmission as well as the location changes of these PRs.

With regards to the transmitting signals, a slotted frequency segmentation model is adopted, in which the whole bandwidth is divided into N non-overlapping narrowband slots centered at $\{f_v\}_{v=1}^{N}$. Each frequency slot can be viewed as a channel for transmission. The power of the transmitted signal of a PR is expanded as:

$$P_s^i = \sum_{v=1}^{N} P_s^i(f_v), \tag{10.24}$$

where $P_s^i(f_v)$ is the power of the transmitted signal of the ith PR at the vth frequency slot, $i = 1, \ldots, N_s$, and $v = 1, \ldots, N$.

Since the transmitted signal will be attenuated according to the distance between the PRs and CRs, in order to estimate the path loss, we assume that any active transmitter locates at a certain finite point out of N_s reference points. The channel can be modeled as AWGN and the channel gain is:[3]

$$H_{i,j} = P_s^i(d_{i,j})^{-\alpha/2}, \tag{10.25}$$

where d_{ij} is the distance between the ith PR and the jth CR and α is the propagation factor. The received power of the transmitted signal at the jth CR is expressed as:

$$P_r^j(f_v) = \sum_{i=1}^{N_s} P_s^i(f_v)H_{i,j}. \tag{10.26}$$

Using the discrete-time Fourier transform, $P_r^j(f_v)$ can be estimated at frequencies $\{f_k = 2\pi k/N\}_{k=0}^{N-1}$ from a number of N time-domain samples. To reduce the number of samples, each CR is equipped with a set of frequency-selective filters that take the linearly random measurements from all frequency components of the signal. The sensing process at each CR can be represented by an $M \times N$ matrix $\boldsymbol{\phi}$ that randomly maps the signal of length N into M random measurements, where $M \ll N$. The entries of matrix $\boldsymbol{\phi}$ can be designed to be random numbers. The frequency-selective filter can be implemented by the frequency-selective surface [406].

At the fusion center, the measurements from N_r CRs can be written as an $MN_r \times 1$ vector:

$$\boldsymbol{M}_{MN_r\times 1} = \boldsymbol{\Phi}_{MN_r\times NN_r}\boldsymbol{H}_{NN_r\times NN_S}\boldsymbol{P_s}_{NN_s\times 1}, \tag{10.27}$$

where $\boldsymbol{P_s}$ is a vector with entries:

$$\boldsymbol{P_s} = [P_s^1(f_1), \ldots, P_s^1(f_N), \ldots, P_s^{N_s}(f_1), \ldots, P_s^{N_s} f_N]^T,$$

[3] For a fading channel, the channel gain can be calculated by $H_{i,j} = P_s^i(d_{i,j})^{-\alpha/2}|h_{i,j}|$, where $h_{i,j}$ is the channel fading gain that can be obtained by averaging out the effect of the channel fading.

and $\boldsymbol{H}$ is a $NN_r \times NN_s$ matrix:

$$\mathbf{H} = \begin{bmatrix} \bar{H}_{11} & \bar{H}_{12} & \dots & \bar{H}_{1N_s} \\ \bar{H}_{21} & \bar{H}_{22} & \dots & \bar{H}_{2N_s} \\ \vdots & \vdots & \ddots & \vdots \\ \bar{H}_{N_r1} & \bar{H}_{N_r2} & \dots & \bar{H}_{N_rN_s} \end{bmatrix},$$

where $\bar{H}_{ji}$ is a diagonal matrix with H_{ji} on its diagonal. $\boldsymbol{\Phi}$ is a $MN_r \times NN_r$ block diagonal matrix where all of the blocks on the main diagonal are $\boldsymbol{\phi}$:

$$\boldsymbol{\Phi} = \begin{bmatrix} \boldsymbol{\phi} & \mathbf{0} & \dots & \mathbf{0} \\ \mathbf{0} & \boldsymbol{\phi} & \dots & \vdots \\ \vdots & \vdots & \ddots & \mathbf{0} \\ \mathbf{0} & \dots & \mathbf{0} & \boldsymbol{\phi} \end{bmatrix}.$$

$\mathbf{0}$ is the matrix whose entries are all zeros.

We rewrite (10.27) in a more compact form:

$$\boldsymbol{M} = \boldsymbol{\Phi} \boldsymbol{H} \boldsymbol{P}_s = \boldsymbol{A} \boldsymbol{P}_s. \tag{10.28}$$

In general, we have significantly fewer measurements $MN_r \ll NN_s$. Solving this kind of underdetermined linear system of equations is usually time consuming and less effective. However, by adopting the recently developed CS technology, we can delicately recover $\boldsymbol{P}_s$ that represents the occupied spectrum and locations of the PRs by exploiting the joint spasity.

In addition, $\boldsymbol{P}_s$ may be different at different time instances due to the mobility of the PRs or the dynamic usage of the spectrum. Yet, applying the CS technique to recover $\boldsymbol{P}_s$ each time is inefficient. We adopt a Kalman filter to predict the location change more accurately according to their moving pattern. As to the dynamic spectrum usage situation, the proposed algorithm enables the CR nodes to respond to spectrum change quickly. Furthermore, joint consideration of the Kalman filter and dynamic spectrum sensing is also proposed to reduce the searching region and complexity.

10.5.2 Joint spectrum sensing and localization algorithm

In our algorithm, we aim to recover $\boldsymbol{P}_s$ to get the spectrum usage and location information of the PRs. Using a Kalman filter and the DCSS algorithm, $\boldsymbol{P}_s$ is constantly updated according to the location change of the PRs and the dynamic spectrum usage environment.

Joint spectrum sensing and location

In (10.27), $\boldsymbol{M}$ is an $M \times N_r$ vector and $\boldsymbol{P}_s$ is an $N \times N_s$ vector. Since $M \times N_r \ll N \times N_s$, directly solving the ill-posed problem is time consuming. Notice that the transmitted signals possess a narrowband nature compared with the broad bandwidth of available spectrum as well as the mobile PRs located sparsely in the operational space.

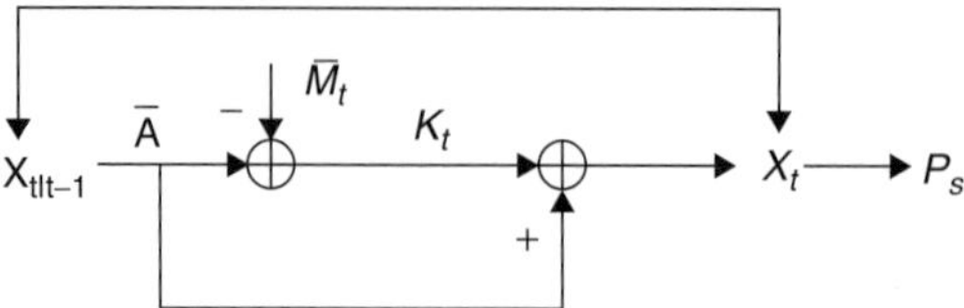

Figure 10.16 Localization using a Kalman filter. $\overline{A}_t$ is the measurement matrix. K^t is the Kalman filter gain updated at every time instance. $\overline{M}_t$ are the measurements at the fusion center at every time instance.

Because of this sparsity, most of the entries of P_s are zeros. To recover it, we can convert the original ill-posed problem to a convex optimization problem, and apply the ℓ_1 norm minimization:

$$\min ||P_s||_1 \quad s.t. \quad M = AP_s. \tag{10.29}$$

We adopt YALL1 [407] to recover P_s, and thus indentify the spectrum usage and localization information of the PRs in the CR network.

Kalman filter for localization

To predict the locations of the mobile users more accurately, we implement a Kalman filter [393] to track the location change of the PRs.

At $t = t_0$, P_s is recovered from the measurements of the CRs. From $t > t_0$, a Kalman filter is used to update P_s instead of using ℓ_1 at every time instance. The Kalman filter prediction and update procedure is shown in Figure 10.16.

The input state X of the Kalman filter is defined as $X = [P_s, \widetilde{L_s}]^T$, where $\widetilde{L_s}$ is the location change information from the PRs inertial navigation systems. We predict the next state as:

$$X_{t|t-1} = FX_{t-1}, \tag{10.30}$$

where F is the state transition matrix:

$$\mathbf{F} = \begin{bmatrix} \mathbf{1} & \mathbf{1} \\ \mathbf{0} & \mathbf{1} \end{bmatrix},$$

$\mathbf{1}$ is an NN_s vector consist of ones and $\mathbf{0}$ is an NN_s vector consist of zeros.

Then, the prediction is updated with the measurement innovation with Kalman filter gain:

$$X_t = X_{t|t-1} + K_t(\overline{M}_t - \overline{A}X_{t|t-1}), \tag{10.31}$$

where $\overline{M}_t$ is a $2MN_r$ measurement vector $\overline{M}_t = [M, \mathbf{0}]^T$. $\overline{A}$ is the $2MN_r \times NN_s$ measurement matrix $\overline{A} = [A; \mathbf{0}]$. $(\overline{M}_t - \overline{A}X_{t|t-1})$ is the measurement innovation. K^t is the Kalman filter gain that is updated at every time instance:

$$K_t = E_{t|t-1}\overline{A}^T(\overline{A}E_{t|t-1}\overline{A}^T + R)^{-1}, \tag{10.32}$$

where

$$E_{t|t-1} = FE_{t-1}F^T + Q, \tag{10.33}$$

R is the measurement noise covariance matrix, and Q is the prediction process noise covariance matrix, which can both be obtained from offline measurements. E is the prior estimate error covariance matrix updated at each iteration by the following formula:

$$E_t = (1 - K_t\overline{A})\, E_{t|t-1}. \tag{10.34}$$

From state X at time t, we can obtain P_s and find out the locations of the PRs.

Dynamic spectrum sensing

At every time instance, we calculate the measurement difference $\Delta M = |M_{t_1}| - |M_{t_2}|$. t_1 and t_2 denote two different time instances. Once $|\Delta M|$ exceeds the preselected threshold τ, it means that the spectrum occupation of the CR network is varied. Either an existing PR stops or starts using a channel, or a new PR joins the network in this situation. To deal with the dynamics in spectrum usage, we adopt dynamic compressive spectrum sensing to solve it as a least-squares problem:

$$\min_{n\in\{1,\dots,N_sN\}} \min_{(P_s)_n} \|A_n(P_s)_n - \Delta M\|_2, \tag{10.35}$$

where A_n is the nth column of sensing matrix A, and $(P_s)_n$ is the nth entity in vector P_s.

In the case of an existing PR releasing its occupied channel, the norm of M_{t_2} is less than that of M_{t_1}. Using the location estimation results of the Kalman filter, we can find out the changing entry $(P_s)^*_n$ by searching in a reduced region Ω:

$$\min_{n\in\Omega} \|A_n(P_s)^*_n - \Delta M\|_2,\ n = 1,\dots,S, \tag{10.36}$$

where Ω denote the set of entries that represent the identified active PRs.

For the situation of a new PR joining the existing CR network, the minimizer of objective function (12) can be written as

$$(P_s)^*_n = \frac{(A_n)^T(\Delta M)}{(A_n)^T A_n}, \quad n = 1,\dots,NN_s. \tag{10.37}$$

We can use a vectorized algorithm [324] to get $(P_s)^*_n$. Then compare

$$\|A_n(P_s)^*_n - \Delta M\|_2,\ n = 1,\dots,NN_s. \tag{10.38}$$

If there is a unique n that gives 0 or a tiny value, then this n is n^*, the entity that has changed since the previous period. Therefore, let

$$P_s \leftarrow (P_s)_{prev} + \begin{cases} (P_s)^*_n, & n = n^*; \\ 0, & \text{otherwise}, \end{cases} \quad n = 1,\dots,NN_s. \tag{10.39}$$

Hence, we can detect the change in spectrum usage quickly with high reliability. Once the P_s is updated, as a byproduct, we can get the location of the newly present and vanishing PR in the CR network as well.

Overall, we adopt a mechanism that keeps a balance between the accuracy and the efficiency. To complete the dynamic spectrum sensing and mobile PRs localization, we recover $\boldsymbol{P}_s$ using CS at regular intervals. Between the two reconstructions, $\boldsymbol{P}_s$ is constantly updated through Kalman filter predictions and dynamic compressive spectrum sensing results. Thus, we can successfully monitor the dynamics of the PRs locations and spectrum usage in the CR network.

10.5.3 Simulations

The simulation setup is described as follows. The operational space is a $1000\,\text{m} \times 1000\,\text{m}$ square field, which contains $N_s = 100$ uniformly distributed reference points. We monitor the dynamic spectrum usage and the locations of PRs for a time interval T. The active primary users randomly locate in the space and move in arbitrary directions in the space, or stop/start using their channels at any time. The time instances t_{stop} and t_{start} are selected according to a uniform distribution $U(0, T)$. We use a simple symmetric lattice random walk model to simulate the mobility of the PRs. Thus, at each time instance, the probabilities of a PR walk to any one of its neighbors are then identical. The transmitted power of PRs is 1 W. The number of available channels N is 64, and $N_r = 20$ CR nodes are randomly deployed in the same field working collaboratively to implement the dynamic sensing, localization and tracking tasks.

To demonstrate the performance of our algorithms, we define the sampling rate r as $r = (M \times N_r)/(N \times N_s)$. The Probability of Detection (POD) and Miss Detection Rate (MDR) are, respectively,

$$POD = \frac{N_{Hit}}{N_{Hit} + N_{Miss}}, \tag{10.40}$$

$$MDR = \frac{N_{Miss}}{N_{Miss} + N_{Correct}}. \tag{10.41}$$

Here, N_{Hit} is the number of successful detections of PRs. N_{Miss} is the number of miss detections, and $N_{correct}$ is the number of correct reports of no appearance of PRs.

Joint compressive spectrum sensing and localization

We validate the proposed joint compressive spectrum sensing and localization algorithm. The number of active PRs ranges from 1 to 3 in the simulation and the sampling rate is from 7.5% to 15%. Figure 10.17 shows the probability of detection and Figure 10.18 shows the miss detection rate.

In Figure 10.17, when the sampling rate is 7.5%, the probability of detection is above 95% in all cases. As the sampling rate increases beyond 15%, spectrum occupation and location information can be recovered with a very high probability. In Figure 10.18, the miss detection probability is very low at a low sampling rate in all the three cases. The MDR decreases as the increment of the sampling rate. Even when three active PRs appear in the operational space, the MDR can be lower than 2% as the sampling rate approaches 15%.

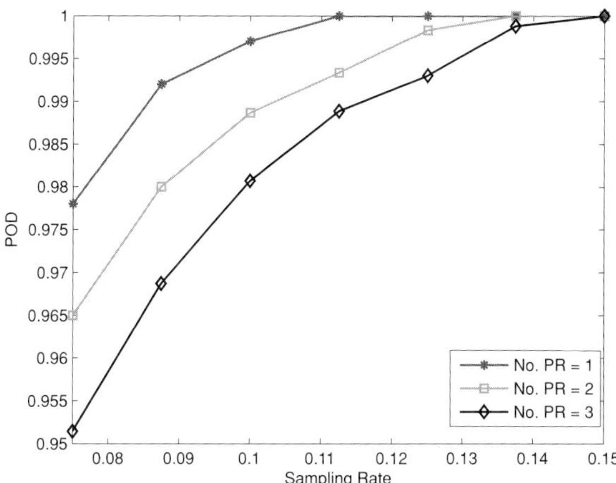

Figure 10.17 POD vs. sampling rate.

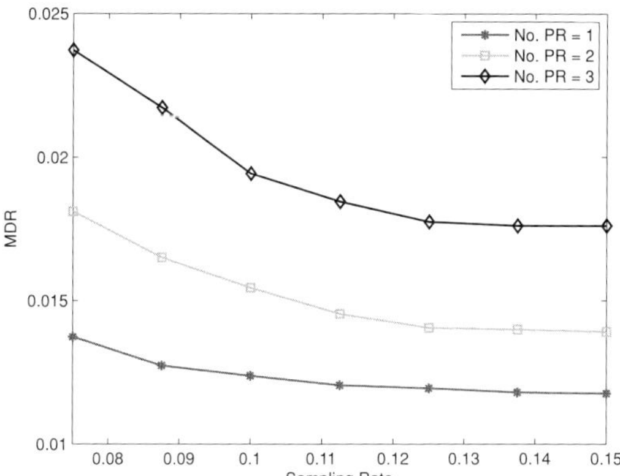

Figure 10.18 MDR vs. sampling rate.

Kalman-filter-based location

Once P_s is reconstructed, the location change of the PRs are tracked by applying the Kalman filter at the fusion center. In the operational space, we define the location change from one reference point to another as a step. Without loss of generality, we define the velocity of the PR as one step between two time instances. We investigate the performance of the Kalman filter, using the averaging distance in the operational field of nine steps. Figure 10.19 shows the POD as the change of the distance.

We can see that the Kalman filter works effectively when there is only one active PR in the operational field. The POD is above 90% when walking a trace of nine steps. The probability of detection decreases as the distance increases. When more active PRs appear in the field, the tracking task becomes more challenging. In the worst case, the performance is still acceptable when all three PRs travel the distance of nine steps.

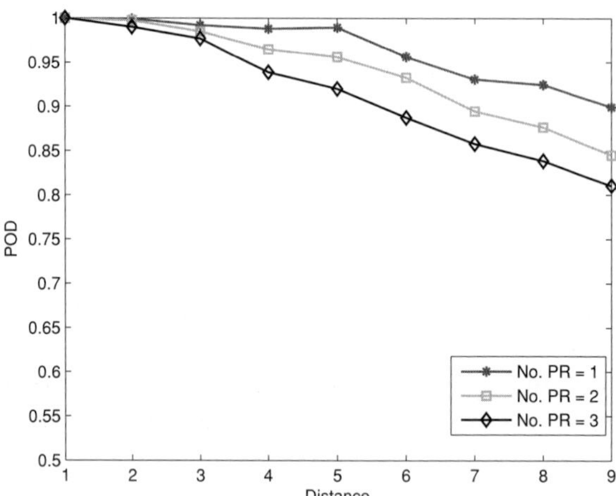

Figure 10.19 POD vs. distance.

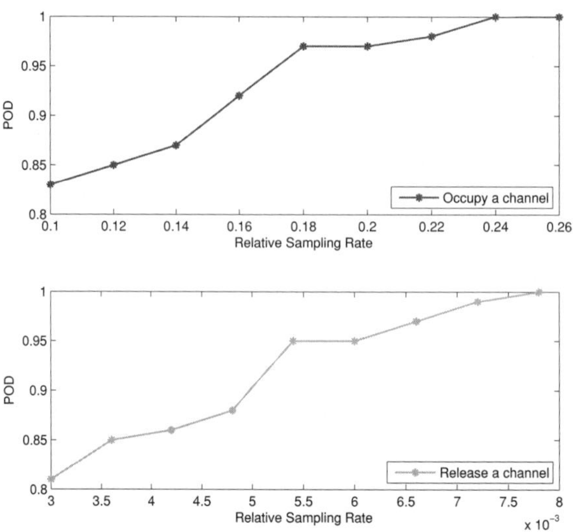

Figure 10.20 POD vs. relative sampling rate.

Dynamic spectrum sensing

In Figure 10.20, we show the probability of detection in dynamic spectrum sensing versus the relative sampling rate. The relative sampling rate is defined as the number of received entries of $\Delta \boldsymbol{M}$ over its total number of entries. Note that the relative sampling rate is referred to the size of $\Delta \boldsymbol{M}$, which is already much compressed compared to $\boldsymbol{P}_s$.

When a new PR occupies a channel, we test the performance of the dynamic spectrum sensing when the relative sampling rate is varied from 10% to 26%. We can see that as the relative sampling rate increases, the POD approaches one. We can see that at the fusion center, only 26% of the entries of $\Delta \boldsymbol{M}$ are needed to guarantee a successful detection. In the case of one existing PR releasing a channel, we can see that the relative sampling rate is quite low. A sampling rate of a mere 0.8% will guarantee a perfect detection.

10.6 Summary

In this chapter, we apply CS for collaborative spectrum sensing in cognitive radio networks to reduce the amount of sensing and transmission overhead of cognitive radio (CR) nodes. We propose to equip each CR node with a frequency-selective filter set, which linearly combines multiple channel information, and lets it send a small number of such linear combinations to the fusion center, where the channel occupancy information is then decoded. Consequently, the amount of channel sensing at the CRs and the number of reports sent from the CRs to the fusion center reduce significantly. Two novel decoding approaches have been proposed – one based on matrix completion and the other based on joint sparsity recovery. The novel matrix completion approach recovers the complete CR-to-center reports from a small number of valid reports and then reconstructs the channel occupancy information. The joint sparsity approach, on the other hand, skips recovering the reports and directly reconstructs channel-occupancy information by exploiting the fact that each occupied channel is observable by multiple CR nodes. Our algorithm enables faster recovery for large-scale cognitive radio networks. The primary user detection performance of the proposed approaches has been evaluated by simulations. The results of random tests show that, in noiseless cases, the number of samples required is no more than 50% of the number of channels in the network to guarantee exact primary user detection for both approaches; while in noisy environments, at low channel occupancy rate, we can still have high probability of detection.

Then, we move from a static environment (in the previous section) to a dynamic one where channel powers and occupancies evolve over time. In the dynamic setting, our method recovers each change quickly for a very small number of measurements. In the simulation, we can see that the proposed method is effective in different SNRs, the number of channels, and the number of received CR reports.

Finally, we exploit the two-fold sparsity in spectrum and location to deal with the spectrum sensing and localization problem in CR networks. To handle the mobility of the PRs and dynamics of the spectrum usage, we implement a Kalman filter to track the location changes of the mobile PRs in an intelligent way, and the proposed spectrum sensing algorithm is applied to update the spectrum usage information quickly and reliably. In addition, the joint concern of Kalman filter and dynamic spectrum sensing is also investigated. Simulation results validate the effectiveness of our algorithm.

References

[1] T. S. Rappaport, *Wireless Communications: Principles and Practice*, 2nd edn. Upper Saddle River, New Jersey, USA: Prentice Hall, 2002.

[2] W. Yu, W. Rhee, S. Boyd, and J. M. Cioffi, "Iterative water-filling for Gaussian vector multiple access channels," *IEEE Trans. on Information Theory*, vol. 50, no. 1, pp. 145–152, Jan. 2004.

[3] R. Cendrillon, M. Moonen, J. Verlinden, T. Bostoen, and W. Yu, "Optimal multiuser spectrum management for digital subscriber lines," in *Proc. of IEEE International Conference on Communications*, Paris, France, June 2004.

[4] R. Cendrillon and M. Moonen, "Iterative spectrum balancing for digital subscriber lines," in *Proc. of IEEE International Conference on Communications*, Seoul, Korea, May 2005.

[5] J. Papandriopoulos and J. S. Evans, "Low-complexity distributed algorithms for spectrum balancing in multi-user DSL networks," in *Proc. of IEEE International Conference on Communications*, Istanbul, Turkey, June 2006.

[6] R. Cendrillon, J. Huang, M. Chiang, and M. Moonen, "Autonomous spectrum balancing for digital subscriber lines," *IEEE Trans. on Signal Processing*, vol. 55, no. 8, pp. 4241–4257, Aug. 2007.

[7] L. Wooyul, K. Youngjae, M. H. Brady, and J. M. Cioffi, "Band-preference dynamic spectrum management in a DSL environment," in *Proc. IEEE Global Telecommunications Conference*, San Francisco, CA, USA, November–December 2006.

[8] J. M. Cioffi, W. Rhee, M. Mohseni, and M. H. Brady, "Band preference in dynamic spectrum management," in *Proc. of Eurasip Conference on Signal Processing*, Vienna, Austria, September 2004.

[9] [Online] available: http://www.umtsworld.com.

[10] [Online] available: http://www.cdg.org.

[11] "IEEE standard for local and metropolitan area networks part 16: Air interface for fixed broadband wireless access systems IEEE Std. 802.16-2004" (Revision of IEEE Std. 802.16-2001), pp. 27–28, 2004.

[12] "Achieving wireless broadband with WiMAX," IEEE Computer Society, Tech. Rep., 2004.

[13] [Online] available: http://www.bluetooth.com/bluetooth/.

[14] "Federal communications commission report," FCC, February 2002, revision of part 15 of the commission's rules regarding ultra-wideband transmission systems, first report and order, available at http://www.naic.edu/$\sim$astro/RXstatus/ Lnarrow/fcc$_{U}$WB.pdf.

[15] A. Batra et al., "Multi-band OFDM physical layer proposal for IEEE 802.15 task group 3a," IEEE P802.15-03/268r3, Tech. Rep., March 2004.

[16] C. Perkins, "Highly dynamic destination-sequenced distance-vector routing (DSDV) for mobile computers," in *Proc. of ACM Conference on Communications Architectures, Protocols and Applications*, London, UK, September 1994.

[17] C.-C. Chiang, H.-K. Wu, W. Liu, and M. Gerla, "Routing in clustered multihop, mobile wireless networks with fading channel," in *Proc. of IEEE Singapore International Conference on Networks*, Singapore, April 1997.
[18] S. Murthy and J. J. Garcia-Luna-Aceves, "An efficient routing protocol for wireless networks," *ACM Mobile Networks and App. J., Special Issue on Routing in Mobile Communication Networks*, vol. 1, no. 2, pp. 183–197, Oct. 1996.
[19] C. Perkins and E. Royer, "Ad-hoc on-demand distance vector routing," in *Proc. of the 2nd IEEE Workshop on Mobile Computing Systems and Applications*, New Orleans, LA, USA, February 1999.
[20] D. B. Johnson, "Routing in ad hoc networks of mobile hosts," in *Proc. of the Workshop on Mobile Computing Systems and Applications*, Santa Cruz, CA, USA, December 1994.
[21] V. D. Park and M. S. Corson, "A highly adaptive distributed routing algorithm for mobile wireless networks," in *Proc. IEEE Conf. on Comp. Comm.*, Kobe, Japan, April 1997.
[22] C. K. Toh, "Associativity-based routing for ad hoc mobile networks," *Wireless Personal Communications: An International Journal*, vol. 4, no. 1, pp. 103–139, Mar. 1997.
[23] R. Dube, C. D. Rais, K. Y. Wang, and S. K. Tripathi, "Signal stability-based adaptive routing (SSA) for ad-hoc mobile networks," *IEEE Personal Communications Magazine*, vol. 4, no. 1, pp. 36–45, Feb. 1997.
[24] K. Menger, "Zur allgemeinen kurventheorie," *Fund. Math.*, vol. 10, pp. 96–115, 1927.
[25] T. Camp, J. Boleng, and V. Davies, "A survey of mobility models for ad hoc network research," *Wireless Communications and Mobile Computing: Special issue on Mobile Ad Hoc Networking: Research, Trends and Applications*, vol. 2, no. 5, pp. 483–502, Sept. 2002.
[26] J. Y. L. Boudec and M. Vojnovic, "Perfect simulation and stationarity of a class of mobility models," in *Proc. IEEE Conf. on Comp. Comm.*, Miami, FL, USA, March 2005.
[27] [Online] available: http://www.antd.nist.gov/wahn_ssn.shtml.
[28] B. Krishnamachari, *Networking Wireless Sensors*. Cambridge, UK: Cambridge University Press, 2005.
[29] S. Meguerdichian, F. Koushanfar, M. Potkonjak, and M. B. Srivastava, "Coverage problems in wireless ad hoc sensor networks," in *Proc. IEEE Conf. on Comp. Comm.* Anchorage, AK, USA, April 2001.
[30] B. Chen and P. K. Varshney, "A Bayesian sampling approach to decision fusion using hierarchical models," *IEEE Trans. on Signal Processing*, vol. 50, no. 8, pp. 1809–1818, Aug. 2002.
[31] M. Ozay and F. T. Y. Vural, "A theoretical analysis of feature fusion in stacked generalization," in *Proc. of Signal Processing and Communications Applications Conference*, Antalya, Turkey, April 2009.
[32] J. P. Walters, Z. Liang, W. Shi, and V. Chaudhary, *Wireless Sensor Network Security: A Survey Security in Distributed, Grid, and Pervasive Computing*. CRC Press, Auerbach Publications, 2006.
[33] K. J. Kerpez, D. L. Waring, S. Galli, J. Dixon, and P. Madon, "Advanced DSL management," *IEEE Communications Magazine*, vol. 41, no. 9, pp. 116–123, Sept. 2003.
[34] D. Tse and P. Viswanath, *Fundamentals of Wireless Communication*. Cambridge, UK: Cambridge University Press, 2005.
[35] L. Zheng and D. N. C. Tse, "Diversity and multiplexing: A fundamental tradeoff in multiple-antenna channels," *IEEE Trans. on Information Theory*, vol. 49, no. 5, pp. 1073–1096, May 2003.

[36] S. M. Alamouti, "A simple transmit diversity technique for wireless communications," *IEEE Journal on Selected Areas in Communications*, vol. 16, no. 8, pp. 1451–1458, Oct. 1998.

[37] S. Haykin, "Cognitive radio: Brain-empowered wireless communications," *IEEE Journal on Selected Areas in Communications*, vol. 23, no. 2, pp. 201–220, Feb. 2005.

[38] T. Issariyakul and E. Hossain, "ORCA-MRT: An optimization-based approach for fair scheduling in multirate TDMA wireless networks," *IEEE Transactions on Wireless Communications*, vol. 4, no. 6, pp. 2823–2835, Nov. 2005.

[39] M. Andrews, "A survey of scheduling theory in wireless data networks," in *IMA Summer Workshop on Wireless Communications*, 2005.

[40] T. Ng, I. Stoica, and H. Zhang, "Packet fair queueing algorithms for wireless networks with location-dependent errors," in *IEEE International Conference on Computer Communications*, San Francisco, CA, USA, March–April 1998.

[41] S. Lu, V. Bharghavan, and R. Srikant, "Fair scheduling in wireless packet networks," *IEEE/ACM Transactions on Networking*, vol. 7, no. 4, pp. 473–337, Aug. 1999.

[42] T. Issariyakul, E. Hossain, and D. I. Kim, "Medium access control protocols for wireless mobile ad hoc networks: Issues and approaches," *Wireless Communications and Mobile Computing*, vol. 3, no. 8, pp. 935–958, Nov. 2003.

[43] A. C. V. Gummalla and J. O. Limb, "Wireless medium access control protocols," *IEEE Communications Surveys and Tutorials*, vol. 3, no. 2, pp. 2–15, 2nd Quarter 2000.

[44] S. Kumar, V. S. Raghavan, and J. Deng, "Medium access control protocols for ad hoc wireless networks: A survey," *Ad Hoc Networks*, vol. 4, no. 3, pp. 326–358, May 2006.

[45] W. Ye and J. Heidemann, *Medium Access Control in Wireless Sensor Networks*. Netherlands: Wireless Sensor Networks, Kluwer Academic Publishers, 2003.

[46] A. Gupta and P. Mohapatra, "A survey on ultra wide band medium access control schemes," *Computer Networks*, vol. 51, no. 11, pp. 2976–2993, Aug. 2007.

[47] I. F. Akyildiz, J. McNair, L. Carrasco, and R. Puigjaner, "Medium access control protocols for multimedia traffic in wireless networks," *IEEE Network*, vol. 13, no. 4, pp. 39–47, July–August 1999.

[48] P. Roshan and J. Leary, *802.11 Wireless LAN Fundamentals*. Cisco Press, Indianapolis, IN, 2004.

[49] M. Satyanarayanan, "Pervasive computing: Vision and challenges," *Personal Communications, IEEE*, vol. 8, no. 4, pp. 10–17, Aug. 2001.

[50] R. Glidden, C. Bockorick, S. Cooper, C. Diorio, D. Dressler, V. Gutnik, C. Hagen, D. Hara, T. Hass, T. Humes, J. Hyde, R. Oliver, O. Onen, A. Pesavento, K. Sundstrom, and M. Thomas, "Design of ultra-low-cost UHF RFID tags for supply chain applications," *Communications Magazine, IEEE*, vol. 42, no. 8, pp. 140–151, Aug. 2004.

[51] L. Mo, Y. He, Y. Liu, J. Zhao, S.-J. Tang, X.-Y. Li, and G. Dai, "Canopy closure estimates with Greenorbs: Sustainable sensing in the forest," in *Proceedings of the 7th ACM Conference on Embedded Networked Sensor Systems*, Berkeley, CA, USA, 2009.

[52] N. B. Priyantha, A. Chakraborty, and H. Balakrishnan, "The cricket location-support system," in *Proceedings of the 6th Annual International Conference on Mobile Computing and Networking*, Boston, MA, USA, 2000.

[53] A. Savvides, C.-C. Han, and M. B. Strivastava, "Dynamic fine-grained localization in adhoc networks of sensors," in *Proceedings of the 7th Annual International Conference on Mobile Computing and Networking*, Rome, Italy, 2001.

[54] D. Niculescu and B. Nath, "Ad hoc positioning system (APS) using AOA," *Twenty-Second Annual Joint Conference of the IEEE Computer and Communications.* San Francisco, CA, USA, March–3 April 2003.
[55] L. Doherty, K. Pister, and L. El Ghaoui, "Convex position estimation in wireless sensor networks," *Twentieth Annual Joint Conference of the IEEE Computer and Communications Societies*. Anchorage, AK, USA, April 2001.
[56] T. He, C. Huang, B. M. Blum, J. A. Stankovic, and T. Abdelzaher, "Range-free localization schemes for large scale sensor networks," in *Proceedings of the 9th Annual International Conference on Mobile Computing and Networking*. San Diego, CA, USA, 2003.
[57] L. Kleinrock and J. Silvester, "Optimum transmission radii for packet radio networks or why six is a magic number," in *NTC '78; National Telecommunications Conference*, Birmingham, AL, USA, Dec. 1978.
[58] L. M. Ni, Y. Liu, Y. C. Lau, and A. P. Patil, "Landmarc: Indoor location sensing using active RFID," *Wirel. Netw.*, vol. 10, no. 6, pp. 701–710, Nov. 2004.
[59] Y. Liu and Z. Yang, *Location, Localization, and Localizability: Location – Awareness Technology for Wireless Networks*. Berlin: Springer, 2011.
[60] Y. Shang, W. Ruml, Y. Zhang, and M. P. J. Fromherz, "Localization from mere connectivity," in *Proceedings of the 4th ACM International Symposium on Mobile Ad Hoc Networking & Computing*, Annapolis, MD, USA, 2003.
[61] D. Niculescu and B. Nath, "Dv based positioning in ad hoc networks," *Telecommunication Systems*, vol. 22, no. 1–4, pp. 267–280, Jan.–Apr. 2003.
[62] D. Moore, J. Leonard, D. Rus, and S. Teller, "Robust distributed network localization with noisy range measurements," in *Proceedings of the 2nd International Conference on Embedded Networked Sensor Systems*, Baltimore, MD, NY, USA, 2004.
[63] H. L. V. Trees, *Detection, Estimation, and Modulation Theory: Radar-Sonar Signal Processing and Gaussian Signals in Noise*. Melbourne, FL, USA: Krieger Publishing Co., Inc., 1992.
[64] J. Liu, Y. Zhang, and F. Zhao, "Robust distributed node localization with error management," in *Proceedings of the 7th ACM International Symposium on Mobile Ad Hoc Networking and Computing*, Florence, Italy.
[65] Z. Yang and Y. Liu, "Quality of trilateration: Confidence based iterative localization," in *Distributed Computing Systems, 2008. ICDCS '08. The 28th International Conference on*, Beijing, China, June 2008, pp. 446–453.
[66] C. Savarese, J. M. Rabaey, and K. Langendoen, "Robust positioning algorithms for distributed ad-hoc wireless sensor networks," in *Proceedings of the General Track of the Annual Conference on USENIX Annual Technical Conference*. Monterey, CA, USA, 2002.
[67] E. Candes, J. Romberg, and T. Tao, "Robust uncertainty principles: Exact signal reconstruction from highly incomplete frequency information," *IEEE Transactions on Information Theory*, vol. 52, pp. 489–509, Feb. 2006.
[68] E. Candes, J. Romberg, and T. Tao, "Stable signal recovery from incomplete and inaccurate information," *Communications on Pure and Applied Mathematics*, vol. 59, no. 8, pp. 1207–1233, Aug. 2006.
[69] E. Candes and T. Tao, "Near optimal signal recovery from random projections: Universal encoding strategies," *IEEE Transactions on Information Theory*, vol. 52, no. 1, pp. 5406–5425, Dec. 2006.

[70] D. Donoho, "Compressed sensing," *IEEE Trans. on Information Theory*, vol. 52, no. 4, pp. 1289–1306, Apr. 2006.
[71] H. Nyquist, "Certain topics in telegraph transmission theory," *IEEE Trans. of the American Institute of Electrical Engineers*, vol. 47, no. 2, pp. 617–644, Apr. 1928.
[72] C. Shannon, "Communication in the presence of noise," *Proc. Institute of Radio Engineers*, vol. 37, no. 1, pp. 10–21, 1949.
[73] B. K. Natarajan, "Sparse approximate solutions to linear systems," *SIAM Journal on Computing*, vol. 24, pp. 227–234, 1995.
[74] M. Elad, *Sparse and Redundant Representations: From Theory to Applications in Signal and Image Processing*. Springer Verlag, Berlin, Heidelberg, Germany, 2010.
[75] J. Starck, F. Murtagh, and J. Fadili, *Sparse Image and Signal Processing: Wavelets, Curvelets, Morphological Diversity*. Cambridge University Press, Cambridge, UK, 2010.
[76] L. Rudin, S. Osher, and E. Fatemi, "Nonlinear total variation based noise removal algorithms," *Physica D: Nonlinear Phenomena*, vol. 60, no. 1–4, pp. 259–268, 1992.
[77] J. Starck, E. Candes, and D. Donoho, "The curvelet transform for image denoising," *IEEE Transactions on Image Processing*, vol. 11, no. 6, pp. 670–684, June 2002.
[78] B. A. Olshausen and D. J. Field, "Emergence of simple-cell receptive field properties by learning a sparse code for natural images," *Nature*, vol. 381, no. 6583, pp. 607–609, 1996.
[79] K. Engan, S. Aase, and J. Husoy, "Multi-frame compression: Theory and design," *Signal Processing*, vol. 80, no. 10, pp. 2121–2140, Oct. 2000.
[80] M. Aharon, M. Elad, and A. Bruckstein, "k-SVD: An algorithm for designing overcomplete dictionaries for sparse representation," *IEEE Transactions on Signal Processing*, vol. 54, no. 11, pp. 4311–4322, Nov. 2006.
[81] M. Yuan and Y. Lin, "Model selection and estimation in regression with grouped variables," *Journal of the Royal Statistical Society: Series B*, vol. 68, no. 1, pp. 49–67, Apr. 2006.
[82] J. Chen and X. Huo, "Theoretical results on sparse representations of multiple-measurement vectors," *IEEE Transactions on Signal Processing*, vol. 54, no. 12, pp. 4634–4643, Dec. 2006.
[83] F. Bach, "Consistency of the group lasso and multiple kernel learning," *The Journal of Machine Learning Research*, vol. 9, pp. 1179–1225, 2008.
[84] D. Malioutov, M. Cetin, and A. Willsky, "A sparse signal reconstruction perspective for source localization with sensor arrays," *IEEE Transactions on Signal Processing*, vol. 53, no. 8, pp. 3010–3022, Aug. 2005.
[85] S. Cotter, B. Rao, K. Engan, and K. Kreutz-Delgado, "Sparse solutions to linear inverse problems with multiple measurement vectors," *IEEE Transactions on Signal Processing*, vol. 53, no. 7, pp. 2477–2488, July 2005.
[86] J. Meng, W. Yin, H. Li, E. Hossain, and Z. Han, "Collaborative spectrum sensing from sparse observations in cognitive radio networks," *IEEE Journal on Selected Topics on Communications, special issue on Advances in Cognitive Radio Networking and Communications*, vol. 29, no. 2, pp. 327–337, Feb. 2011.
[87] M. Fazel, "Matrix rank minimization with applications," *Department of Electrical Engineering, Stanford University*, Doctoral Thesis, 2002.
[88] E. J. Candès and B. Recht, "Exact low-rank matrix completion via convex optimization," in *Proc. of IEEE Annual Allerton Conference on Communication, Control, and Computing*, Monticello, IL, USA, September 2008, pp. 806–812.

[89] B. Recht, M. Fazel, and P. Parrilo, "Guaranteed minimum-rank solutions of linear matrix equations via nuclear norm minimization," *SIAM Review*, vol. 52, no. 3, pp. 471–501, 2010.

[90] Z. Liu and L. Vandenberghe, "Interior-point method for nuclear norm approximation with application to system identification," *SIAM Journal on Matrix Analysis and Applications*, vol. 31, no. 3, pp. 1235–1256, 2009.

[91] A. So and Y. Ye, "Theory of semidefinite programming for sensor network localization," *Mathematical Programming*, vol. 109, no. 2, pp. 367–384, 2007.

[92] C. Tomasi and T. Kanade, "Shape and motion from image streams under orthography: A factorization method," *International Journal of Computer Vision*, vol. 9, no. 2, pp. 137–154, 1992.

[93] T. Morita and T. Kanade, "A sequential factorization method for recovering shape and motion from image streams," *IEEE Transactions on Pattern Analysis and Machine Intelligence*, vol. 19, no. 8, pp. 858–867, 1997.

[94] D. Goldberg, D. Nichols, B. Oki, and D. Terry, "Using collaborative filtering to weave an information tapestry," *Communications of the ACM*, vol. 35, no. 12, pp. 61–70, 1992.

[95] E. Candes, X. Li, Y. Ma, and J. Wright, "Robust principal component analysis?" *Journal of the ACM*, vol. 58, no. 3, pp. 1–37, May 2011.

[96] R. Baraniuk, V. Cevher, M. Duarte, and C. Hegde, "Model-based compressive sensing," *IEEE Transactions on Information Theory*, vol. 56, no. 4, pp. 1982–2001, Apr. 2010.

[97] Y. Eldar and M. Mishali, "Robust recovery of signals from a structured union of subspaces," *IEEE Transactions on Information Theory*, vol. 55, no. 11, pp. 5302–5316, Nov. 2009.

[98] Y. Lu and M. Do, "Sampling signals from a union of subspaces," *Signal Processing Magazine, IEEE*, vol. 25, no. 2, pp. 41–47, Mar. 2008.

[99] E. Candes and J. Romberg, "Sparsity and incoherence in compressive sampling," *Inverse Problems*, vol. 23, no. 3, pp. 969–985, 2007.

[100] P. Feng and Y. Bresler, "Spectrum-blind minimum-rate sampling and reconstruction of multiband signals," IEEE, International Conference on Acoustics, Speech and Signal Processing, Atlanta, GA, USA, 1996.

[101] M. Vetterli, P. Marziliano, and T. Blu, "Sampling signals with finite rate of innovation," *IEEE Trans. on Signal Processing*, vol. 50, no. 6, pp. 1417–1428, June 2002.

[102] E. Candes and T. Tao, "Decoding by linear programming," *IEEE Transactions on Information Theory*, vol. 51, no. 12, pp. 4203–4215, Dec. 2005.

[103] Y. Zhang, "Theory of compressive sensing via l1-minimization: A non-RIP analysis and extensions," *Rice University CAAM Technical Report TR08-11*, 2008.

[104] E. Candes and Y. Plan, "A probabilistic and RIPless theory of compressed sensing," *IEEE Transactions on Information Theory*, vol. 57, no. 11, pp. 7235–7254, Nov. 2010.

[105] D. Donoho and X. Huo, "Uncertainty principles and ideal atomic decompositions," *IEEE Transactions on Information Theory*, vol. 47, no. 7, pp. 2845–2862, Nov. 2001.

[106] R. Gribonval and M. Nielsen, "Sparse representations in unions of bases," *IEEE Transactions on Information Theory*, vol. 49, no. 12, pp. 3320–3325, Dec. 2003.

[107] Y. Zhang, "A simple proof for recoverability of ℓ_1-minimization: Go over or under?" *Rice University CAAM Technical Report TR05-09*, 2005.

[108] A. Cohen, W. Dahmen, and R. A. DeVore, "Compressed sensing and best k-term approximation," *Journal of the American Mathematical Society*, vol. 22, no. 1, pp. 211–231, Jan. 2009.

[109] E. Candes, "The restricted isometry property and its implications for compressed sensing," *Comptes Rendus Mathematique*, vol. 346, no. 9–10, pp. 589–592, 2008.

[110] S. Foucart and M. Lai, "Sparsest solutions of underdetermined linear systems via lq-minimization for $0 < q \leq 1$," *Applied and Computational Harmonic Analysis*, vol. 26, no. 3, pp. 395–407, 2009.

[111] S. Foucart, "A note on guaranteed sparse recovery via ℓ_1-minimization," *Applied and Computational Harmonic Analysis*, vol. 29, no. 1, pp. 97–103, July 2010.

[112] T. Cai, L. Wang, and G. Xu, "Shifting inequality and recovery of sparse signals," *IEEE Transactions on Signal Processing*, vol. 58, no. 3, pp. 1300–1308, Mar. 2010.

[113] Q. Mo and S. Li, "New bounds on the restricted isometry constant δ_{2k}," *Applied and Computational Harmonic Analysis*, vol. 31, no. 3, pp. 460–468, 2011.

[114] M. Davenport, "PhD thesis: Random observations on random observations: Sparse signal acquisition and processing," Rice University, Houston, TX, USA, 2010.

[115] R. Baraniuk, M. Davenport, R. Devore, and M. Wakin, "A simple proof of the restricted isometry property for random matrices," *Constructive Approximation*, vol. 28, no. 3, pp. 253–263, 2007.

[116] S. Mendelson, A. Pajor, and N. Tomczak-Jaegermann, "Uniform uncertainty principle for Bernoulli and subgaussian ensembles," *Constructive Approximation*, vol. 28, no. 3, pp. 277–289, 2008.

[117] H. Rauhut, "Compressive sensing and structured random matrices," *Theoretical Foundations and Numerical Methods for Sparse Recovery*, vol. 9, pp. 1–92, 2010.

[118] J. Bourgain, S. Dilworth, K. Ford, S. Konyagin, and D. Kutzarova, "Explicit constructions of RIP matrices and related problems," *Duke Mathematical Journal*, vol. 159, no. 1, pp. 145–185, 2011.

[119] J. Haupt, L. Applebaum, and R. Nowak, "On the restricted isometry of deterministically subsampled Fourier matrices," *IEEE Conference on Information Sciences and Systems (CISS)*, Princeton, NJ, USA, 2010.

[120] P. Indyk, "Explicit constructions for compressed sensing of sparse signals," *SIAM Symposium on Discrete Algorithms (SODA)*, pp. 30–33, 2008.

[121] S. Vavasis, "Derivation of compressive sensing theorems from the spherical section property," [online] available: www.student.math.uwaterloo.ca/∼co769/simplif.pdf.

[122] B. S. Kashin, "Diameters of some finite-dimensional sets and classes of smooth functions," *Mathematics of the USSR-Izvestiya*, vol. 11, p. 317, 1977.

[123] A. Garnaev and E. D. Gluskin, "The widths of a euclidean ball," *Dokl. Akad. Nauk SSSR*, vol. 277, no. 5, pp. 1048–1052, 1984.

[124] J. Tropp and A. Gilbert, "Signal recovery from random measurements via orthogonal matching pursuit," *IEEE Transactions on Information Theory*, vol. 53, no. 12, pp. 4655–4666, Dec. 2007.

[125] D. Donoho, Y. Tsaig, I. Drori, and J.-C. Starck, "Sparse solution of underdetermined linear equations by stagewise orthogonal matching pursuit," *IEEE Transactions on Information Theory*, vol. 58, no. 2, pp. 1094–1121, Feb. 2012.

[126] D. Needell and R. Vershynin, "Signal recovery from incomplete and inaccurate measurements via regularized orthogonal matching pursuit," *IEEE Journal of Selected Topics in Signal Processing*, vol. 4, no. 2, pp. 310–316, Apr. 2010.

[127] W. Dai and O. Milenkovic, "Subspace pursuit for compressive sensing: Closing the gap between performance and complexity," *IEEE Transactions on Information Theory*, vol. 55, no. 5, pp. 2230–2249, May 2009.

[128] D. Needell and J. Tropp, "Cosamp: Iterative signal recovery from incomplete and inaccurate samples," *Communications of the ACM*, vol. 53, no. 12, pp. 93–100, Dec. 2010.

[129] T. Blumensath and M. Davies, "Iterative hard thresholding for compressed sensing," *Applied and Computational Harmonic Analysis*, vol. 27, no. 3, pp. 265–274, 2009.

[130] T. Blumensath and M. Davies, "Normalized iterative hard thresholding: Guaranteed stability and performance," *Selected Topics in Signal Processing, IEEE Journal of*, vol. 4, no. 2, pp. 298–309, Apr. 2010.

[131] D. Du and F. Hwang, *Combinatorial Group Testing and Its Applications*. World Scientific Pub. Co. Inc., 2000.

[132] R. Berinde, A. Gilbert, P. Indyk, H. Karloff, and M. Strauss, "Combining geometry and combinatorics: A unified approach to sparse signal recovery," *46th Annual Allerton Conference on Communication, Control and Computing*, Monticello, IL, USA, Sept. 2008.

[133] A. Gilbert and P. Indyk, "Sparse recovery using sparse matrices," *Proceedings of the IEEE*, vol. 98, no. 6, pp. 937–947, June 2010.

[134] A. Gilbert, M. Strauss, J. Tropp, and R. Vershynin, "One sketch for all: Fast algorithms for compressed sensing," *Proceedings 39th ACM Symp. Theory of Computing, San Diego*, 2007.

[135] A. Gilbert, Y. Li, E. Porat, and M. Strauss, "Approximate sparse recovery: Optimizing time and measurements," *Proceedings of the 42nd Symposium on Theory of Computing*, Cambridge, MA, USA, 2010.

[136] Y. Zhang, "When is missing data recoverable?" *Rice University CAAM Technical Report TR06-15*, 2006.

[137] W. P. Ziemer, *Weakly Differentiable Functions: Sobolev Spaces and Functions of Bounded Variation*, ser. Graduate Texts in Mathematics. Springer, 1989.

[138] S. Kullback, "The Kullback–Leibler distance," *The American Statistician*, vol. 41, no. 4, pp. 340–341, 1987.

[139] S. Kullback, *Information Theory and Statistics*. Dover Pubns, Mineola, N.Y., USA, 1997.

[140] D. Hosmer and S. Lemeshow, *Applied Logistic Regression*. Wiley-Interscience, New York, NY, USA, 2000.

[141] J. Duchi, S. Shalev-Shwartz, Y. Singer, and T. Chandra, "Efficient projections onto the l1-ball for learning in high dimensions," *Proceedings of the 25th, International Conference on Machine Learning*, Helsinki, Finland, 2008.

[142] A. Quattoni, X. Carreras, M. Collins, and T. Darrell, "An efficient projection for $\ell_{1,\infty}$ regularization," in *Proceedings of the 26th Annual International Conference on Machine Learning*, Montreal, Quebec, Canada, June 2009.

[143] J. Liu and J. Ye, "Efficient euclidean projections in linear time," in *Proceedings of the 26th Annual International Conference on Machine Learning*, Montreal, Quebec, Canada, June 2009.

[144] E. van den Berg and M. Friedlander, "Probing the pareto frontier for basis pursuit solutions," *SIAM Journal on Scientific Computing*, vol. 31, no. 2, pp. 890–912, 2008.

[145] E. van den Berg, M. Schmidt, M. Friedlander, and K. Murphy, "Group sparsity via linear-time projection," Technical Report TR-2008-09, University of British Columbia, 2008.

[146] S. Boyd and L. Vandenberghe, *Convex Optimization*. Cambridge, UK: Cambridge University Press, 2004.

[147] D. P. Bertsekas, *Nonlinear Programming, 2nd edn*. Belmont, Massachusetts: Athena Scientific, 1999.

[148] D. Donoho, "De-noising by soft-thresholding," *IEEE Transactions on Information Theory*, vol. 41, no. 3, pp. 613–627, May 1995.

[149] J. Cai, E. J. Candès, and Z. Shen, "A singular value thresholding algorithm for matrix completion," Stanford University, Tech. Rep., October 2008, [online] available: http://arxiv.org/abs/0810.3286.

[150] S. Ma, D. Goldfarb, and L. Chen, "Fixed point and Bregman iterative methods for matrix rank minimization," *Mathematical Programming Series A and B*, vol. 128, no. 1–2, pp. 321–353, June 2009.

[151] J. Cai and S. Osher, "Fast singular value thresholding without singular value decomposition," *UCLA CAM Report 10-24*, 2010.

[152] J. Darbon and M. Sigelle, "Image restoration with discrete constrained total variation, Part I: fast and exact optimization," *Journal of Mathematical Imaging and Vision*, vol. 26, no. 3, pp. 261–276, 2006.

[153] D. Goldfarb and W. Yin, "Parametric maximum flow algorithms for fast total variation minimization," *SIAM Journal on Scientific Computing*, vol. 31, no. 5, pp. 3712–3743, 2009.

[154] T. F. Chan, H. M. Zhou, and R. H. Chan, "Continuation method for total variation denoising problems," *Advanced Signal Processing Algorithms*, vol. 2563, no. 1, pp. 314–325, 1995.

[155] A. Chambolle, "An algorithm for total variation minimization and applications," *Journal of Mathematical Imaging and Vision*, vol. 20, no. 1–2, pp. 89–97, 2004.

[156] A. Chambolle, "Total variation minimization and a class of binary MRF models," *Energy Minimization Methods in Computer Vision and Pattern Recognition, Lecture Notes in Computer Science 3757*, pp. 136–152, 2005.

[157] B. Wohlberg and P. Rodriguez, "An iteratively reweighted norm algorithm for minimization of total variation functionals," *IEEE Signal Processing Letters*, vol. 14, no. 12, pp. 948–951, December 2007.

[158] Y. Wang, J. Yang, W. Yin, and Y. Zhang, "A new alternating minimization algorithm for total variation image reconstruction," *SIAM Journal on Imaging Sciences*, vol. 1, no. 3, pp. 248–272, 2008.

[159] T. Goldstein and S. Osher, "The split Bregman method for l1 regularized problems," *SIAM Journal on Imaging Sciences*, vol. 2, no. 2, pp. 323–343, 2009.

[160] M. Figueiredo and R. Nowak, "An EM algorithm for wavelet-based image restoration," *IEEE Transactions on Image Processing*, vol. 12, no. 8, pp. 906–916, Aug. 2003.

[161] C. De Mol and M. Defrise, "A note on wavelet-based inversion algorithms," *Contemporary Mathematics*, vol. 313, pp. 85–96, 2002.

[162] J. Bect, L. Blanc-Feraud, G. Aubert, and A. Chambolle, "A ℓ_1-unified variational framework for image restoration," *European Conference on Computer Vision*, Slovansky Ostrov, Prague 1, Czech Republic, May 2004.

[163] J. Douglas and H. H. Rachford, "On the numerical solution of the heat conduction problem in 2 and 3 space variables," *Transactions of the American Mathematical Society*, vol. 82, no. 2, pp. 421–439, 1956.

[164] D. H. Peaceman and H. H. Rachford, "The numerical solution of parabolic elliptic differential equations," *SIAM Journal on Applied Mathematics*, vol. 3, no. 1, pp. 28–41, 1955.

[165] D. Gabay and B. Mercier, "A dual algorithm for the solution of nonlinear variational problems via finite-element approximations," *Computers and Mathematics with Applications*, vol. 2, pp. 17–40, 1976.

[166] J. Eckstein and D. P. Bertsekas, "On the Douglas-Rachford splitting method and the proximal point algorithm for maximal monotone operators," *Mathematical Programming*, vol. 55, no. 3, pp. 293–318, June 1992.

[167] M. Fortin and R. Glowinski, *Augmented Lagrangian Methods*. Amsterdam, The Netherlands: North-Holland, 1983.

[168] P. L. Combettes and V. R. Wajs, "Signal recovery by proximal forward-backward splitting," *SIAM Journal on Multiscale Modeling and Simulation*, vol. 4, no. 4, pp. 1168–1200, 2005.

[169] R. Rockafellar, *Convex Analysis*. Princeton, NJ: Princeton University Press, 1970.

[170] S. Ma, W. Yin, Y. Zhang, and A. Chakraborty, "An efficient algorithm for compressed MR imaging using total variation and wavelets," *IEEE International Conference on Computer Vision and Pattern Recognition (CVPR'08)*, Anchorage, AK, USA, June 2008.

[171] R. Tibshirani, M. Saunders, S. Rosset, J. Zhu, and K. Knight, "Sparsity and smoothness via the fused lasso," *Journal of the Royal Statistical Society: Series B (Statistical Methodology)*, vol. 67, no. 1, pp. 91–108, 2005.

[172] Z.-Q. Luo and P. Tseng, "On the linear convergence of descent methods for convex essentially smooth minimization," *SIAM Journal on Control and Optimization*, vol. 30, no. 2, pp. 408–425, 1990.

[173] E. T. Hale, W. Yin, and Y. Zhang, "Fixed-point continuation for l1-minimization: Methodology and convergence," *SIAM Journal on Optimization*, vol. 19, no. 3, pp. 1107–1130, 2008.

[174] Y. Nesterov, "A method of solving a convex programming problem with convergence rate $O(1/k^2)$," *Soviet Mathematics Doklady*, vol. 27, no. 2, pp. 372–376, 1983.

[175] Y. Nesterov, "Gradient methods for minimizing composite objective function," www.optimization-online.org, *CORE Discussion Paper 2007/76*, 2007.

[176] P. Tseng, "On accelerated proximal gradient methods for convex-concave optimization," *submitted to SIAM Journal on Optimization*, 2008.

[177] A. Beck and M. Teboulle, "A fast iterative shrinkage-thresholding algorithm for linear inverse problems," *SIAM Journal on Imaging Sciences*, vol. 2, no. 1, pp. 183–202, 2009.

[178] S. Becker, J. Bobin, and E. Candès, "NESTA: A fast and accurate first-order method for sparse recovery," *SIAM Journal of Imaging Science*, vol. 4, no. 1, pp. 1–39, 2011.

[179] W. Deng, W. Yin, and Y. Zhang, "Group sparse optimization by alternating direction method," *Rice University CAAM Technical Report TR11-06*, 2011.

[180] M. R. Hestenes, "Multiplier and gradient methods," *Journal of Optimization Theory and Applications*, vol. 4, no. 5, pp. 303–320, 1969.

[181] M. J. D. Powell, "A method for nonlinear constraints in minimization problems," in *Optimization*, R. Fletcher, Ed. New York: Academic Press, 1972, pp. 283–298.

[182] R. Glowinski and P. Le Tallec, *Augmented Lagrangian and Operator-Splitting Methods in Nonlinear Mechanics*. Philadelphia, Pennsylvania: SIAM, 1989.

[183] J. Eckstein, "Nonlinear proximal point algorithms using Bregman functions, with applications to convex programming," *Mathematics of Operations Research*, vol. 18, no. 1, pp. 202–226, 1993.

[184] L. Bregman, "The relaxation method of finding the common points of convex sets and its application to the solution of problems in convex programming," *USSR Computational Mathematics and Mathematical Physics*, vol. 7, no. 3, pp. 200–217, 1967.

[185] W. Yin and S. Osher, "Error forgetting of Bregman iteration," *Rice University CAAM Technical Report TR12-03. Submitted to Journal of Scientific Computing, Special issue in honor of Stanley Osher's 70th birthday*, 2012.

[186] S. Osher, M. Burger, D. Goldfarb, J. Xu, and W. Yin, "An iterative regularization method for total variation-based image restoration," *SIAM Journal on Multiscale Modeling and Simulation*, vol. 4, no. 2, pp. 460–489, 2005.

[187] J. Yang and Y. Zhang, "Alternating direction algorithms for ℓ_1-problems in compressive sensing," *Rice University CAAM Technical Report TR09-37*, 2009.

[188] K. J. Arrow, L. Hurwicz, and H. Uzawa, *Studies in Nonlinear Programming*. Stanford, CA: Stanford University Press, 1958.

[189] W. Yin, S. Osher, D. Goldfarb, and J. Darbon, "Bregman iterative algorithms for l1-minimization with applications to compressed sensing," *SIAM Journal on Imaging Sciences*, vol. 1, no. 1, pp. 143–168, 2008.

[190] S. Osher, Y. Mao, B. Dong, and W. Yin, "Fast linearized Bregman iteration for compressive sensing and sparse denoising," *Communications in Mathematical Sciences*, vol. 8, no. 1, pp. 93–111, 2010.

[191] J.-F. Cai, S. Osher, and Z. Shen, "Convergence of the linearized Bregman iteration for ℓ_1-norm minimization," *Mathematics of Computation*, vol. 78, no. 268, pp. 2127–2136, 2009.

[192] M. Friedlander and P. Tseng, "Exact regularization of convex programs," *SIAM Journal on Optimization*, vol. 18, no. 4, pp. 1326–1350, 2007.

[193] W. Yin, "Analysis and generalizations of the linearized Bregman method," *SIAM Journal on Imaging Sciences*, vol. 3, no. 4, pp. 856–877, 2010.

[194] M.-J. Lai and W. Yin, "Augmented ℓ_1 and nuclear-norm models with a globally linearly convergent algorithm," *Rice University CAAM Technical Report TR12-02. Submitted to SIAM Journal on Imaging Sciences*, 2012.

[195] H. Zhang, J. Cai, L. Cheng, and J. Zhu, "Strongly convex programming for exact matrix completion and robust principal component analysis," *Preprint*, 2012.

[196] R. T. Rockafellar, "The multiplier method of Hestenes and Powell applied to convex programming," *Journal of Optimization Theory and Applications*, vol. 12, no. 6, pp. 555–562, 1973.

[197] B. He, L. Liao, D. Han, and H. Yang, "A new inexact alternating directions method for monotone variational inequalities," *Mathematical Programming*, vol. 92, no. 1, pp. 103–118, 2002.

[198] W. Deng and W. Yin, "On the global linear convergence of the alternating direction method of multipliers," *Rice University CAAM Technical Report TR12-14*, 2012.

[199] G. Chen and M. Teboulle, "Convergence analysis of a proximal-like minimization algorithm using Bregman functions," *SIAM Journal on Optimization*, vol. 3, no. 3, pp. 538–543, 1993.

[200] X. Zhang, M. Burger, and S. Osher, "A unified primal-dual algorithm framework based on Bregman iteration," *Journal of Scientific Computing*, vol. 46, no. 1, pp. 20–46, 2011.

[201] S. Boyd, N. Parikh, E. Chu, B. Peleato, and J. Eckstein, "Distributed optimization and statistical learning via the alternating direction method of multipliers," *Foundations and Trends in Machine Learning*, vol. 3, no. 1, pp. 1–122, 2010.

[202] D. Bertsekas and J. Tsitsiklis, *Parallel and Distributed Computation: Numerical Methods*, 2nd edn. Nashua, NH, Athena Scientific, 1997.

[203] D. Goldfarb, S. Ma, and K. Scheinberg, "Fast alternating linearization methods for minimizing the sum of two convex functions," Mathematical Programming, March 1997.

[204] B. He and X. Yuan, "On non-ergodic convergence rate of Douglas-Rachford alternating direction method of multipliers," *Optimization Online*, 2012.

[205] T. Goldstein, B. OòDonoghue, and S. Setzer, "Fast alternating direction optimization methods," *UCLA CAM Report 12-35*, 2012.

[206] J. Eckstein and D. Bertsekas, "An alternating direction method for linear programming," *Laboratory for Information and Decision Systems, MIT,* [online] available: http://hdl.handle.net/1721.1/3197, 1990.

[207] J. Ortega and W. Rheinboldt, *Iterative Solution of Nonlinear Equations in Several Variables*. Amsterdam, The Netherlands: Academic Press, 1970.

[208] C. Hildreth, "A quadratic programming procedure," *Naval Research Logistics Quarterly*, vol. 4, no. 1, pp. 79–85, 1957.

[209] J. Warga, "Minimizing certain convex functions," *Journal of the Society for Industrial and Applied Mathematics*, vol. 11, no. 3, pp. 588–593, 1963.

[210] A. Auslender, *Optimisation: Méthodes Numériques*. Masson, 1976.

[211] R. Sargent and D. Sebastian, "On the convergence of sequential minimization algorithms," *Journal of Optimization Theory and Applications*, vol. 12, no. 6, pp. 567–575, 1973.

[212] S. Han, "A successive projection method," *Mathematical Programming*, vol. 40, no. 1, pp. 1–14, 1988.

[213] O. Mangasarian and R. Leone, "Parallel successive overrelaxation methods for symmetric linear complementarity problems and linear programs," *Journal of Optimization Theory and Applications*, vol. 54, no. 3, pp. 437–446, 1987.

[214] P. Tseng, "Dual coordinate ascent methods for non-strictly convex minimization," *Mathematical Programming*, vol. 59, no. 1, pp. 231–247, 1993.

[215] P. Tseng, "Convergence of a block coordinate descent method for nondifferentiable minimization," *Journal of Optimization Theory and Applications*, vol. 109, no. 3, pp. 475–494, 2001.

[216] P. Tseng and S. Yun, "A coordinate gradient descent method for nonsmooth separable minimization," *Mathematical Programming*, vol. 117, no. 1, pp. 387–423, 2009.

[217] Y. Li and S. Osher, "Coordinate descent optimization for l1 minimization with applications to compressed sensing: A greedy algorithm," *UCLA CAM Report 09-17*, 2009.

[218] M. Razaviyayn, M. Hong, and Z. Luo, "A unified convergence analysis of coordinatewise successive minimization methods for nonsmooth optimization," *Report of University of Minnesota, Twin Cities*, 2012.

[219] X. Wei, Y. Yuan, and Q. Ling, "DOA estimation using a greedy block coordinate descent algorithm," *Report of University of Science and Techonolgy of China*, 2012.

[220] B. Efron, T. Hastie, I. Johnstone, and R. Tibshirani, "Least angle regression," *Annals of Statistics*, vol. 32, no. 2, pp. 407–499, 2004.

[221] M. Best, "An algorithm for the solution of the parametric quadratic programming problem," *CORR Report 82-84, University of Waterloo*, 1982.

[222] L. Ghaoui, V. Viallon, and T. Rabbani, "Safe feature elimination in sparse supervised learning," *Arxiv preprint arXiv:1009.4219*, 2010.

[223] R. Tibshirani, J. Bien, J. Friedman, T. Hastie, N. Simon, J. Taylor, and R. Tibshirani, "Strong rules for discarding predictors in lasso-type problems," *Journal of the Royal Statistical Society: Series B*, vol. 57, no. 7, pp. 245–266, 2012.

[224] S. Wright, R. Nowak, and M. Figueiredo, "Sparse reconstruction by separable approximation," *IEEE Transactions on Signal Processing*, vol. 57, no. 7, pp. 2479–2493, 2009.

[225] Z. Wen, W. Yin, H. Zhang, and D. Goldfarb, "On the convergence of an active set method for l1-minimization," *Optimization Methods and Software*, vol. 27, no. 6, pp. 1–20, 2011.

[226] J. Nocedal and S. J. Wright, *Numerical Optimization*. New York: Springer-Verlag, 1999.

[227] E. Candes, M. Wakin, and S. Boyd, "Enhancing sparsity by reweighted 1 minimization," *Journal of Fourier Analysis and Applications*, vol. 14, no. 5, pp. 877–905, 2008.

[228] R. Chartrand and W. Yin, "Iteratively reweighted algorithms for compressive sensing," *ICASSP'08*, Las Vegas, NV, USA, March–April 2008.

[229] I. Daubechies, R. DeVore, M. Fornasier, and C. Güntürk, "Iteratively reweighted least squares minimization for sparse recovery," *Communications on Pure and Applied Mathematics*, vol. 63, no. 1, pp. 1–38, 2010.

[230] K. Mohan and M. Fazel, "Iterative reweighted least squares for matrix rank minimization," *48th Annual Allerton Conference on Communication, Control, and Computing (Allerton)*, Monticello, IL, USA, October 2010.

[231] M. Fornasier, H. Rauhut, and R. Ward, "Low-rank matrix recovery via iteratively reweighted least squares minimization," *SIAM Journal on Optimization*, vol. 21, no. 4, pp. 1614–1640, 2011.

[232] M.-J. Lai, Y. Xu, and W. Yin, "On unconstrained nonconvex minimization for sparse vector and low-rank matrix recovery," *Submitted to SIAM Journal on Numerical Analysis*, 2011.

[233] R. Chartrand and V. Staneva, "Restricted isometry properties and nonconvex compressive sensing," *Inverse Problems*, vol. 24, no. 3, pp. 1–14, 2008.

[234] M. Lai and L. Liu, "The null space property for sparse recovery from multiple measurement vectors," *Applied and Computational Harmonic Analysis*, vol. 30, no. 3, pp. 402–406, 2011.

[235] Q. Sun, "Recovery of sparsest signals via ℓ_q-minimization," *Arxiv preprint:1005.0267*, 2011.

[236] D. Ge, X. Jiang, and Y. Ye, "A note on complexity of L_p minimization," *Stanford University Technical Report 2010*, 2010.

[237] S. Foucart, "Hard thresholding pursuit: An algorithm for compressive sensing," *SIAM Journal on Numerical Analysis*, vol. 49, no. 6, pp. 2543–2563, 2011.

[238] Y. Wang and W. Yin, "Sparse signal reconstruction via iterative support detection," *SIAM Journal on Imaging Sciences*, vol. 3, no. 3, pp. 462–491, 2010.

[239] K. Toh and S. Yun, "An accelerated proximal gradient algorithm for nuclear norm regularized linear least squares problems," *Pacific Journal of Optimization*, vol. 6, no. 3, pp. 615–640, 2010.

[240] J. Yang and X. Yuan, "Linearized augmented Lagrangian and alternating direction methods for nuclear norm minimization," *Optimization Online*, 2011.

[241] R. Keshavan, A. Montanari, and S. Oh, "Matrix completion from a few entries," *Information Theory, IEEE Transactions on*, vol. 56, no. 6, pp. 2980–2998, 2010.

[242] Z. Wen, W. Yin, and Y. Zhang, "Solving a low-rank factorization model for matrix completion by a nonlinear successive over-relaxation algorithm," *Rice University CAAM Report TR10-07*, 2010.

[243] [Online] available: http://en.wikipedia.org/wiki/Analog-to-digital_converter.

[244] J. A. Tropp, J. N. Laska, M. F. Duarte, J. K. Romberg, and R. G. Baraniuk, "Beyond Nyquist: Efficient sampling of sparse bandlimited signals," *IEEE Transactions on Information Theory*, vol. 56, no. 1, pp. 520–544, January 2010.

[245] S. Kirolos, J. Laska, M. Wakin, M. Duarte, D. Baron, T. Ragheb, Y. Massoud, and R. Baraniuk, "Analog-to-information conversion via random demodulation," in *IEEE Dallas Circuits and Systems Workshop (DCAS)*, Dallas, TX, 2006.

[246] J. N. Laska, S. Kirolos, M. F. Duarte, T. S. Ragheb, R. G. Baraniuk, and Y. Massoud, "Analog-to-information conversion via random demodulation," in *IEEE International Symposium on Circuits and Systems, ISCAS*, New Orleans, LA, May 2007.

[247] M. Mishali and Y. C. Eldar, "From theory to practice: Sub-Nyquist sampling of sparse wideband analog signals," *IEEE Journal of Selected Topics in Signal Processing*, vol. 4, no. 2, pp. 375–391, April 2010.

[248] M. Mishali and Y. C. Eldar, "Expected rip: Conditioning of the modulated wideband converter," in *2009 IEEE Information Theory Workshop*, Sicily, Italy, 2009.

[249] M. Mishali, Y. C. Eldar, and J. A. Tropp, "Efficient sampling of sparse wideband analog signals." IEEE 25th Convention of Electrical and Electronics in Israel, Israel, 2008.

[250] M. Mishali, Y. C. Eldar, and A. Elron, "Xampling: Signal acquisition and processing in union of subspaces," *IEEE Transactions on Signal Processing*, vol. 59, no. 10, pp. 4719–4734, Oct. 2011.

[251] T. Michaeli and Y. C. Eldar, "Xampling at the rate of innovation," *IEEE Transactions on Signal Processing*, vol. 60, no. 3, pp. 1121–1133, May 2012.

[252] K. Gedalyahu and Y. Eldar, "Time-delay estimation from low-rate samples: A union of subspaces approach," *IEEE Transactions on Signal Processing*, vol. 58, no. 6, pp. 3017–3031, June 2011.

[253] E. Matusiak and Y. Eldar, "Sub-Nyquist sampling of short pulses," *IEEE Transactions on Signal Processing*, vol. 60, no. 3, pp. 1134–1148, Mar. 2012.

[254] M. Mishali and Y. C. Eldar, "Xampling: Compressed sensing of analog signals," *Book Chapter, Compressed Sensing Theory and Applications*. Cambridge, UK: Cambridge University Press, 2012.

[255] M. Mishali, Y. Eldar, O. Dounaevsky, and E. Shoshan, "Xampling: Analog to digital at sub-Nyquist rates," *IET Circuits Devices System*, vol. 5, no. 1, pp. 8–20, 2011.

[256] M. A. Davenport, P. T. Boufounos, M. B. Wakin, and R. G. Baraniuk, "Signal processing with compressive measurements," *IEEE J. Sel. Topics Signal Process.*, vol. 4, no. 2, pp. 445–460, 2010.

[257] T. Ragheb, S. Kirolos, J. Laska, A. Gilbert, M. Strauss, R. Baraniuk, and Y. Massoud, "Implementation models for analog-to-information conversion via random sampling," in *Midwest Symposium on Circuits and Systems*, Montreal, Canada, 2007.

[258] J. Laska, S. Kirolos, Y. Massoud, R. Baraniuk, A. Gilbert, M. Iwen, and M. Strauss, "Random sampling for analog-to-information conversion of wideband signals," in *IEEE Dallas/CAS Workshop on Design, Applications, Integration and Software*, Dallas, TX, 2006.

[259] J. A. Tropp, M. B. Wakin, M. F. Duarte, D. Baron, and R. G. Baraniuk, "Random filters for compressive sampling and reconstruction," in *IEEE International Conference on Acoustics, Speech and Signal Processing, 2006. ICASSP 2006 Proceedings*, Toulouse, France, May 2006.

[260] J. Meng, J. Ahmadi-Shokouh, H. Li, E. J. Charlson, Z. Han, S. Noghanian, and E. Hossain, "Sampling rate reduction for 60 GHZ UWB communication using compressive sensing," in *Asilomar Conference on Signals, Systems & Computers*, Pacific Grove, CA, 2009.

[261] M. Mishali and Y. C. Eldar, "Blind multiband signal reconstruction: Compressed sensing for analog signals," *IEEE Transactions on Signal Processing*, vol. 57, no. 3, pp. 993–1009, March 2009.

[262] S. R. Schnelle, J. P. Slavinsky, P. T. Boufounos, M. A. Davenport, and R. G. Baraniuks, "A compressive phase-locked loop," in *IEEE International Conference on Acoustics, Speech, and Signal Processing (ICASSP)*, Kyoto, Japan, 2012.

[263] M. Davenport, S. Schnelle, J. Slavinsky, R. Baraniuk, M. Wakin, and P. Boufounos, "A wideband compressive radio receiver," in *The 2010 Military Communications Conference*, San Jose, CA, 2010.

[264] M. A. Davenport, J. N. Laska, J. R. Treichler, and R. G. Baraniuk, "The pros and cons of compressive sensing for wideband signal acquisition: Noise folding vs. dynamic range," *IEEE Transactions on Signal Processing*, vol. 60, no. 9, pp. 4628–4642, Sept. 2012.

[265] J. Treichler, M. Davenport, J. Laska, and R. Baraniuk, "Dynamic range and compressive sensing acquisition receivers," in *7th U.S./Australia Joint Workshop on Defense Applications of Signal Processing (DASP)*, Coolum, Australia, 2011.

[266] M. Lexa, M. Davies, and J. Thompson, "Reconciling compressive sampling systems for spectrally-sparse continuous-time signals," *IEEE Transactions on Signal Processing*, vol. 60, no. 1, pp. 151–171, Jan. 2012.

[267] P. Maechler, C. Studer, D. Bellasi, A. Maleki, A. Burg, N. Felber, H. Kaeslin, and R. G. Baraniuk, "VLSI implementation of approximate message passing for signal restoration and compressive sensing," *IEEE Journal on Emerging and Selected Topics in Circuits and Systems*, vol. 2, no. 3, pp. 1–11, Oct. 2012.

[268] X. Chen, Z. Yu, S. Hoyos, B. M. Sadler, and J. Silva-Martinez, "A sub-Nyquist rate sampling receiver exploiting compressive sensing," *IEEE Transactions on Circuits and Systems I: Regular Papers*, vol. 58, no. 3, pp. 507–520, Mar. 2011.

[269] M. A. Davenport and M. B. Wakin, "Compressive sensing of analog signals using discrete prolate spheroidal sequences," *Applied and Computational Harmonic Analysis*, vol. 33, no. 3, pp. 438–472, Nov. 2012.

[270] A. Goldsmith, *Wireless Communications*. Cambridge, UK: Cambridge University Press, 2005.

[271] R. Baraniuk, "Compressive sensing," *IEEE Signal Processing Magazine*, vol. 24, no. 4, pp. 118–121, July 2007.

[272] W. U. Bajwa, J. Haupt, A. M. Sayeed, and R. Nowak, "Compressed channel sensing: A new approach to estimating sparse multipath channels," *Proceedings of the IEEE*, vol. 98, no. 6, pp. 1058–1076, June 2010.

[273] J. L. Paredes, G. R. Arce, and Z. Wang, "Ultra-wideband compressed sensing channel estimation," *IEEE Journal on Selected Topics in Signal Processing*, vol. 1, no. 3, pp. 383–395, Oct. 2007.

[274] C. R. Berger, Z. Wang, Z. Huang, and S. Zhou, "Application of compressive sensing to sparse channel estimation," *IEEE Communications Magazine*, vol. 48, no. 11, pp. 164–174, Nov. 2010.

[275] P. Zhang, Z. Hu, R. C. Qiu, and B. M. Sadler, "Compressive sensing based ultra-wideband communication system," in *Proc. of IEEE International Conference on Communications (ICC)*, Kyoto, Japan, June 2009.

[276] J. Romberg, "Multiple channel estimation using spectrally random probes," in *Proc. SPIE Wavelets XIII*, San Diego, CA, USA, 2009.

[277] W. U. Bajwa, A. Sayeed, and R. Nowak, "Compressed sensing of wireless channels in time, frequency, and space," in *Proc. of 42nd Asilomar Conf. Signals, Systems, and Computers*, Pacific Grove, CA, 2008.

[278] J. Meng, Y. Li, N. Nguyen, W. Yin, and Z. Han, "Compressive sensing based high resolution channel estimation for OFDM system," *IEEE Journal of Selected Topics in Signal Processing, special issue on Robust Measures and Tests Using Sparse Data for Detection and Estimation*, vol. 6, no. 1, pp. 15–25, Feb. 2012.

[279] C. R. Berger, J. Gomes, and J. M. F. Moura, "Study of pilot designs for cyclic-prefix OFDM on time-varying and sparse underwater acoustic channels," in *Proc. of IEEE Oceans*, Spain, June 2011.

[280] J. Huang, C. R. Berger, S. Zhou, and P. Willett, "Iterative sparse channel estimation and decoding for underwater MIMO-OFDM," *EURASIP Journal on Advances in Signal Processing*, vol. 2010, no. 460379, 2010.

[281] S. Gleichman and Y. C. Eldar, "Blind compressed sensing," *IEEE Transactions on Information Theory*, vol. 57, no. 10, pp. 6958–6975, Oct. 2011.

[282] S. Gleichman and Y. C. Eldar, "Multichannel blind compressed sensing," in *IEEE Sensor Array and Multichannel Signal Processing Workshop*, Hoboken, NJ, June 2010.

[283] G. Mileounis, B. Babadi, N. Kalouptsidis, and V. Tarokh, "An adaptive greedy algorithm with application to nonlinear communications," *IEEE Transactions on Signal Processing*, vol. 58, no. 6, pp. 2998–3007, June 2010.

[284] Y. C. Eldar, P. Kuppinger, and H. Bolcskei, "Block-sparse signals: Uncertainty relations and efficient recovery," *IEEE Trans. Sig. Process.*, vol. 58, no. 6, pp. 3042–3054, June 2010.

[285] D. Eiwena, G. Tauböck, F. Hlawatsch, and H. G. Feichtinger, "Group sparsity methods for compressive channel estimation in doubly dispersive multicarrier systems," in *2010 IEEE Eleventh International Workshop on Signal Processing Advances in Wireless Communications (SPAWC)*, Vienna, Austria, June 2010.

[286] J. Haupt, W. U. Bajwa, G. Raz, and R. Nowak, "Toeplitz compressed sensing matrices with application to sparse channel estimation," *IEEE Transactions on Information Theory*, vol. 56, no. 11, pp. 5862–5875, Nov. 2010.

[287] G. Tauböck, F. Hlawatsch, D. Eiwen, and H. Rauhut, "Compressive estimation of doubly selective channels in multicarrier systems: Leakage effects and sparsity-ehancing processing," *IEEE Journal on Selected Topics in Signal Processing*, vol. 4, no. 2, pp. 255–271, Apr. 2010.

[288] G. Tauböck and F. Hlawatsch, "Compressed sensing based estimation of doubly selective channels using a sparsity-optimized basis expansion," in *Proceedings of the 16th European Signal Processing Conference (EUSIPCO 2008)*, Lousanne, Switzerland, August 2008.

[289] W. U. Bajwa, A. Sayeed, and R. Nowak, "Learning sparse doubly-selective channels," in *in Proc. of 46th Allerton conf. Communication, Control, and Computing*, Monticello, IL, USA, October 2008.

[290] O. Edfors, M. Sandell, J. J. V. de Beek, D. Landström, and F. Sjöberg, "An introduction to orthogonal frequency division multiplexing," Luleå Sweden: Luleå Tekniska Universitet, Tech. Rep., 1996.

[291] S. Takaoka and F. Adachi, "Pilot-assisted adaptive interpolation channel estimation for OFDM signal reception," in *Proc. of IEEE Vehicular Technology Conference (VTC)*, Milan, Italy, May 2004.

[292] J. Byun and N. P. Natarajan, "Adaptive pilot utilization for OFDM channel estimation in a time varying channel," in *Proc. of IEEE Wireless and Microwave Technology Conference (WAMICON)*, Clearwater, FL, USA, April 2009.

[293] W. U. Bajwa, J. D. Haupt, G. M. Raz, S. J. Wright, and R. D. Nowak, "Toeplitz-structured compressed sensing matrices," in *Proc. of IEEE/SP Workshop on Statistical Signal Processing (SSP)*, Madison, WI, USA, August 2007.

[294] J. Romberg, "Compressive sensing by random convolution," *SIAM J. Imaging Sci.*, vol. 2, no. 4, pp. 1098–1128, 2009.

[295] J. K. Romberg and R. Neelamani, "Sparse channel separation using random probes," *Inverse Problem*, vol. 26, no. 11, pp. 1–25, Nov. 2010.

[296] C. R. Berger, S. Zhou, P. Willett, B. Demissie, and J. Heckenbach, "Compressed sensing for OFDM/MIMO radar," in *Proc. of IEEE Annual Asilomar Conference on Signals, Systems and Computers*, Pacific Grove, CA, USA, October 2008.

[297] C. R. Berger, S. Zhou, and P. Willett, "Signal extraction using compressed sensing for passive radar with OFDM signals," in *Proc. of IEEE International Conference on Information Fusion*, Cologne Germany, June–July 2008.

[298] G. Tauböck and F. Hlawatsch, "A compressed sensing technique for OFDM channel estimation in mobile environments: Exploiting channel sparsity for reducing pilots," in *Proc. of IEEE International Conferenc on Acoustics, Speech, and Signal Processing (ICASSP)*, Las Vegas, NV, USA, April 2008.

[299] C. R. Berger, S. Zhou, W. Chen, and P. Willett, "Sparse channel estimation for OFDM: Over-complete dictionaries and super-resolution methods," in *Proc. of IEEE International Workshop on Signal Process. Advances in Wireless Communications (SPAWC)*, Perugia, Italy, June 2009.

[300] M. Mohammadnia-Avval, A. Ghassemi, and L. Lampe, "Compressive sensing recovery of nonlinearly distorted OFDM signals," in *Proc. of IEEE International Conference on Communications (ICC)*, Kyoto, Japan, June 2011.

[301] P. Schniter, "A message-passing receiver for BICM-OFDM over unknown clustered-sparse channels," *IEEE Journal on Selected Topics in Signal Processing*, vol. 5, no. 8, pp. 1462–1474, Dec. 2011.

[302] A. Gomaa and N. Al-Dhahir, "A sparsity-aware approach for NBI estimation in MIMO-OFDM," *Wireless Communication, IEEE Transactions on*, vol. 10, no. 6, pp. 1854–1862, June 2011.

[303] M. Ledoux, Ed., *The Concentration of Measure Phenomenon*. American Mathematical Society, 2001.

[304] W. Guo and W. Yin, "Edgecs: Edge guided compressive sensing reconstruction," Rice University, Tech. Rep., 2009.

[305] M. Stojanovic, "On the relationship between capacity and distance in an underwater acoustic communication channel," in *Proceedings of the First ACM International Workshop on UnderWater Networks (WUWNet)*, Los Angeles, CA, USA, Sept. 2006.

[306] E. M. Sozer, M. Stojanovic, and J. G. Proakis, "Underwater acoustic networks," *IEEE Journal of Oceanic Engineering*, vol. 25, no. 1, pp. 72–83, Jan. 2000.

[307] J. Preisig, "Acoustic propagation considerations for underwater acoustic communications network development," in *Proceedings of the 1st ACM International Workshop on Underwater Networks*, Los Angeles, CA, USA, Sept. 2006.

[308] P. C. Etter, *Underwater Acoustic Modeling and Simulation, 3rd edn.* UK: Spon Press, Taylor & Francis Group, 2003.

[309] R. J. Urick, *Principles of Underwater Sound, 3rd edn.* New York: McGraw-Hill, 1983.

[310] "Technical guides – speed of sound in sea-water," in *the National Physical Laboratory,* [online] available: http://www.npl.co.uk/acoustics/techguides/soundseawater/content.html.

[311] E. Hossain, D. Niyato, and Z. Han, *Dynamic Spectrum Access in Cognitive Radio Networks*. Cambridge, UK: Cambridge University Press, 2009.

[312] L. Lai, H. E. Gamal, H. Jiang, and H. V. Poor, "Cognitive medium access: Exploration, exploitation and competition," *IEEE Transactions on Mobile Computing*, vol. 10, no. 2, pp. 239–253, Feb. 2011.

[313] Q. Zhao and A. Swami, "A decision-theoretic framework for opportunistic spectrum access," *IEEE Wireless Communications Magazine*, vol. 14, no. 4, pp. 14–20, Aug. 2007.

[314] D. Chen, S. Yin, Q. Zhang, M. Liu, and S. Li, "Mining spectrum usage data: A large-scale spectrum measurement study," in *Proceedings of the 15th Annual International Conference on Mobile Computing and Networking*, Beijing, China, September 2009.
[315] J. Riihijärvi, P. Mähönen, M. Wellens, and M. Gordziel, "Characterization and modelling of spectrum for dynamic spectrum access with spatial statistics and random fields," in *Proc. of IEEE 19th International Symposium on Personal, Indoor and Mobile Radio Communications (PIMRC)*, Cannes, France, 2008.
[316] M. Wellens, J. Riihijarvi, M. Gordziel, and P. Mähönen, "Spatial statistics of spectrum usage: From measurements to spectrum models," in *Proc. of IEEE International Conference on Communications (ICC)*, Bresden, Germany, June 2009.
[317] H. Li, Z. Han, and Z. Zhang, "Communication over random fields: A statistical framework for cognitive radio networks," in *Proc. of IEEE Globe Communication Conference (Globecom)*, Houston, TX, USA, December 2011.
[318] M. Wellens, J. Riihijärvi, and P. Mähönen, "Spatial statistics and models of spectrum use," *Elsevier Computer Communications*, vol. 32, no. 18, pp. 1998–2011, Aug. 2009.
[319] H. Li, "Reconstructing geographical-spectral pattern in cognitive radio networks," in *Proc. of the Fifth International Conference on Cognitive Radio Oriented Wireless Networks Communications (CROWNCOM)*, Cannes, France, June 2010.
[320] E. J. Candés and B. Recht, "Exact matrix completion via convex optimization," *Magazine Communications of the ACM*, vol. 55, no. 6, pp. 111–119, June 2012.
[321] A. Oka and L. Lampe, "Compressed sensing of Gauss-Markov random field with wireless sensor networks," in *Proc. of Sensor Array and Multichannel Signal Processing Workshop, 2008. SAM 2008. 5th IEEE*, Vancouver, Canada, July 2008.
[322] V. Cevher, M. F. Duarte, C. Hegde, and R. G. Baraniuk, "Sparse signal recovery using Markov random fields," in *Proc. of the Workshop on Neural Information Processing Systems*, Vancouver, Canada, Dec. 2008.
[323] R. Kindermann and J. L. Snell, *Markov Random Fields and Their Applications*. American Mathematical Society, 1980.
[324] W. Yin, Z. Wen, S. Li, J. Meng, and Z. Han, "Dynamic compressive spectrum sensing for cognitive radio networks," in *Proc. of Conference on Information Sciences and Systems (CISS)*, Baltimore, MD, USA, March 2011.
[325] S. Burer and R. Monteiro, "Local mimima and convergence in low-rank semidefinite programming," *Mathematical Programming*, vol. 103, no. 3, pp. 427–444, 2005.
[326] N. Kalouptsidis, G. Mileounis, B. Babadi, and V. Tarokh, "Adaptive algorithms for sparse nonlinear channel estimation," in *IEEE/SP 15th Workshop on Statistical Signal Processing*, Cardiff, United Kingdom, Sept. 2009.
[327] H. F. Harmuth, *Transmission of Information by Orthogonal Functions*. New York, NY, USA, Springer, 1969.
[328] X. Shen, R. C. Q. M. Guizani, and T. Le-Ngoc, *Ultra-Wideband Wireless Communications and Networks*. Hoboken, NJ, USA, Wiley, 2006.
[329] J. Zhang, P. Orlik, Z. Sahinoglu, A. F. Molisch, and P. Kinney, "UWB systems for wireless sensor networks," *Proceedings of IEEE*, vol. 97, no. 2, pp. 313–331, Feb. 2009.
[330] D. Yang, H. Li, G. D. Peterson, and A. Fathy, "Compressive sensing TDOA for UWB positioning systems," in *Proc. of IEEE Radio and Wireless Symposium (RWS)*, Santa Clara, CA, USA, January 2011.
[331] P. Zhang, H. Zhen, R. C. Qiu, and B. M. Sadler, "A compressed sensing based ultra-wideband communication system," in *Proc. of IEEE International Conference on Communications (ICC)*, Dresden, Germany, June 2009.

[332] D. Yang, H. Li, G. D. Peterson, and A. Fathy, "Compressed sensing based UWB receiver: Hardware compressing and FPGA reconstruction," in *Proc. of Conference on Information Sciences and Systems (CISS)*, Baltimore, MD, USA, March 2009.

[333] D. Pozar, *Microwave Engineering*. Hoboken, NY, USA: Wiley, 1998.

[334] Y. Zhu, J. D. Zuegel, J. R. Marciante, and H. Wu, "Downlink scheduling using compressed sensing a reconfigurable, multi-gigahertz pulse shaping circuit based on distributed transversal filters," in *Proc. of IEEE International Symposium on Circuits and Systems*, Island of Kos, Greece, May 2006.

[335] M. E. Tipping, "Sparse Bayesian learning and the relevance vector machine," *Journal of Machine Learning Research*, vol. 1, pp. 211–244, 2001.

[336] S. Ji, Y. Xue, and L. Carin, "Bayesian compressive sensing," *IEEE Trans. Signal Processing*, vol. 56, no. 6, pp. 1094–1121, June 2008.

[337] A. Oka and L. Lampe, "A compressed sensing receiver for bursty communication with UWB impulse radio," in *Proc. of IEEE International Conference on Ultra-Wideband (ICUWB)*, Vancouver, Canada, Sept. 2009.

[338] Z. Sahinoglu, S. Gezici, and I. Guvenc, *Ultra-wideband Positioning Systems: Theoretical Limits, Ranging Algorithms and Protocols*. Cambridge, UK: Cambridge University Press, 2011.

[339] C. Feng, W. S. A. Au, S. Valaee, and Z. Tan, "Compressive sensing based positioning using rss of wlan access points," in *Proc. of IEEE Conference on Computer Communications (Infocom)*, San Diego, CA, USA, March 2010.

[340] D. Milioris, G. Tzagkarakis, P. Jacquet, and P. Tsakalides, "Indoor positioning in wireless lans using compressinve sensing signal-strength fingerprints," in *Proc. of EUSIPCO*, Barcelona, Spain, August–September 2011.

[341] X. Li, Q. Han, V. Chakravarthy, and Z. Wu, "Joint spectrum sensing and primary user localization for cognitive radio via compressed sensing," in *Proc. of IEEE Military Communication Conference (MILCOM)*, San Jose, CA, USA, October–November 2010.

[342] L. Liu, Z. Han, Z. Wu, and L. Qian, "Collaborative compressive sensing based dynamic spectrum sensing and mobile primary user localization in cognitive radio networks," in *Proc. of IEEE Global Telecommunications Conference (Globecom)*, Houston, TX, USA, December 2011.

[343] C. Berrou, A. Glavieux, and P. Thitimajshima, "Near Shannon limit error-correcting coding and decoding: Turbo codes," in *Proc. of IEEE International Conference on Communications (ICC)*, Geneva, Switzerland, May 1993.

[344] A. J. Viterbi, *CDMA: Principles of Spread Spectrum Communication*. Addison-Wesley, Boston, MA, USA, 1995.

[345] S. Verdu, *Multiuser Detection*. Cambridge University Press, Cambridge, UK, 1998.

[346] D. L. Donoho, Y. Tsaig, I. Drori, and J. Starck, "Sparse solution of underdetermined systems of linear equations by stagewise orthogonal matching pursuit," *IEEE Trans. on Information Theory*, vol. 58, no. 2, pp. 1094–1121, Feb. 2012.

[347] D. Divsalar, M. K. Simon, and D. Raphaeli, "Improved parallel interference cancellation for CDMA," *IEEE Trans. on Communications*, vol. 46, no. 2, pp. 258–268, Feb. 1998.

[348] A. L. C. Hui and K. B. Letaief, "Successive interference cancellation for multiuser asynchronous DS/CDMA detectors in multipath fading links," *IEEE Trans. on Communications*, vol. 46, no. 3, pp. 384–391, Mar. 1998.

[349] A. K. Fletcher, S. Rangan, and V. K. Goyal, "On-off random access channels: A compressed sensing framework," *preprint*, 2009.

[350] M. J. Wainwright, *Sharp thresholds for high-dimensional and noisy recovery of sparsity*. Univ. of California, Berkeley, Dept. of Statistics, Tech. Rep., 2008.

[351] H. Wang and C. Leng, "Unified lasso estimation via least square approximation," *Journal of American Statistical Association*, vol. 102, no. 479, pp. 1039–1048, Sept. 2007.

[352] X. Liu, E. K. P. Chong, and N. B. Shroff, "A framework for opportunistic scheduling in wireless networks," *Computer Networks*, vol. 41, no. 4, pp. 451–474, Mar. 2003.

[353] S. R. Bhaskaran, L. Davis, A. Grant, S. Hanly, and P. Tune, "Downlink scheduling using compressed sensing," in *Proc. of IEEE Workshop on Information Theory*, Taormina, Sicily, Italy, December 2009.

[354] F. Fazel, M. Fazel, and M. Stojanovic, "Random access compressed sensing for energy-efficient underwater sensor networks," *IEEE Journal on Selected Areas in Communications*, vol. 29, no. 8, pp. 1660–1670, Aug. 2011.

[355] M. S. B. K. S. Lee, S.Pattern and A. Ortega, "Spatially-locallized compressed sensing and rotuing in multi-hop sensor networks," in *Proc. of 3rd International Conference on Geosensor Networks (GSN)*, Oxford, UK, 2009.

[356] G. Staple and K. Werbach, "The end of spectrum scarcity," *IEEE Spectrum Archive*, vol. 41, no. 3, pp. 48–52, Mar. 2004.

[357] H. Kim and K. G. Shin, "Efficient discovery of spectrum opportunities with MAC-layer sensing in cognitive radio networks," *IEEE Trans. on Mobile Computing*, vol. 7, no. 5, pp. 533–545, May 2008.

[358] "Longley-rice methodology for evaluating TV coverage and interference," Office of Engineering and Technology (OET), Federal Communications Commision, Tech. Rep., July 1997, OET Bulletin NO. 69.

[359] S. M. Mishra, A. Sahai, and R. Brodersen, "Cooperative sensing among cognitive radios," in *Proc. of IEEE International Conference on Communications (ICC)*, Istanbul, Turkey, June 2006.

[360] A. Ghasemi and E. S. Sousa, "Collaborative spectrum sensing for opportunistic access in fading environments," in *Proc. of IEEE International Symposium on New Frontiers in Dynamic Spectrum Access Networks*, Baltimore, MD, USA, November 2005.

[361] A. Ghasemi and E. S. Sousa, "Opportunistic spectrum access in fading channels through collaborative sensing," *Journal of Communications (JCM)*, vol. 2, no. 2, pp. 71–82, Mar. 2007.

[362] W. Saad, Z. Han, M. Debbah, A. Hjørungnes, and T. Başar, "Coalitional games for distributed collaborative spectrum sensing in cognitive radio networks," in *Proc. of IEEE Conference on Computer Communications (INFOCOM)*, Rio de Janeiro, Brazil, April 2009.

[363] G. Ghurumuruhan and Y. Li, "Cooperative spectrum sensing in cognitive radio: Part I: Two user networks," *IEEE Transactions on Wireless Communications*, vol. 6, no. 6, pp. 2204–2213, June 2007.

[364] G. Ghurumuruhan and Y. Li, "Cooperative spectrum sensing in cognitive radio: Part II: Multiuser networks," *IEEE Transactions on Wireless Communications*, vol. 6, no. 6, pp. 2214–2222, June 2007.

[365] J. Unnikrishnan and V. V. Veeravalli, "Cooperative sensing for primary detection in cognitive radio," *IEEE Journal of Selected Topics in Signal Processing*, vol. 2, no. 1, pp. 18–27, Feb. 2008.

[366] S. Cui, Z. Quan, and A. Sayed, "Optimal linear cooperation for spectrum sensing in cognitive radio networks," *IEEE Journal of Selected Topics in Signal Processing*, vol. 2, no. 1, pp. 28–40, Feb. 2008.

[367] W. Zhang, C. Sun, and K. B. Letaief, "Cooperative spectrum sensing for cognitive radios under bandwidth constraints," in *Proc. of IEEE Wireless Communications and Networking Conference*, Hong Kong, China, March 2007.

[368] C. H. Lee and W. Wolf, "Energy efficient techniques for cooperative spectrum sensing in cognitive radios," in *Proc. of IEEE Consumer Communications and Networking Conference (CCNC)*, Las Vegas, NV, USA, January 2008.

[369] J. Meng, H. Li, and Z. Han, "Sparse event detection in wireless sensor networks using compressive sensing," in *Proc. of IEEE Conference on Information Sciences and Systems (CISS)*, Baltimore, MD, USA, March 2009.

[370] R. H. Keshavan and S. Oh, "Matrix completion from a few entries," in *Proc. of International Symposium on Information Theory (ISIT)*, Seoul, Korea, July 2009.

[371] D. Goldfarb and S. Ma, "Convergence of fixed point continuation algorithms for matrix rank minimization," Department of IEOR, Columbia University, Tech. Rep., June 2009, [online] available: http://arxiv.org/abs/0906.3499.

[372] M. F. Duarte, S. Sarvotham, M. B. Wakin, D. Baron, and R. Baraniuk, "Joint sparsity models for distributed compressed sensing," in *Proc. of Workshop on Signal Processing with Adaptive Sparse Structured Representations*, Rennes, France, November 2005.

[373] J. Tropp, A. C. Gilbert, and M. J. Strauss, "Simultaneous sparse approximation via greedy pursuit," in *Proc. of IEEE International Conference on Acoustics, Speech, and Signal Processing (ICASSP)*, Philadelphia, PA, USA, March 2005.

[374] M. Fornasier and H. Rauhut, "Recovery algorithms for vector-valued data with joint sparsity constraints," *SIAM Journal on Numerical Analysis*, vol. 46, no. 2, pp. 577–613, March 2008.

[375] C. La and M. N. Do, "Signal reconstruction using sparse tree representation," in *Proc. SPIE Conference on Wavelet Applications in Signal and Image Processing*, San Diego, CA, USA, August 2005.

[376] E. Visotsky, S. Kuffner, and R. Peterson, "On collaborative detection of TV transmissions in support of dynamic spectrum sensing," in *Proc. of IEEE International Symposium on New Frontiers in Dynamic Spectrum Access Networks*, Baltimore, MD, USA, November 2005.

[377] W. Zhang and K. B. Letaief, "Cooperative spectrum sensing with transmit and relay diversity in cognitive networks," *IEEE Transactions on Wireless Communications*, vol. 7, no. 12, pp. 4761–4766, Dec. 2008.

[378] L. S. Cardoso, M. Debbah, P. Bianchi, and J. Najim, "Cooperative spectrum sensing using random matrix theory," in *Proc. of IEEE International Symposium on Wireless Pervasive Computing (ISWPC)*, Santorini, Greece, May 2008.

[379] B. Wang, K. J. R. Liu, and T. Clancy, "Evolutionary game framework for behavior dynamics in cooperative spectrum sensing," in *IEEE Global Communication Conference*, New Orleans, LA, USA, December 2008.

[380] C. Sun, W. Zhang, and K. B. Letaief, "Cluster-based cooperative spectrum sensing in cognitive radio systems," in *Proc. of IEEE International Conference on Communications (ICC)*, Glasgow, Scotland, June 2007.

[381] L. Cao and H. Zheng, "Distributed rule-regulated spectrum sharing," *IEEE Journal on Selected Areas in Communications: Special Issue on Cognitive Radio: Theory and Applications*, vol. 26, no. 1, pp. 130–145, Jan. 2008.

[382] Y. Liang, Y. Zeng, E. Peh, and A. T. Hoang, "Sensing-throughput tradeoff for cognitive radio networks," *IEEE Trans. on Wireless Communications*, vol. 7, no. 4, pp. 1326–1337, Apr. 2008.

[383] X. Zhou, J. Ma, Y. Li, Y. H. Kwon, and A. C. K. Soong, "Cooperative spectrum sensing with transmit and relay diversity in cognitive networks," *IEEE Trans. on Communications*, vol. 58, no. 2, pp. 463–466, Feb. 2010.

[384] Z. Tian, "Compressed wideband sensing in cooperative cognitive radio networks," in *Proc. of IEEE Global Telecommunications Conference (GLOBECOM)*, New Orleans, LO, USA, December 2008.

[385] Y. Wang, Z. Tian, and C. Feng, "A two-step compressed spectrum sensing scheme for wideband cognitive radios," in *IEEE Global Telecommunications Conference*, Miami, FL, USA, December 2010.

[386] S. Hong, "Multi-resolution bayesian compressive sensing for cognitive radio primary user detection," in *IEEE Global Telecommunications Conference*, December 2010.

[387] F. Zeng, C. Li, and Z. Tian, "Distributed compressive spectrum sensing in cooperative multihop cognitive networks," *IEEE Journal of Selected Topics in Signal Processing*, vol. 5, no. 1, pp. 37–38, Feb. 2010.

[388] Z. Tian, E. Blasch, W. Li, G. Chen, and X. Li, "Performance evaluation of distributed compressed wideband sensing for cognitive radio networks," in *11th International Conference on Information Fusion*, Cologne, Germany, June–July 2008.

[389] Z. Tian, G. Leus, and V. Lottici, "Joint dynamic resource allocation and waveform adaptation for cognitive networks," *IEEE Journal on Selected Areas in Communications*, vol. 29, no. 2, pp. 443 –454, Feb. 2011.

[390] M. S. Asif and J. Romberg, "Dynamic updating for l_1 minimization," *IEEE Journal of Selected Topics in Signal Processing*, vol. 4, no. 2, pp. 421–434, Apr. 2010.

[391] M. S. Asif and J. Romberg, "Dynamic updating for sparse time varying signals," in *43rd Annual Conference on Information Sciences and Systems (CISS)*, Baltimore, MD, USA, March 2009.

[392] M. S. Asif and J. Romberg, "Streaming measurements in compressive sensing: l1 filtering," in *42nd Asilomar Conference on Signals, Systems and Computers*, Pacific Grove, CA, USA, October 2008.

[393] N. Vaswani, "Kalman filtered compressed sensing," in *International Conference on Image Processing*, San Diego, CA, USA, October 2008.

[394] X. Li, V. Chakravarthy, and Z. Wu, "Joint spectrum sensing and primary user localization for cognitive radio via compressed sensing," in *IEEE MILCOM*, San Jose, CA, USA, October 2010.

[395] J. A. Bazerque and G. B. Giannakis, "Distributed spectrum sensing for cognitive radio networks by exploiting sparsity," *IEEE Transations on Signal Processing*, vol. 58, no. 3, pp. 1847–1862, Mar. 2010.

[396] P. Drineas, R. Kannan, and M. Mahoney, "Fast Monte Carlo algorithms for matrices II: Computing low-rank approximations to a matrix," *SIAM Journal on Computing*, vol. 36, no. 1, pp. 158–183, 2006.

[397] J. Meng, W. Yin, H. Li, E. Hossain, and Z. Han, "Collaborative spectrum sensing from sparse observations using matrix completion for cognitive radio networks," in *Proc. of IEEE International Conference on Acoustics, Speech, and Signal Processing (ICASSP)*, Dallas, TX, USA, March 2010.

[398] Mosek ApS Inc., "The Mosek optimization tools, ver 4." 2006.

[399] R. Bixby, Z. Gu, and E. Rothberg, "Gurobi optimization, www.gurobi.com," 2009-2011.

[400] W. Yin, "Gurobi Mex: A MATLAB interface for Gurobi," 2009–2010, [online] available: http://www.caam.rice.edu/∼ wy1/gurobimex.

[401] W. Yin, S. P. Morgan, J. Yang, and Y. Zhang, "Practical compressive sensing with Toeplitz and circulant matrices," in *Proc. of Visual Communications and Image Processing*, Huang Shan, An Hui, China, July 2010.

[402] NTIA, "FCC frequency allocation chart," 2003, [online] available: http://www.ntia.doc.gov/osmhome/allochrt.pdf.

[403] "Spectrum policy task force report," Federal Communication Commission, Tech. Rep. 135, November 2002, eT Docket.

[404] E. van den Berg and M. Friedlander, "Joint-sparse recovery from multiple measurements," Department of Computer Science, University of British Columbia, Technical Report TR-2009-07, Apr. 2009.

[405] D. Coppersmith and S. Winograd, "Matrix multiplication via arithmetic progressions," *Journal of Symbolic Computation*, vol. 9, no. 3, pp. 251–280, March 1990.

[406] B. A. Munk, *Frequency Selective Surfaces: Theory and Design*. Wiley, Hoboken, NJ, USA, 2000.

[407] Y. Zhang, J. Yang, and W. Yin, *YALL1: your algorithm for L1*, 2009, [online] available: http://yall1.blogs.rice.edu/.

Index